Theory of Edge Diffraction in Electromagnetics:

Origination and Validation of the Physical Theory of Diffraction

Revised Printing

Theory of Edge Diffraction in Electromagnetics:

Origination and Validation of the Physical Theory of Diffraction

Revised Printing

Pyotr Yakovlevich Ufimtsev

Edited by Andrew J. Terzuoli, Jr.
Air Force Institute of Technology (AFIT)

Translated by Richard D. Moore
(Formerly with USAF Foreign Technology Division)

Raleigh, NC
www.scitechpub.com

Published by SciTech Publishing, Inc.
911 Paverstone Drive, Suite B
Raleigh, NC 27615
(919) 847-2434, fax (919) 847-2568
scitechpublishing.com

This is a revised reissue of the book originally published by Tech Science Press in 2003.

Editor: Dudley R. Kay
Production Director: Susan Manning
Production Coordinator: Robert Lawless
Typesetting: SNP Best-set Typesetter Ltd., Hong Kong
Cover Design: Aaron Lawhon
Printer: Edwards Brothers, Inc.

Printed in the United States of America

ISBN: 9781891121661

Foreword
by K. M. Mitzner

The Lockheed F-117 Stealth Fighter and the Northrop B-2 Stealth Bomber play key roles in today's United States Air Force.

These were the first two major aircraft designs to employ the principles of Pyotr Ufimtsev's Physical Theory of Diffraction (PTD). Ben Rich, who oversaw the F-117 project as head of Lockheed's fabled Skunk Works, refers to Professor Ufimtsev's work as "the Rosetta Stone breakthrough for stealth technology." At Northrop, where I worked on the B-2 project, we were so enthusiastic about PTD that a co-worker and I sometimes broke into choruses of "Go Ufimtsev" to the tune of "On, Wisconsin."

And so today the rather abstract physics and mathematics developed by this charming and unassuming old-world gentleman are influencing military strategy and tactics and thus helping shape history — not just through the F-117 and the B-2, but through the many military systems of many kinds that now incorporate stealth technology based on PTD.

It can indeed be argued that the PTD-based B-2, even before it was operational, played a major role in ending the Cold War. The Soviet Union was ringed with a "fence" of radar installations, spaced so that their detection ranges for conventional aircraft overlapped to prevent any vehicle from sneaking through. But if stealth technology could cut the detection range by a factor of n, then the Soviets would need n times as many radars to maintain the fence, and the cost could not be borne by their economy. This may well have been an important consideration in the recognition under Gorbachev that the Soviet Union could no longer compete militarily with the West.

Professor Ufimtsev, a native of Russia who now lives in Los Angeles, was working at a government institute in Moscow at the time he was publishing his theory, but apparently the Soviet Union did not take full advantage of PTD.

~

Those of us who were developing design techniques for stealth vehicles at places like Northrop and Lockheed and Grumman (which, when it was still separate from Northrop, gave Professor Ufimtsev a medal) needed a method of evaluating the residual scattering from a body that had already been configured to be stealthy. Professor Ufimtsev's PTD turned out to be that method — and also gave us guidance as to how to make the body even more stealthy.

Most radar systems are monostatic or near-monostatic — that is, the radar transmitter and receiver are located along the same line of sight to the target or along lines of sight that differ by only a small angle — and thus the typical system sees only the energy that is returned from the target at or near the backscatter direction. For a given direction of incidence to a surface, the strongest backscatter is associated with surface points where the line of sight is normal to the surface or to an edge — if such surface points exist. The stealth designer therefore seeks to minimize the range of incidence directions for which such points do exist and to position those directions so that they are unlikely to be the line of sight from a radar. Then he uses special materials to further reduce the scattering.

To this end, the surface of the F-117 is composed primarily of polygonal facets, with two adjacent facets meeting to form a wedge with a straight edge and flat faces. Wedges with straight edges — but with curved faces — are also key to the B-2 design, where the wedges are formed by the union of the shallow upper and lower surfaces along the polygonal planform. Both designs incorporate special materials.

So PTD had to — and did — prove itself as a computational tool for problems involving flat or curved surfaces, straight edges, materials other than perfect conductors, and backscatter for lines of sight far off normal to the surfaces and edges.

Perhaps even more important, PTD was a driving force in the design of both vehicles. The primary motivation for the flat facet design of the F-117 was, in fact, to be able to do a PTD analysis of the vehicle using then-existing computer capability.

As to the B–2, we were well into the design process when the government made a major change in the mission specification that required a major change in the vehicle design. If we had not been ready to respond, this could have added billions to the cost of the project or even killed it. But we had already worked out the method of redesign as an abstract exercise in applying the principles of PTD, we obtained experimental proof of concept within a few days, and the project went on smoothly.

~

Professor Ufimtsev first introduced PTD in a series of Soviet journal articles beginning in the late 1950s. In 1962 he consolidated his results in book form as *Metod kraevykh voln v fizicheskoi teorii diffraktsii*, which was translated into English in 1971 by the U.S. Air Force Foreign Technology Division[1] as *Method of Edge Waves in the Physical Theory of Diffraction*.

The present book contains an improved and updated development of the most significant material in the 1962 book, augmented by results and insights from Professor Ufimtsev's more recent research.

Professor Ufimtsev initially developed PTD as an improvement on the physical optics (PO) approximation to scattering from a body. Both approaches are based on the observations that:

a) Physically, the field scattered from a body is exactly the same as the radiation field produced by *equivalent electric and magnetic surface currents* on the body. These equivalent currents are obtained respectively by rotating the tangential magnetic field 90° in the plane of the surface (in accordance with the right-hand rule for current) and rotating the tangential electric field 90° in the opposite sense. In the case of a perfectly conducting body, the equivalent electric surface current is the true surface current and there is no equivalent magnetic surface current.

b) Mathematically, the scattered field can be expressed as an integral over the body surface of a linear function of the equivalent surface currents.

---

1. The USAF Foreign Technology Division was redesignated the National Air Intelligence Center in 1993. Its primary responsibility is the production of scientific and technical aerospace intelligence for the United States Air Force.

Solution of a scattering problem can thus be reduced conceptually to determination of the equivalent surface currents and then evaluation of the integral. Both steps can be quite challenging, especially for large bodies of complicated shape and material structure.

Both PTD and PO are approximate methods that employ the results of simpler scattering problems, often called *canonical problems,*[1] to determine the field scattered by the body of interest. Conceptually both methods involve first approximating the surface currents and then evaluating the surface integral exactly or approximately. Often, however, the two steps can be merged, so that explicit use of the surface currents is bypassed, by making use of pre-calculated *diffraction coefficients* and of established formulas for the results of exact and asymptotic integrations.

The surface currents in PTD are expressed as the sum of a first approximation called the *uniform* part plus a correction term called the *nonuniform* part. Usually the uniform part is defined as identical with the PO approximation. But PTD is also applicable to problems, including the thin conductor configurations introduced in Chapter 6, where a different choice of uniform part is appropriate.

The canonical problem for physical optics is simply scattering from an infinite flat plate. At a point on the surface of a perfect conductor, for example, the PO surface current is the current that the source field would produce on a perfectly conducting plate tangent at the point. A coated surface can often be characterized by an *impedance boundary condition* (IBC), that is, a *surface impedance* (ohms) at each point that fixes the ratio of the equivalent magnetic current (which can be expressed in volts/meter) in one direction to the equivalent electric current (amperes/meter) in the normal direction; for this case, the PO surface currents at a point are the currents for a tangent plane with uniform surface impedance equal to the impedance at the point.

---

1. Strictly speaking, a canonical problem is one of a limited *canon* of basic problems — such as scattering from an infinite perfectly conducting wedge — with results that can be used in constructing solutions to more complicated problems. But it is quite natural and very convenient to broaden the definition to include any problem with results that can be so used.

The nonuniform part often arises from truncations of the geometry, such as the end of a wire or an edge along which smooth sections of the surface meet. The nonuniform currents that arise from an edge tend to be concentrated near the edge and are thus called *fringe currents.* Fringe currents are also created at less drastic discontinuities, for example, lines across which there is a jump in surface curvature or in surface impedance. Other examples of nonuniform contributions include creeping waves launched from a shadow boundary into the shadowed region and correction terms to PO, for example, due to surface curvature. The details of the problem at hand determine which nonuniform contributions are important and which can safely be neglected.

For a body with edges, a first-order approximation to the fringe current near a point on an edge can be found from the canonical problem of an infinite wedge with edge and faces tangent to the true edge and faces at the point. The scattered field is then found as the sum of a contribution from the physical optics currents on the body surface plus contributions from the fringe currents associated with each edge.

For high enough frequency and an observation point far enough from the body, the surface integral describing the first-order fringe current scattering can be reduced to a line integral along the edges, and the corresponding PO surface integral can be reduced to an analogous line integral plus possible additional contributions from discrete points on the body surface. The two line integrals can then be summed to give an expression for the edge-scattered field in terms of contributions, called *elementary edge waves,* from each point on the edge. An elementary edge wave can furthermore be interpreted as the field of a geometrical ray that is diffracted from the corresponding edge point, that is, one of the rays that are launched from the point in various directions because of the presence of the edge.

The ray description of the edge scattering can frequently be simplified. For far-field scattering of a plane wave by a flat perfectly conducting plate with a convex polygonal edge, for example, the edge integral can be reduced to ray field contributions emanating from the corners. If the plate has instead a convex curved edge, the far-field backscatter at high enough frequency reduces to ray contributions from the two points at which the line of sight is normal to the edge.

Additional terms can be added to the fringe current to characterize higher-order effects such as the current disturbance generated at a corner where two edges meet or the currents associated with interactions between edges. Evaluating these higher-order effects is not, however, usually an easy job.

The discussion above shows that use of current-based PTD does not sacrifice the insight that comes with a geometrical ray description of the scattering. Where a ray description is meaningful, it can indeed always be extracted by applying appropriate integration techniques. The ray description is especially valuable when PTD is being used to drive a design rather than just evaluate it.

The PTD approach is a high frequency technique in the sense that both the approximations to the surface currents and the resulting approximations to the scattered field tend to become more accurate as frequency increases. As is usual with methods that work well in one part of the spectrum, much effort has gone into refinements that extend the frequency range.

All in all, PTD is a robust theory of proven value and amenable to further development. No wonder it became known both at Lockheed and at Northrop as "industrial-strength" diffraction theory!

KENNETH M. MITZNER

APRIL 17, 2003

Preface from the Editor

Over the last twenty-five years, the Air Force Institute of Technology (AFIT) has educated numerous people in the area of radar observability. This included graduate education at the masters and Ph.D. level, as well as continuing education in the form of short courses. These programs involved those that literally wrote the book in the area; people like the chief engineers and program managers of various operational platforms, as well as academics working in the forefront of electromagnetic scattering phenomena. We were fortunate to perform this education during a time of relative world peace. A previous effort during the not so fortunate time of WWII represented an earlier major initiative in radar and its uses. That period resulted in an excellent documentation known as the *Massachusetts Institute of Technology Radiation Laboratory Series,* commonly known as the MIT "Rad Lab" Series.

Slightly over ten years ago, a number of us who were working in this field thought it would be worthwhile to create a series of documents, similar to the Rad Lab Series, to record the thrust in radar observability that had occurred more recently. With the almost concurrent arrival of Professor Pyotr ("Peter") Ya. Ufimtsev into the USA, we felt that he would be one of the most appropriate, if not the most appropriate person, to author the first book in the series. Most of us were intimately familiar with his book on the Physical Theory of Diffraction (PTD), which played a significant role in this effort. About three years ago during one of Professor Ufimtsev's visits to AFIT as a guest lecturer, I discussed this possibility with him and he very excitedly told me he had started a second book on edge diffraction in Russian, but he had never translated it into English. We decided to embark on the task of translating his original work from Russian to English, then update and edit the work to produce a book that could be of significant value in the continuing education process.

So we needed to locate someone who not only could faithfully translate Russian into English but also was familiar with the technical area at hand. I had known Richard Moore from the days when he worked for the Low Observables Office at Foreign Technology Division and immediately thought he would be the perfect person. When I finally caught up with him in the Washington, D.C. area, he said he had been looking for material to translate in this subject area for the longest time and would be most honored to work on such an important project. My administration at AFIT was very supportive of this endeavor, which essentially had to be accomplished on my own time, because of the necessity to protect the copyright for a future publisher. When all was said and done, the project fell together naturally and seamlessly. We sent Peter's handwritten Russian manuscript to Richard and over the subsequent time the three of us interacted and iterated to produce the document that you now see.

It has been my distinct pleasure to work with two people as talented as Peter and Richard. At this time, because of the other priorities that we all face, I do not know if there will ever be another series like the Rad Lab Series, but I do know that if it does come to pass, this volume will be the most fitting first volume that I could ever imagine. I wish to thank Peter, Richard, my family, and my administration at AFIT for giving me the opportunity, time, and support to contribute to this endeavor. I also thank God for the spirit, fortitude, persistence, and determination to want to accomplish this project. We all hope that you will enjoy this volume, and gain as much from it technically, as we have in the process of creating it. Please feel free to send me your comments via the e-mail below.

ANDREW J. TERZUOLI, JR.
a.j.terzuoli@ieee.org

DAYTON, OHIO USA
MARCH, 2003

Translator's Note

It wasn't long after I had begun working at FTD's Low Observables Office that I heard bits and pieces of a story that a Soviet scientist was actually responsible for the theoretical developments that ultimately led to the design and production of the yet-to-be-unveiled F–117. Somehow, Bill Schroeder and Denys Overholser of Lockheed's Skunk Works had acquired an FTD machine translation of a Russian document entitled, *Metod kraevykh voln v fizicheskoi teorii diffraktsii.* After deciphering its contents (machine translations were never particularly easy to read, but an amazing accomplishment at the time, nevertheless) these two gentlemen understood the significance of what they held in their hands — a blueprint for the straightforward design of radar evading aircraft. Thus, Project Harvey and Have Blue were born and so was an intelligence legend that can hold its own against the best Tom Clancy can offer.

In the Intelligence Community Ufimtsev's name became almost mythic. I say his *name* because no one at the time really knew exactly who this mysterious man was or what secret passageways we imagined he must have taken to work each morning as he donned the same trench coat that we too were always so stereotypically pictured as wearing. But regardless of the personas tacked onto all involved, a critical question remained. How far ahead of us were they? Most everyone was quick to admit that the Russian physicists and mathematicians were tops, and given the critical success the U.S. military was experiencing with PTD almost a decade after this text had been published in Moscow, many of us figured we were in for some fierce competition. Perhaps it was time for us to burn the midnight oil.

In the face of a rapidly changing world, I caught sight of this man at an AFIT seminar years later, and if he owned a trench coat, he certainly wasn't wearing it now. Brushing politics aside, Ufimtsev expressed no interest in dark sunglasses, fedoras, and secret passageways. What seemed to excite this man was instructing students about the properties of black bodies and describing how the F-117 scattered radar waves off its fuselage and planform like combed back spines on a well-groomed hedgehog. It was unmistakable. Here, Ufimtsev's animation expressed what he obviously loved most — learning, and in turn, teaching. And there I sat, thinking I had as much to learn from this gentleman as the students huddled over the desks in front of me. Thankfully, I slept a little more soundly that night.

Almost fifteen years after my story begins I still find it hard to believe that today I'm sitting here collaborating with this remarkable man. I say this not only because his development of PTD revolutionized our ability to model electromagnetic scattering, but also for the reason that his development of PTD in a very real, direct, and tangible way played an unexpected role in changing the political landscape of our world. Since that time quietly whispered terms such as *stealth* and *low observables* have practically become household words, the Berlin Wall and Iron Curtain have become relics of the past, and Pyotr Ufimtsev, who played an unwitting role in all of this, took up a new residency in Los Angeles where he worked for a time, to the irony of all ironies, at Northrop's B-2 Division.

When Andy Terzuoli succeeded in tracking me down three years ago with the hope that translating Ufimtsev's follow-on manuscript on edge diffraction would personally be of interest, I could hardly believe such an opportunity so close to home could have befallen me. It was an incredible honor to work with these gentlemen in the production of this book, and I wish to thank them for their insight, advice, and most admirable patience as I worked on the translation and assembly of this updated text. Although a human translation of *Metod kraevykh voln* has yet to be completed, it is my hope that this offering can satisfy the interests of the technically and historically curious until an English edition of it, too, can perhaps one day be made available.

In addition to thanking Andy and Pyotr for this unique opportunity, most sincere thanks are due to Peter F. H. Priest, Benedikt A. Munk, and Kirby H. Williams, who not only pointed me down some roads less traveled but also encouraged me to pave a few of my own, and in so doing did much to steer my life toward a juncture with this translation.

RICHARD D. MOORE

MEMORIAL DAY, 2003

Acknowledgments

The author appreciates the interest and valuable comments provided by V. A. Fock in support of this research. He also recommended publishing a book containing this material. As examination will show, this material stood the test of the time and has not lost its scientific significance. The author also benefited from technical discussions with L. A. Vainshtein. Experimental verification of the theoretical calculations was provided by E. N. Maizel's and L. S. Chugunova. Numerical calculations presented in this book were completed and originally plotted by the computer support staff. The author is very thankful to all of the above mentioned colleagues.

The author is indebted to K. M. Mitzner, a leading expert in the theoretical analysis of electromagnetic scattering and one of the designers of the B-2 stealth bomber, for sharing his very valuable and interesting thoughts on the development and impact of PTD. His introduction of the incremental length diffraction coefficient (ILDC) concept was a significant contribution to scattering theory and facilitated greatly the application of PTD to practical design problems.

The author also wishes to thank R. D. Moore for his translation of the original Russian manuscript and his assitance in the preparation of the revised printing of this book. Thanks are also due to A. J. Terzuoli, Jr. of the Air Force Institute of Technology for sponsoring this effort.

This book includes in revised form materials from certain articles by the author and Chapters 1, 4, 6, and 7 of his book on PTD [13]. The author thanks the editorial boards of the related journals and the *Radio i Svyaz* publishing house for their permission to use these materials.

Contents

Introduction

Documentation on edge waves first appeared in diffraction theory as early as the beginning of the 19th century. At that time, Young examined the field in the shadow region behind an aperture in an opaque plane and found it can be represented in the form of spherical waves centered on the aperture edge [1, 2]. This physical representation of the diffracted field was later developed and mathematically formulated in the theory of diffraction from thin black screens by Kirchhoff, Maggi, and Kottler between 1882 and 1923. Near the middle of the 20th century, diffraction theory focused on exact solutions to different boundary value problems of the wave equation and the study of their properties. These solutions further served as a basis for the asymptotic representation of the diffracted field in the limiting case as $\lambda \to 0$ (where λ is the wavelength).

The research that continued in the following decades showed that one of the central phenomena of diffraction is the formation of edge waves. These waves arise near vertices and edges on the surface of a body and become significant components of diffracted field when the linear dimension of the scattering body is greater than the wavelength, λ.

This text categorizes and summarizes the work of the author conducted in the 1950s–70s and is mainly connected with research on edge diffracted waves.This research stemmed from practical considerations and focused on the development of approximate methods to calculate the scattering of electromagnetic waves from real objects.

The first chapter generalizes Kirchhoff-Kottler theory to volumetric black bodies. This theory is of interest for two reasons. First, from a historical point of view, it was the first theory (Maggi, Kottler) that successfully extracted edge waves from the diffracted field. Second, this theory also relates to the problem of coating a scatterer with radar absorbing materials.

It provides a limit on the radar cross section reduction that can be achieved with the use of perfectly absorbing coatings.

Subsequent chapters examine the diffraction of electromagnetic waves from perfectly conducting bodies. Chapters 2 and 3 discuss edge waves diffracted from convex and concave bodies of revolution possessing edges. The Physical Theory of Diffraction (PTD) in its original form is applied here for calculation of radar cross section. A central and original idea of PTD is introduction of the concept of a "**nonuniform**" current which is caused by diffraction at sharp edges, vertices, small protrusions, and other inhomogeneities at the object surface. In Chapter 4, it is shown that the radiation created by this nonuniform current is the physical nature of the depolarization phenomenon in backscattering.

The theory developed in Chapters 2 and 3 addresses only primary edge waves, which provide the main contributions to the scattered field. However, a more thorough description of the field requires the examination of higher-order edge waves resulting from multiple diffraction. These waves are investigated in Chapters 6 and 7. Current edge waves due to multiple diffraction at dipoles and strips are studied in Chapter 6 using the parabolic equation method which is first introduced in Chapter 5. A physical approach is developed in Chapter 7 to examine the radiation of multiple edge waves.

A more rigorous analysis of higher-order edge waves is presented in Chapters 8–11. This analysis is based on the solution of functional equations and provides new exact and asymptotic expressions for the current on a strip and for the far field scattered from a strip. This mathematical theory is extended in Chapter 12 to investigate diffraction at an open resonator consisting of two parallel plates. Two basic topics of interest are associated with this problem. What is the nature of the resonance phenomenon and what is its influence on the scattered field? It is demonstrated here that the resonance is the result of constructive interference of multiple waveguide modes reflected from the ends of the plates. New physical phenomena are discovered while investigating the scattered field.

The results obtained in Chapters 8–12 complete our research on edge waves and validate the approximate methods developed in previous chapters.

The theory presented in this book can be considered as consisting of two parts. Chapters 1–4, and 7 constitute the physical theory of diffraction, which is based on a series of physical assumptions. The remaining chapters

(Chapters 5, 6, 8–11, and 12) present the mathematical theory based on the solution of the parabolic equation and on the rigorous wave equation.

Some results of this work were previously published in a series of articles and were presented at All-Union Symposiums and School-Seminars on the Diffraction of Waves [3–12, 14–26]. In 1962, the author published the book, the *Method of Edge Waves in the Physical Theory of Diffraction* [13]. Chapters 1, 4, 6, and 7 of that publication are incorporated in revised form in Chapters 2, 4, and 7 of this text.

To show the place this present book might occupy in the existing professional literature, it begins with a short review of the known edge diffraction techniques and briefly explains basic distinguish features of the present book. Other related references are given and discussed concurrently. During preparation of this revised printing, the book was updated with the inclusion of additional comments and references. Notice among them the recent author's monograph *Fundamentals of the Physical Theory of Diffraction* (Wiley & Sons, Inc., 2007) devoted to the modern PTD and based on the concept of elementary edge waves.

While reading this book, one should keep in mind that some notations and symbols are often used throughout the book. Their physical meaning can be the same or similar. However, their mathematical definitions can be completely different in different chapters.

The Gaussian system of units and an $e^{-i\omega t}$ time dependence are used in this book. The relationships between the Gaussian system of units and the International system (SI) are presented in the Appendix 1.

Review of Edge Diffraction Techniques

Edge diffraction has been investigated in many books. Here, we provide brief comments on these publications and then explain what is new in the present book on the Theory of Edge Diffraction (TED) in Electromagnetics.

1. J. R. Mentzer. *Scattering and Diffraction of Radio Waves.* Pergamon Press, London, New York, 1955.

   This book contains:

   - Physical Optics results for thin plates.
   - The results for a thin wire of finite length found by using the variation technique. Simple expressions are provided without physical interpretation.
   - Approximations for scattering from a thick circular cylinder of finite length. The current on the cylindrical part of the surface is assumed to be the same as that on the infinitely long cylinder. Scattering from the bases (disks) is ignored. Edge/fringe currents are not taken into account.

2. R. W. P. King and T. T. Wu. *The Scattering and Diffraction of Waves.* Harvard University Press, Cambridge, Massachusetts, 1959.

   - Only one chapter here relates to edge diffraction. The strip and disk diffraction problems are considered. It collects the known results for the total scattering cross section. Several first terms of the high frequency asymptotic expansions are presented.

3. H. Honl, A. W. Maue, and K. Westpfahl. *Theorie der Beugung.* Springer-Verlag, Berlin, Gottingen, Heidelberg, 1961.

   - Integral equations for the current on perfectly conducting plates are formulated.

- For the field scattered from a strip and a disk, several first terms of the high frequency asymptotic expansions are presented. These are nonuniform asymptotics which are not valid for grazing incidence and observation directions.

4. P. Ya. Ufimtsev. *Method of Edge Waves in the Physical Theory of Diffraction.* Moscow, Soviet Radio Publishing House, 1962.

    - The Physical Theory of Diffraction (PTD) is developed which provides the first term of the high frequency asymptotic expansions for edge waves. Objects with convex illuminated surface are considered.
    - Asymptotic solution for the thin wire of finite length is found which includes all multiple edge waves.
    - The nature of depolarization in backscattering is investigated. A technique to measure the scattered field generated by the diffraction component of the surface current is described.

5. V. A. Borovikov. *Diffraction at Polygons and Polyhedrons*. Nauka, Moscow, 1966.

    - High frequency asymptotic expansions are constructed for edge waves far from the scattering edges. Different types of asymptotics are presented for the field in the ray regions and in the vicinity of geometrical optics boundaries. Diffraction of waves incident under grazing directions is considered. Some basic results are included in publication [13] cited below.

6. J. J. Bowman, T. B. A. Senior, and P. L. E. Uslenghi, Eds. *Electromagnetic and Acoustic Scattering by Simple Shapes*. Hemisphere Publishing Corporation, New York, 1969.

    - This book collects known results in the diffraction theory related to simple objects. It contains asymptotic solutions for edge diffraction from a wedge, half-plane, disk, and wire. In particular, it includes some PTD results of [4].
    - Several first terms of asymptotic expansions are presented.

7. B. Noble. *Methods Based on the Wiener-Hopf Technique.* Pergamon Press, New York, 1958.

    - Edge diffraction problems are investigated.

- Exact solutions are presented for the scattering and radiation problems related to the semi-infinite parallel plate waveguide and circular waveguide.
- The general asymptotic expressions are complex. They contain infinite series.
- Simple asymptotic expressions are derived for the far field scattered from a strip that take into account diffraction interaction between edges. They contain several first terms of the nonuniform asymptotic expansion that are not applicable for grazing directions.

8. L. A. Vainshtein. *Theory of Diffraction and the Factorization Method.* Golem Press, 1969.
    - In this book the same exact solutions as those in [7] are found for the radiation from a semi-infinite parallel plate waveguide and circular waveguide.
    - Among other investigated problems are scattering from an infinite periodic structure of parallel half-planes and from diaphragms in waveguides.
    - A special asymptotic function is introduced which takes into account all multiple interactions between edges.
9. L. A. Vainshtein. *Open Resonators and Open Waveguides*. Soviet Radio, Moscow, 1966.
    - Characteristic oscillations in open resonators are investigated. The field outside the resonators is not considered. The radiation losses are evaluated through the imaginary part of characteristic frequencies. Asymptotic expressions for eigen modes and characteristic frequencies are found. This theory is based on the parabolic equation.
10. P. L. E. Uslenghi, (Editor), *Electromagnetic Scattering*, Academic Press, New York, 1978.
    - This book is a collection that includes review articles related to the edge diffraction. They mainly consider the Geometrical Theory of Diffraction (GTD) and its modification, Uniform Asymptotic Theory of Diffraction (UAT).
    - The Spectral Theory of Diffraction (STD) is exposed in the article by R. Mitra and Y. Rahmat-Samii. This theory exploits the

PTD concept of fringe (nonuniform) current and can be interpreted as a spectral version of PTD. This theory is demonstrated for a half-plane illuminated by an arbitrary incident wave and for two parallel half-planes. High frequency asymptotics for the primary and secondary edge diffraction are constructed.

11. G. L. James. *Geometrical Theory of Diffraction for Electromagnetic Waves.* London, Peregrinus, 1986.

12. D. A. McNamara, C. W. I. Pistorius, and J. A. G. Malherbe. *Introduction to the Uniform Geometrical Theory of Diffraction.* Boston-London: Artech House, 1990.

13. V. A. Borovikov and B. E. Kinber. *Geometrical Theory of Diffraction,* Bath, U.K.: IEE/Bookcraft, 1994.

   - The original GTD fails near shadow boundaries, caustics, and foci. Some modifications of GTD are considered in [11–13] to overcome these shortcomings.

This book on the Theory of Edge Diffraction in Electromagnetics differs from the above publications in the following aspects:

- It contains the theory of black bodies which is important for understanding of physics of diffraction, especially scattering at large bistatic angles. In particular, this theory is of interest for development of future stealth technology. This topic is not covered in the above publications[1–13].
- It extends PTD [4] to concave objects with the horn type cavities. Application to the problem of practical interest is demonstrated. This topic is not covered in the above publications [1–13].
- The GTD/UTD/UAT approaches described in the above books enable the determination of the first term in the asymptotic expansion for 3-D edge waves. However, they are ray based techniques and fail near caustics and foci where the diffraction field does not have a ray structure.
- In contrast, current and wave-based techniques are developed in this book on TED. That is why they are valid at caustics and foci. Some problems of practical interest related to these phenomena are investigated here.

- Chapter 4 of this book is a revised form of Chapter 6 of the above book [4] by the author. This Chapter describes the depolarization nature and measurement of the field radiated by the nonuniform/diffraction component of the surface current. These topics are not covered in the above books[1–3, 5–13].
- This book analyzes the nature of edge diffraction by investigating primary shadow radiation and its interaction with edge faces. This topic is not considered in the above books [1–13].
- In this book a solution to the wire problem is presented that is based on the parabolic equation. It is the first application of this technique to the wire problem.
- Direct integration of approximate expressions for the current leads to field expressions that do not satisfy the reciprocity principle. A special technique is developed in this book to overcome this shortcoming. No similar results exist in the above books [1–3, 5–13].
- Many asymptotic expressions for the current on a strip and for the far scattered field are presented in the above books [1–3, 5–13]. However, these asymptotic expressions are not uniform. They are not applicable for arbitrary points of observation. In particular, most of them are not valid for grazing angles. Furthermore, those expressions which were derived for grazing angles are not valid for other directions.
- Unique uniform asymptotics are presented in this book. They allow one to calculate the current and far field (with high precision!) for an arbitrary observation point on the strip and in the far zone. They are also valid for an arbitrary direction of the incident wave. Similar results do not exist in the above books.
- Books [8, 9] examine the scattering/radiation from semi-infinite waveguides. This book investigates a more complex problem, the scattering from a finite length waveguide, which actually represents an open resonator.
- Properties of characteristic oscillations in open resonators were studied in book [9]. However, their influence on the scattered field was not considered there. This important and interesting topic is investigated here.

1

Diffraction of Electromagnetic Waves at Black Bodies: Generalization of Kirchhoff-Kottler Theory

§1.1 Black Bodies

As was mentioned earlier, this chapter serves as a historical introduction to the study of edge waves. No less important, this chapter also introduces Kirchhoff-Kottler theory, which is shown to be useful in the analysis of some actual problems. Specifically, it is interesting to determine whether it is possible to use layers of radar absorbing material to eliminate the electromagnetic field scattered by an object. If this is impossible, then to what extent can one decrease the field scattered from an ideal surface that completely absorbs all energy incident upon it? These questions lead us to the well known "black body" problem.

Although from an electrodynamic point of view, it is principally impossible to formulate the conditions that model the black body concept, such an abstract idea is, nevertheless, useful in some cases. Physically, a black body can be approximated as a medium with very slowly changing electromagnetic parameters, for which the coefficient of reflection is zero (media with an ideal impedance matched condition).

The application of the black body concept to diffraction was first formulated at the end of the 19th century by Kirchhoff while he was conducting research on scalar wave diffraction from thin opaque screens. Later, Kottler extended the concept of a black screen to electromagnetic waves. In particular, he showed that Kirchhoff's solution for black screens was an exact solution to the problem with discontinuous boundary values — the Sprungwertproblem [27, 28]. The model for black screens proposed by Kirchhoff led to a series of other models (Sommerfeld, Voigt, Kottler). A clear description of these models can be found, for instance, in Bakker and Copson [29]. However, the Kirchhoff-Kottler model is more appropriate for our research.

The theory developed by Kirchhoff and Kottler for thin black screens can be generalized for arbitrary volumetric black bodies. However, there was no literature that provided a complete description of the field scattered by these objects. In 1963 the following research was conducted to fill this void and was later published in article [20].[1] It should be noted that Neugebauer examined the diffraction of plane electromagnetic waves from thin black and semitransparent screens with apertures [30, 31]. However, his application of the rigorous form of Babinet's principle in the complementary problem is not consistent with the approximations he makes in the original problem.

§1.2 Vector Analog of Helmholtz Theorems

First, we will derive some useful theorems that are the vector analogs of the well known Helmholtz theorems that apply to scalar wave fields [29].

Using the vector analog of Green's theorem [32] one can state the following:[2]

THEOREM 1. In a homogeneous volume V bounded by a surface S (Fig. 1.2.1), let the vectors **E** and **H** satisfy Maxwell's equations and let the scalar Green's function e^{ikr}/r and both its first and second derivatives be continuous and finite, then

$$\mathcal{E}_V + \mathcal{E}_S = 0, \quad \mathcal{H}_V + \mathcal{H}_S = 0 \tag{1.2.1}$$

[1]The material of this article is included in this chapter with the permission of Kluwer Academic/Plenum Publishers.

[2]See Appendix 2 for detail.

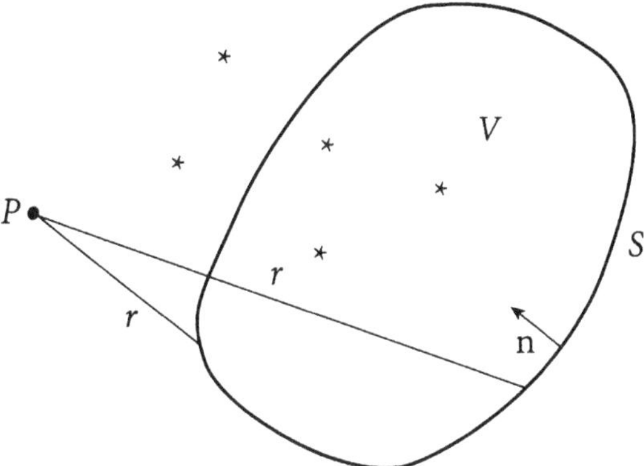

Figure 1.2.1: Geometry of Theorem 1. Reprinted with the permission of Kluwer Academic/Plenum Publishers.

$$\left.\begin{aligned}\mathcal{E}_{V,S} &= -\frac{1}{ik}\left(\nabla\nabla\cdot\mathbf{A}^{e}_{V,S} + k^2\mathbf{A}^{e}_{V,S}\right) - \nabla\times\mathbf{A}^{m}_{V,S}\\ \mathcal{H}_{V,S} &= -\frac{1}{ik}\left(\nabla\nabla\cdot\mathbf{A}^{m}_{V,S} + k^2\mathbf{A}^{m}_{V,S}\right) + \nabla\times\mathbf{A}^{e}_{V,S}\end{aligned}\right\} \tag{1.2.2}$$

$$\mathbf{A}^{e,m}_{V} = \frac{1}{c}\int_V \mathbf{j}^{e,m}\frac{e^{ikr}}{r}dV\,, \tag{1.2.3}$$

$$4\pi\mathbf{A}^{e}_{S} = \int_S \mathbf{n}\times\mathbf{H}\frac{e^{ikr}}{r}dS\,,\quad 4\pi\mathbf{A}^{m}_{S} = -\int_S \mathbf{n}\times\mathbf{E}\frac{e^{ikr}}{r}dS \tag{1.2.4}$$

Here, $\mathbf{j}^e(\mathbf{j}^m)$ is the volume density of the external electric (magnetic) currents; **E**, **H** are the electric and magnetic vector intensities of the total field, and r is the distance between the source and observation points located outside the volume V. The cross products in the vector potentials $\mathbf{A}^e_S$

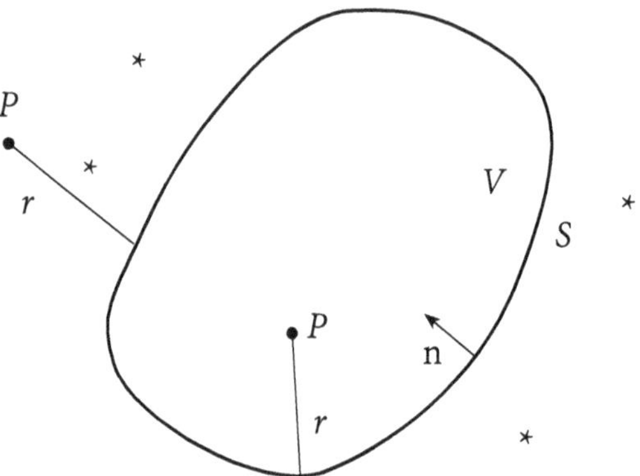

Figure 1.2.2: Geometry of Theorem 2. Reprinted with the permission of Kluwer Academic/Plenum Publishers.

and $\mathbf{A}_S^m$ can be interpreted as the equivalent electric and magnetic surface currents, respectively. It should be noted that this text uses an absolute system of units (Gaussian) and an $e^{-i\omega t}$ time dependence.

Using Eqn. (1.2.1) it is not difficult to establish the following theorems, which are the vector analogs of the well known Helmholtz theorems.

THEOREM 2. Let all sources be located outside a volume V. If an observation point is located at P (see Fig. 1.2.2), then

$$\mathcal{E}_s(P) = \begin{cases} \mathbf{E}(P), & \text{if } P \text{ is inside } V \\ 0, & \text{if } P \text{ is outside } V \end{cases} \tag{1.2.5}$$

$$\mathcal{H}_s(P) = \begin{cases} \mathbf{H}(P), & \text{if } P \text{ is inside } V \\ 0, & \text{if } P \text{ is outside } V \end{cases} \tag{1.2.6}$$

THEOREM 3. Let all sources be located inside a volume V (Fig. 1.2.3), and let the field satisfy the radiation condition. Then,

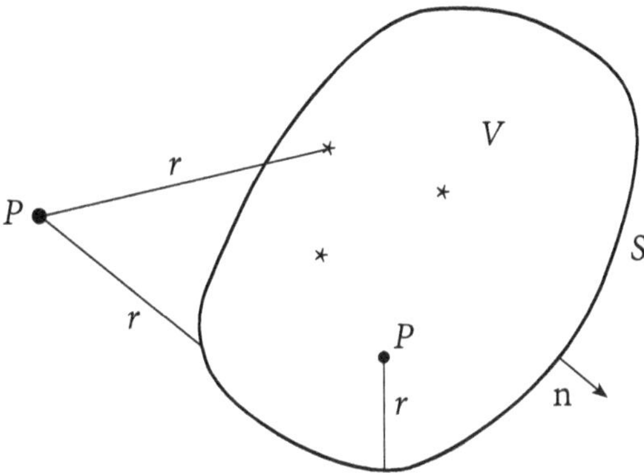

Figure 1.2.3: Geometry of Theorem 3. Reprinted with the permission of Kluwer Academic/Plenum Publishers.

$$\mathcal{E}_s(P)=\begin{cases}\mathbf{E}(P), & \text{if } P \text{ is outside } V\\ 0, & \text{if } P \text{ is inside } V\end{cases} \tag{1.2.7}$$

$$\mathcal{H}_s(P)=\begin{cases}\mathbf{H}(P), & \text{if } P \text{ is outside } V\\ 0, & \text{if } P \text{ is inside } V\end{cases} \tag{1.2.8}$$

Theorems 2 and 3 are well known and can be presented in slightly different forms [32, 33]. However, the portion of these theorems addressing the absence of reverse waves, which propagate back to the sources, is usually proved using a cumbersome method employing the uniqueness theorem to examine the discontinuity of the field on S. Using our formulation of Theorem 1, the absence of reverse waves is completely trivial.

§1.3 Definition of the Black Body and the Shadow Contour Theorem

Let us calculate the field at a point P due to the diffraction of electromagnetic waves from a body (Fig. 1.3.1). To do this, Theorem 2 will be applied to a volume V bounded by the surface Σ and the surface of the body (S is

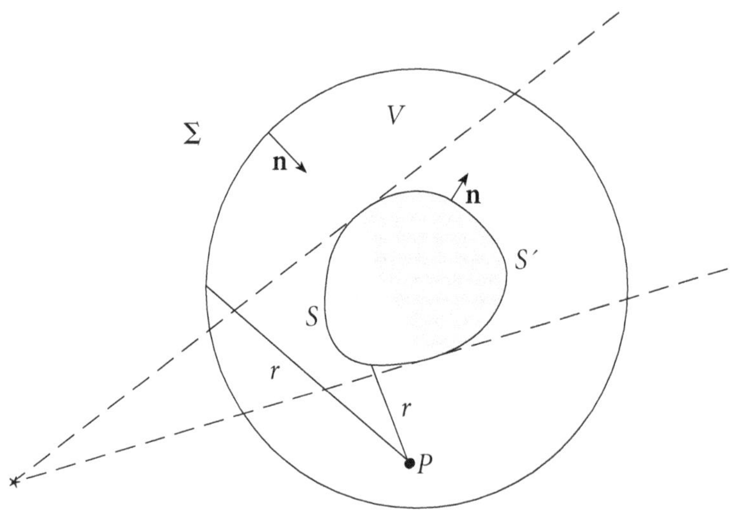

Figure 1.3.1: Diffraction of waves from a black body, bounded by the surfaces S and S'. Reprinted with the permission of Kluwer Academic/Plenum Publishers.

the illuminated surface of the body, S' is the shadowed surface). Given that all of the sources are located outside the volume, we obtain

$$\left.\begin{aligned}\mathbf{E}(P)&=\mathcal{E}_{\Sigma}+\mathcal{E}_{S}+\mathcal{E}_{S'},\\ \mathbf{H}(P)&=\mathcal{H}_{\Sigma}+\mathcal{H}_{S}+\mathcal{H}_{S'},\end{aligned}\right\}\tag{1.3.1}$$

where $\mathcal{E}$ and $\mathcal{H}$ are determined by Eqns. (1.2.2) and (1.2.4).

On the surface Σ,

$$\mathbf{E}=\mathbf{E}_0+\mathbf{E}_P,\quad \mathbf{H}=\mathbf{H}_0+\mathbf{H}_P.\tag{1.3.2}$$

Here, $\mathbf{E}_0$ and $\mathbf{H}_0$ are the source fields located outside the volume bounded by the surface Σ and $\mathbf{E}_P$ and $\mathbf{H}_P$ are the fields scattered by the body.

As a result of Theorem 3, the equivalent currents induced by $\mathbf{E}_P$ and $\mathbf{H}_P$ on the surface Σ do not generate a field interior to V. Therefore,

$$\mathcal{E}_\Sigma = \mathbf{E}_0(P), \quad \mathcal{H}_\Sigma = \mathbf{H}_0(P), \tag{1.3.3}$$

which leads to

$$\left.\begin{aligned} \mathbf{E}(P) &= \mathbf{E}_0(P) + \mathcal{E}_S + \mathcal{E}_{S'}, \\ \mathbf{H}(P) &= \mathbf{H}_0(P) + \mathcal{H}_S + \mathcal{H}_{S'}, \end{aligned}\right\} \tag{1.3.4}$$

if P is located interior to V.

Let us consider the body to be perfectly black, i.e., it absorbs all incident energy. Then, according to Kirchhoff-Kottler theory,

$$\mathbf{E} = \mathbf{E}_0, \quad \mathbf{H} = \mathbf{H}_0 \tag{1.3.5}$$

on the illuminated surface of the body (S), and

$$\mathbf{E} = \mathbf{H} = 0 \tag{1.3.6}$$

on the shadowed surface (S'). Then, the total field produced at P by the diffraction of the electromagnetic waves from the black body is equal to

$$\left.\begin{aligned} \mathbf{E}(P) &= \mathbf{E}_0(P) + \mathcal{E}_S, \\ \mathbf{H}(P) &= \mathbf{H}_0(P) + \mathcal{H}_S. \end{aligned}\right\} \tag{1.3.7}$$

The second terms on the right side of Eqn. (1.3.7) describe the field scattered by the black body and are determined by

$$\left.\begin{aligned} \mathcal{E}_S &= -\frac{1}{ik}\left(\nabla\nabla\cdot\mathbf{A}_S^e + k^2\mathbf{A}_S^e\right) - \nabla\times\mathbf{A}_S^m, \\ \mathcal{H}_S &= -\frac{1}{ik}\left(\nabla\nabla\cdot\mathbf{A}_S^m + k^2\mathbf{A}_S^m\right) + \nabla\times\mathbf{A}_S^e, \end{aligned}\right\} \tag{1.3.8}$$

where

$$4\pi \mathbf{A}_S^e = \int_S \mathbf{n}\times\mathbf{H}_0 \frac{e^{ikr}}{r} dS, \quad 4\pi \mathbf{A}_S^m = -\int_S \mathbf{n}\times\mathbf{E}_0 \frac{e^{ikr}}{r} dS \tag{1.3.9}$$

and **n** is the external surface normal of the body.

One should note that the above definition of a black body leads to the scattered field expressions which are not self-consistent. The reason is that one cannot specify simultaneously independent boundary conditions for **E** and **H**, since they are interrelated by the Maxwell's equations. Because of this, Eqn. (1.3.7) is not self-consistent: as the observation point approaches the black body surface, the values of **E** and **H**, determined by this equation do not agree with the initial assumptions of Eqns. (1.3.5) and (1.3.6).

Kottler [27, 28] mentioned that for black infinitely thin screens, Eqn. (1.3.7) can be considered as a rigorous solution to the problem of a discontinuous boundary values (Sprungwertproblem) where the tangential components E_t and H_t satisfy the conditions

$$E_t\big|_S - E_t\big|_{S'} = E_{0t}, \quad H_t\big|_S - H_t\big|_{S'} = H_{0t}. \tag{1.3.10}$$

However, this definition of a black body, in contrast to Eqns. (1.3.5) and (1.3.6), is applicable only for thin screens and not for volumetric bodies.

Our attention will now be focused on Eqn. (1.3.7). *The shadow boundary on the body's surface will be referred to as the shadow contour.* Let us compare the field scattered by two different black bodies that possess the same shadow contour. The illuminated region of the first body will be referred to as S_1, and the illuminated region of the second body will be referred to as S_2. According to Eqn. (1.3.7), the field scattered by these bodies will be $\mathcal{E}_{S_1}$, $\mathcal{H}_{S_1}$ and $\mathcal{E}_{S_2}$, $\mathcal{H}_{S_2}$ respectively. Theorem 2 will now be applied to the volume V bounded by the imaginary surfaces S_1 and S_2 (Fig. 1.3.2) and located in the external field $\mathbf{E}_0$, $\mathbf{H}_0$. Then,

$$\mathcal{E}_{S_1} - \mathcal{E}_{S_2} = -\mathbf{E}_0(P), \quad \mathcal{H}_{S_1} - \mathcal{H}_{S_2} = -\mathbf{H}_0(P), \tag{1.3.11}$$

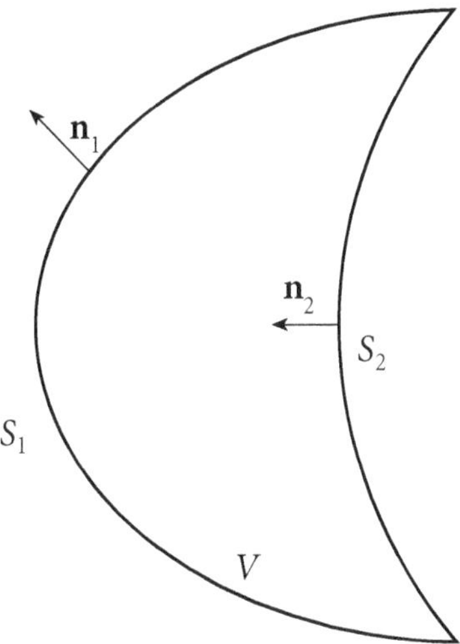

Figure 1.3.2: The shadow contour theorem. Reprinted with the permission of Kluwer Academic/Plenum Publishers.

if the observation point is located within V, and

$$\mathcal{E}_{S_1} - \mathcal{E}_{S_2} = 0, \quad \mathcal{H}_{S_1} - \mathcal{H}_{S_2} = 0, \tag{1.3.12}$$

if the observation point is located outside V. Therefore, Eqn. (1.3.12) may be used to formulate the following *shadow contour* theorem.

> *The field scattered by a black body (for a given mutual orientation between the source and body) does not depend on the surface of the body and is completely determined by its shadow contour (boundary between the illuminated and shadowed regions on the body surface).*

It should be noted that the theorem given here is for an arbitrary incident field $\mathbf{E}_0$, $\mathbf{H}_0$ and for arbitrary volumetric black bodies. A similar conclusion in a simpler problem was obtained by Kottler [28]. He showed that the diffracted field behind a black planar screen with an aperture is completely determined only by the aperture edge. He proved this for the special cases of the excitation by a plane wave and by an elementary dipole.

It is possible to give the following physical interpretation of the shadow contour theorem. Different black bodies possessing the same shadow contour (for a given direction of the incident field!) create identical shadows. However, the shadow can be considered as the result of the superposition of the shadow radiation generated by a black body and the incident field, which cancel each other. Thus, in this region the shadow radiation can be interpreted as the wave beam equal to minus the incident field. Because of transverse diffusion, this shadow radiation penetrates into the remainder of space and appears as the sole source of the scattered field, since the illuminated surface of a black body does not radiate and does not reflect. Therefore, different black bodies possessing identical shadow contours produce identical scattered fields.

Let us note that Eqns. (1.3.8) and (1.3.9) provide the following formal interpretation of electromagnetic waves diffracted from a black body. The black body is both a perfect electric conductor ($\epsilon = i\infty$, $\mu = 1$) and a perfect magnetic conductor ($\epsilon = 1$, $\mu = i\infty$). Half of the energy incident upon the body diffracts as though it were a perfect electric conductor, and the other half diffracts as though it were a perfect magnetic conductor. This concept about black bodies was previously developed in more detail by Lemanskii and Zakhar'ev [34, 35].

Thus, we have shown that according to Kirchhoff-Kottler theory, scattered fields possess an "edge" behavior. This will now be studied in more detail to provide answers to the practical questions presented at the beginning of this chapter.

§1.4 Complementary Principle for Thin Black Screens

In free space let a volume V_0 encompass the external sources of an electromagnetic field. The two following problems will now be examined:

1. Diffraction of electromagnetic waves from a thin black plate S (planar or curved). See, Fig. 1.4.1*a*.
2. Diffraction of the same waves from a complementary black screen Σ (planar or curved), which together with the surface S separates space into two regions. See, Fig. 1.4.1*b*.

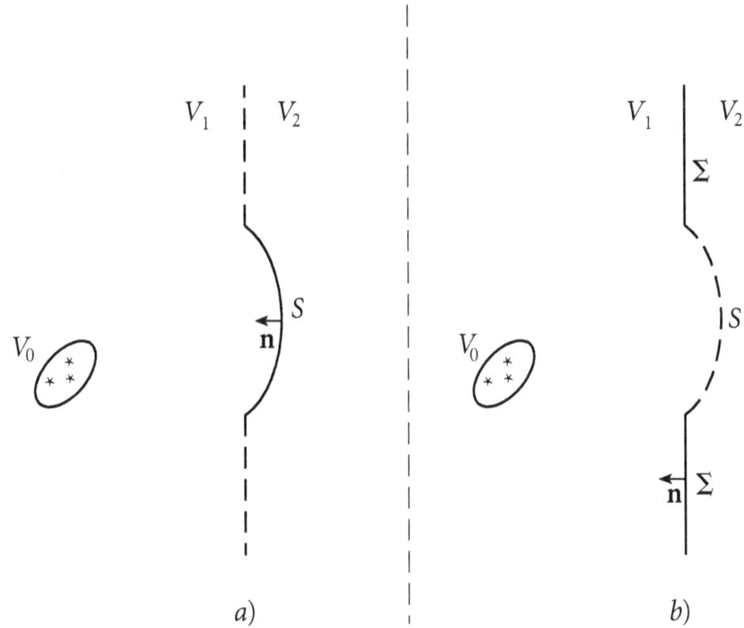

Figure 1.4.1: Diffraction from complementary screens. Reprinted with the permission of Kluwer Academic/Plenum Publishers.

The connection between the solutions to these problems establishes the following addition theorem. Let $\mathcal{E}_S$, $\mathcal{H}_S$ be the field scattered by the black plate S and $\mathcal{E}_\Sigma$, $\mathcal{H}_\Sigma$ be the field scattered by the black screen Σ. According to Theorem 3 of § 1.2:

$$\mathcal{E}_S = -\mathcal{E}_\Sigma\,, \quad \mathcal{H}_S = -\mathcal{H}_\Sigma\,, \tag{1.4.1}$$

if the point of observation P is located in volume V_1 and

$$\mathcal{E}_S = -\mathcal{E}_\Sigma - \mathbf{E}_0\,, \quad \mathcal{H}_S = -\mathcal{H}_\Sigma - \mathbf{H}_0\,, \tag{1.4.2}$$

if P is lies within V_2 (see, Fig. 1.4.1).

§1.5 Total Scattering Cross Section for Black Bodies

The total scattering cross section is determined by

$$\sigma = \frac{\mathcal{P}}{\mathcal{P}_0}, \tag{1.5.1}$$

where

$$\mathcal{P}_0 = \frac{c}{8\pi}\left|\mathbf{E}_0\right|^2, \tag{1.5.2}$$

which represents the time-average power flux density of the incident wave, and

$$\mathcal{P} = \frac{c}{8\pi}\Re e\left\{\int \mathbf{n}\cdot\left[\mathbf{E}\times\mathbf{H}^*\right]dS\right\}, \tag{1.5.3}$$

which is the time-average total power of the field scattered into surrounding space. The integration in Eqn. (1.5.3) can be evaluated over any closed surface containing the body. Choosing to integrate over the surface of the black body, itself, we obtain

$$\mathcal{P} = \frac{c}{8\pi}\Re e\left\{\int_{S'} \mathbf{n}\cdot\left[\mathbf{E}_0\times\mathbf{H}_0^*\right]dS\right\}, \tag{1.5.4}$$

because the scattered field is equal to zero on the illuminated surface (S) of the black body, but on the shadowed side, $\mathbf{E} = -\mathbf{E}_0$, $\mathbf{H} = -\mathbf{H}_0$.

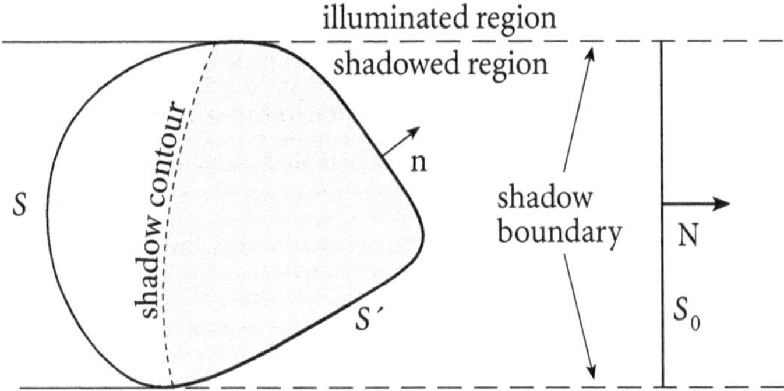

Figure 1.5.1: Depiction of the total scattering cross section.

Assuming that the body is illuminated by a plane wave, we obtain

$$\mathcal{P} = \frac{c}{8\pi}|\mathbf{E}_0|^2 \int_{S'} \cos(\mathbf{n}\cdot\mathbf{N})dS = \frac{c}{8\pi}|\mathbf{E}_0|^2 S_0 . \tag{1.5.5}$$

Here, S_0 is the cross sectional area of the shadow (Fig. 1.5.1). It follows that the total scattering cross section of the black body is equal to

$$\sigma = S_0 . \tag{1.5.6}$$

It should be noted that for convex perfectly conducting bodies, the linear dimensions of which are large relative to a wavelength, the total scattering cross section is equal to $2S_0$. Thus, the total scattering cross section of a black body is half the total cross-section of a metallic body possessing an identical shadow contour.

Next, the analysis of diffraction from a half-plane, strip, and disk, will provide a more detailed understanding of the characteristics of the field scattered by black bodies.

§1.6 Black Half-Plane

The diffraction of a plane wave from a black half-plane was examined in [28, 29], which calculated the scattered field near an edge ($kr \ll 1$) and in the region $kr \gg 1$ far from the shadow boundary. This problem possesses two interesting objectives that are not addressed in these references. The first objective is to represent the field when $kr \gg 1$ far from the edge by a uniform asymptotic expression that is applicable both far from and near the shadow boundary behind the half-plane. The second objective is the following. It was shown earlier that the field scattered by a black body can be considered as a field radiated by equivalent electric and magnetic currents. However, it is known that the illumination of a half plane by an incident wave excites electric and magnetic currents each of which separately creates a reflected wave. It is interesting to determine the nature of reflected waves from a black half-plane. Let us examine the diffraction of the plane wave given by

$$E_z = E_{0z} e^{ik(x\cos\alpha + y\sin\alpha)}, \quad H_z = 0 \tag{1.6.1}$$

from a black half-plane S (Fig. 1.6.1). The scattered field is described by Eqns. (1.3.8) and (1.3.9), where the integration is evaluated over the region occupied by the half-plane ($x = 0, -\infty \leqslant y \leqslant 0, -\infty \leqslant z \leqslant \infty$). It is possible to show that (see, for example [13]),

$$\left.\begin{aligned} A_z^e &= \frac{i}{4} E_{0z} I \cos\alpha, & A_x^e &= A_y^e = 0, \\ A_y^m &= -\frac{i}{4} E_{0z} I, & A_x^m &= A_z^m = 0, \end{aligned}\right\} \tag{1.6.2}$$

where

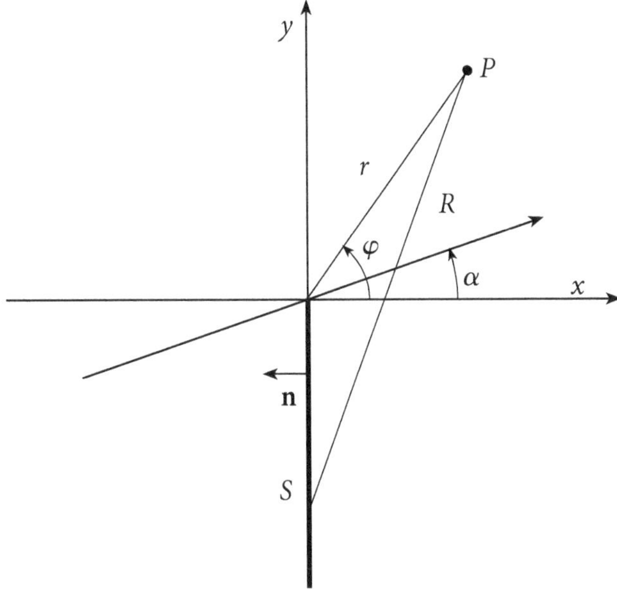

Figure 1.6.1: Diffraction from a black half-plane $x = 0, -\infty \leq y \leq 0$.

$$I = \frac{i}{\pi} \int_{-\infty}^{\infty} \frac{e^{i(v|x|+wy)}}{v(w-\frown{p})} dw, \tag{1.6.3}$$

$$p = k\sin\alpha, \quad v = \sqrt{k^2 - w^2}, \quad \Im\mathfrak{m}\{v\} \geqslant 0, \quad v = k \text{ for } w = 0.$$

and the $\frown$ symbol indicates the contour of integration skirts above the pole at $w = p$. In addition the branch points at $w = \pm k$ will be displaced from the real axis, making $\Im\mathfrak{m}\{k\} > 0$. In the right half space ($x > 0$), Eqn. (1.6.3) may be written as

$$I = \frac{1}{ik\pi} J + \begin{cases} 0 & \text{for} \quad \varphi > \alpha \\ \dfrac{2}{k\cos\alpha} e^{ikr\cos(\alpha-\varphi)} & \text{for} \quad \varphi < \alpha \end{cases}, \tag{1.6.4}$$

where

$$J = \int_{0^- + i\infty}^{0^+ - i\infty} \frac{e^{ikr\cos\zeta}\, d\zeta}{\sin\alpha - \sin(\zeta + \varphi)}. \tag{1.6.5}$$

Here and throughout this book, we will use the notation $x^+ = \lim(x+\epsilon)$ and $x^- = \lim(x-\epsilon)$ where ϵ is a positive value tending to zero ($\epsilon \to 0$). Utilizing a method proposed by Pauli [36], one can obtain the following asymptotic expression for J when $kr \gg 1$

$$J = 2\sqrt{\pi} e^{i\pi/4} \frac{e^{ikr\cos(\alpha-\varphi)}}{\cos[(\alpha+\varphi)/2]} \int_{\infty \sin[(\alpha-\varphi)/2]}^{\sqrt{2kr}\sin[(\alpha-\varphi)/2]} e^{it^2}\, dt. \tag{1.6.6}$$

The remaining terms of the asymptotic expansion are on the order of $(kr)^{-1/2}$ and less in comparison with Eqn. (1.6.6). Some details of Pauli's method are presented below in §5.4.

As a result, the total field in the right half-space ($x > 0, kr \gg 1$) is equal to

$$\mathbf{E} = \boldsymbol{\mathcal{E}}^{(d)} + \begin{cases} \mathbf{E}_0 & \text{for} \quad \varphi > \alpha \\ 0 & \text{for} \quad \varphi < \alpha \end{cases}, \tag{1.6.7}$$

$$\mathbf{H} = \boldsymbol{\mathcal{H}}^{(d)} + \begin{cases} \mathbf{H}_0 & \text{for} \quad \varphi > \alpha \\ 0 & \text{for} \quad \varphi < \alpha \end{cases}, \tag{1.6.8}$$

where $\boldsymbol{\mathcal{E}}^{(d)}$, $\boldsymbol{\mathcal{H}}^{(d)}$ is the diffracted component of the scattered field that does not include plane waves. In polar coordinates, $\boldsymbol{\mathcal{E}}^{(d)}$ and $\boldsymbol{\mathcal{H}}^{(d)}$ are comprised of the following components:

$$\left.\begin{aligned}
\mathcal{E}_z^{(d)} &= \left[2i\phi(r,\alpha-\varphi)\cos\alpha + \frac{e^{ikr}}{\sqrt{2kr}}\sin\frac{\alpha+\varphi}{2}\right]\beta,\\
\mathcal{H}_\varphi^{(d)} &= -\left[2i\phi(r,\alpha-\varphi)\cos\alpha\cos(\alpha-\varphi) + \frac{e^{ikr}}{\sqrt{2kr}}\sin\frac{3\alpha-\varphi}{2}\right]\beta,\\
\mathcal{H}_r^{(d)} &= 2\left[i\phi(r,\alpha-\varphi)\sin(\alpha-\varphi) - \frac{e^{ikr}}{\sqrt{2kr}}\cos\frac{\alpha-\varphi}{2}\right]\beta\cos\alpha,
\end{aligned}\right\} \tag{1.6.9}$$

$$\mathcal{E}_r^{(d)} = \mathcal{E}_\varphi^{(d)} = \mathcal{H}_z^{(d)} = 0, \tag{1.6.10}$$

where

$$\beta = \frac{E_{0z}e^{i\pi/4}}{2\sqrt{\pi}\cos\dfrac{\alpha+\varphi}{2}}, \quad \phi(r,\psi) = e^{ikr\cos\psi}\int_{\infty\sin(\psi/2)}^{\sqrt{2kr}\sin(\psi/2)} e^{it^2}\,dt. \tag{1.6.11}$$

Far from the shadow boundary (i.e., when $(kr)^{1/2}|\sin[(\alpha-\varphi)/2]| \gg 1$) this field appears as a cylindrical wave

$$\left.\begin{aligned}
\mathcal{E}_z^{(d)} &= -\mathcal{H}_\varphi^{(d)} = \frac{1}{2}E_{0z}\cot\frac{\alpha-\varphi}{2}\frac{e^{i(kr+\pi/4)}}{\sqrt{2\pi kr}},\\
\mathcal{E}_r^{(d)} &= \mathcal{E}_\varphi^{(d)} = \mathcal{H}_r^{(d)} = \mathcal{H}_z^{(d)} = 0,
\end{aligned}\right\} \tag{1.6.12}$$

the amplitude of which is independent of the orientation of the half-plane relative to the angle of incidence of the plane wave and is determined only by the value of $\alpha-\varphi$. This is the principal difference between a wave scattered by a black half-plane, from the diffracted wave,

$$\mathcal{E}_z^{(d)} = -\mathcal{H}_\varphi^{(d)} = \frac{1}{2} E_{0z} \frac{\cos\dfrac{\alpha+\varphi}{2} - \sin\dfrac{\alpha-\varphi}{2}}{\cos\dfrac{\alpha+\varphi}{2}\sin\dfrac{\alpha-\varphi}{2}} \frac{e^{i(kr+\pi/4)}}{\sqrt{2\pi kr}},$$

scattered by a perfectly conducting half-plane (see, for example, [13]).

On the shadow boudary (for $\varphi = \alpha$), Eqn. (1.6.9) indicates that

$$\mathcal{E}_z^{(d)} = \begin{cases} -\dfrac{1}{2} E_{0z} e^{ikr} & \text{for} \quad \varphi = \alpha^+, \\ +\dfrac{1}{2} E_{0z} e^{ikr} & \text{for} \quad \varphi = \alpha^-. \end{cases} \tag{1.6.13}$$

Thus, the total field is continuous across the shadow boundary and at the boundary it is equal to one-half the incident wave.

The field for left half-space ($x < 0$) will now be examined. In this case, Eqn. (1.6.3) becomes

$$I = \frac{1}{ik\pi} \int_{0^- + i\infty}^{0^+ - i\infty} \frac{e^{ikr\cos\zeta} d\zeta}{\sin\alpha + \sin(\zeta - \varphi)} + \begin{cases} 0 \quad \text{for} \quad \varphi < \pi - \alpha, \\ \dfrac{2}{k\cos\alpha} e^{-ikr\cos(\alpha+\varphi)} \quad \text{for} \quad \varphi > \pi - \alpha. \end{cases} \tag{1.6.14}$$

The second term in Eqn. (1.6.14) is a reflected plane wave. This means that the equivalent electric and magnetic currents, flowing on the black half-plane create the reflected plane waves. The field of these waves is equal to

$$\boldsymbol{\mathcal{E}}_0 = \boldsymbol{\mathcal{E}}_0^e + \boldsymbol{\mathcal{E}}_0^m, \quad \boldsymbol{\mathcal{H}}_0 = \boldsymbol{\mathcal{H}}_0^e + \boldsymbol{\mathcal{H}}_0^m, \tag{1.6.15}$$

where

$$\left.\begin{aligned} \boldsymbol{\mathcal{E}}_0^e &= -\frac{1}{ik}\left(\nabla\nabla\cdot\mathbf{A}_S^e + k^2\mathbf{A}_S^e\right), \quad \boldsymbol{\mathcal{E}}_0^m = -\nabla\times\mathbf{A}_0^m, \\ \boldsymbol{\mathcal{H}}_0^e &= \nabla\times\mathbf{A}_S^e, \quad \boldsymbol{\mathcal{H}}_0^m = -\frac{1}{ik}\left(\nabla\nabla\cdot\mathbf{A}_S^m + k^2\mathbf{A}_S^m\right), \end{aligned}\right\} \tag{1.6.16}$$

and the values of $\mathbf{A}_S^e$ and $\mathbf{A}_S^m$ are determined according to Eqn. (1.6.2) and the second term of Eqn. (1.6.14). It is possible to show that

$$\left.\begin{aligned} \mathcal{E}_{0z}^e &= -\mathcal{E}_{0z}^m = -\frac{1}{2}E_{0z}e^{-ikr\cos(\alpha+\varphi)}\,, \\ \mathcal{H}_{0\varphi}^e &= -\mathcal{H}_{0\varphi}^m = -\frac{1}{2}E_{0z}\cos(\alpha+\varphi)e^{-ikr\cos(\alpha+\varphi)}\,, \\ \mathcal{H}_{0r}^e &= -\mathcal{H}_{0r}^m = -\frac{1}{2}E_{0z}\sin(\alpha+\varphi)e^{-ikr\cos(\alpha+\varphi)}\,, \end{aligned}\right\} \tag{1.6.17}$$

$$\mathcal{H}_{0z}^e = \mathcal{H}_{0z}^m = \mathcal{E}_{0\varphi}^e = \mathcal{E}_{0\varphi}^m = \mathcal{E}_{0r}^e = \mathcal{E}_{0r}^m = 0\,. \qquad \text{(1.6.17, cont.)}$$

It follows that the reflected plane waves being separately radiated by the electric and magnetic currents completely cancel one another:

$$\boldsymbol{\mathcal{E}}_0^e = -\boldsymbol{\mathcal{E}}_0^m\,,\quad \boldsymbol{\mathcal{H}}_0^e = -\boldsymbol{\mathcal{H}}_0^m \tag{1.6.18}$$

and

$$\boldsymbol{\mathcal{E}}_0 = \boldsymbol{\mathcal{H}}_0 = 0\,. \tag{1.6.19}$$

As a result, the scattered field $\boldsymbol{\mathcal{E}}$, $\boldsymbol{\mathcal{H}}$ in left half-space ($x < 0$) is determined only by the first term in Eqn. (1.6.14), and is also equal to Eqn. (1.6.12) when $(kr)^{1/2}|\cos[(\alpha+\varphi)/2]| \gg 1$. When $\varphi = \pi-\alpha$, which is the boundary of the reflected wave in the case of a perfectly conducting half-plane, the field scattered by the black half-plane is equal to

$$\mathcal{E}_z = -\mathcal{H}_\varphi = -\frac{1}{2}E_{0z}\tan\alpha\frac{e^{i(kr+\pi/4)}}{\sqrt{2\pi kr}}\,. \tag{1.6.20}$$

For the case of diffraction of the plane wave

$$H_z = H_{0z}e^{ik(x\cos\alpha+y\sin\alpha)}\,,\quad E_z = 0\,, \tag{1.6.21}$$

a black half-plane creates a scattered field in the region $kr \gg 1$ that can be described by Eqns. (1.6.9) and (1.6.12) if we replace E_z with H_z, $-H_{\varphi,r}$ with $E_{\varphi,r}$ and E_{0z} with H_{0z}. In other words, the amplitude of the field scattered by a black half-plane is independent of the polarization of the incident wave.

In conclusion, it should be noted that the solution for the black half-plane, according to the shadow contour theorem, is also a solution to the diffraction of a plane wave from a black arbitrary cylindrical surface occupying the half-space $-\infty \leqslant y \leqslant 0$.

§1.7 Black Strip and Black Disk

Let a plane wave given by

$$\mathbf{E} = \mathbf{E}_0 e^{ik(x\cos\alpha + y\sin\alpha)}, \quad \mathbf{H} = \mathbf{H}_0 e^{ik(x\cos\alpha + y\sin\alpha)}, \tag{1.7.1}$$

illuminate a black strip of infinite length (Fig. 1.7.1). The scattered field is determined by Eqns. (1.3.8) and (1.3.9). It is possible to show that the far field ($r \gg ka^2$) is equal to

$$\left.\begin{aligned} \mathcal{E}_z = -\mathcal{H}_\varphi = E_{0z}(\cos\alpha + \cos\varphi)\frac{\sin[ka(\sin\alpha - \sin\varphi)]}{\sin\alpha - \sin\varphi}\frac{e^{i(kr+3\pi/4)}}{\sqrt{2\pi kr}}, \\ \mathcal{E}_\varphi = \mathcal{H}_z = H_{0z}(\cos\alpha + \cos\varphi)\frac{\sin[ka(\sin\alpha - \sin\varphi)]}{\sin\alpha - \sin\varphi}\frac{e^{i(kr+3\pi/4)}}{\sqrt{2\pi kr}}, \end{aligned}\right\} \tag{1.7.2}$$

In the direction of the main shadow lobe ($\varphi = \alpha$) we find that

$$\left.\begin{aligned} \mathcal{E}_z = -\mathcal{H}_\varphi = 2kaE_{0z}\cos\alpha\frac{e^{i(kr+3\pi/4)}}{\sqrt{2\pi kr}}, \\ \mathcal{E}_\varphi = \mathcal{H}_z = 2kaH_{0z}\cos\alpha\frac{e^{i(kr+3\pi/4)}}{\sqrt{2\pi kr}}, \end{aligned}\right\} \tag{1.7.3}$$

which is equivalent to a field scattered by a perfectly conducting strip (for the condition $ka \gg 1, |\alpha| \ll \pi/2$). However, a black strip, in contrast to a perfectly conducting strip, does not radiate either in the direction of specular reflection ($\varphi = \pi - \alpha$) or in the direction of the source ($\varphi = \pi + \alpha$). Furthermore, Eqn. (1.7.2) can be used to show that the intensity of the field scattered from the black strip is independent of the polarization of the incident wave.

With the aid of Eqn. (1.7.2) and the relationships

$$\left.\begin{aligned} \mathcal{E}_z = -\mathcal{H}_\varphi = E_{0z} ka\, e(\alpha,\varphi)\sqrt{\frac{2}{\pi kr}}e^{i(kr+3\pi/4)}, \\ \mathcal{E}_\varphi = \mathcal{H}_z = H_{0z} ka\, h(\alpha,\varphi)\sqrt{\frac{2}{\pi kr}}e^{i(kr+3\pi/4)}, \end{aligned}\right\} \tag{1.7.4}$$

the values of $e(\alpha, \varphi)$ and $h(\alpha, \varphi)$, which characterize the scattering pattern of a black strip, were numerically calculated. Calculations were performed for $ka = \sqrt{28}$ with $\alpha = 0°$ and $-45°$. These results were presented in Figs. 1.7.2 and 1.7.3 using a solid line. For comparison, graphs of $e_0(\alpha, \varphi)$ (dashed curve) and $h_0(\alpha, \varphi)$ (dash-dot curve) show (within the physical optics approximation) the scattering pattern for a perfectly conducting strip. It can be seen that in the region of the main shadow lobe, the black strip radiates in a manner identical to the perfectly conducting strip. It is very interesting to note that in some other directions, the black strip radiates energy more than the perfectly conducting strip.

For a plane wave

$$\mathbf{E} = \mathbf{E}_0 e^{ik(z\cos\gamma + y\sin\gamma)}, \quad \mathbf{H} = \mathbf{H}_0 e^{ik(z\cos\gamma + y\sin\gamma)}, \tag{1.7.5}$$

diffracting from a black disk (see Fig. 1.7.4), the scattered field in the far zone ($r \gg ka^2$) is

$$\begin{aligned} \mathcal{E}_\varphi = -\mathcal{H}_\vartheta = -\frac{ia}{2}\big[E_{0x}(\cos\gamma + \cos\vartheta)\sin\varphi + \\ + H_{0x}(1 + \cos\gamma\cos\vartheta)\cos\varphi\big]\mathcal{V}, \end{aligned} \tag{1.7.6}$$

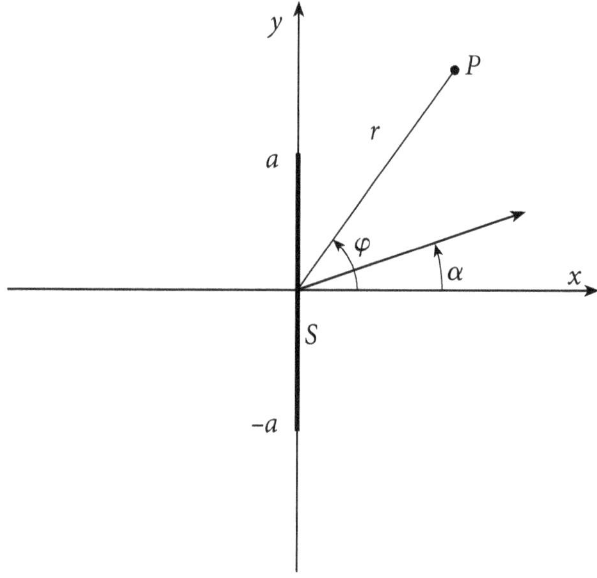

Figure 1.7.1: Segment of the y-axis ($-a \le y \le a$) depicting the cross section of a strip in the $x = 0$ plane, where α is the angle of incidence. Reprinted with the permission of Kluwer Academic/Plenum Publishers.

$$\mathcal{E}_\vartheta = \mathcal{H}_\varphi = \frac{ia}{2}\left[E_{0x}(1+\cos\gamma\cos\vartheta)\cos\varphi - \right.$$

$$\left. -H_{0x}(\cos\gamma + \cos\vartheta)\sin\varphi\right]\mathcal{V}, \tag{1.7.6, cont.}$$

where

$$\mathcal{V} = \frac{J_1\left(ka\sqrt{\lambda^2+\mu^2}\right)}{\sqrt{\lambda^2+\mu^2}}\frac{e^{ikR}}{R}, \tag{1.7.7}$$

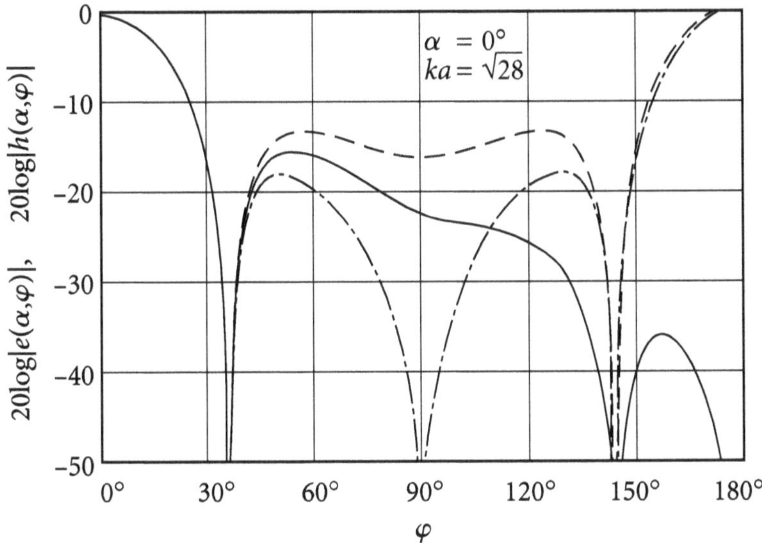

Figure 1.7.2: Field scattered by a strip. The solid line depicts the field scattered by a black strip. The dashed and dashed-dotted curves correspond to the field scattered by a perfectly conducting strip for different polarizations of the incident wave.

$$\left.\begin{aligned} \lambda &= -\sin\vartheta\cos\varphi, \\ \mu &= \sin\gamma - \sin\vartheta\sin\varphi, \end{aligned}\right\} \qquad (1.7.7,\ \text{cont.})$$

and J_1 is the Bessel function. In the direction of the main shadow lobe ($\varphi = \pi/2$, $\vartheta = \gamma$) the scattered field is equal to

$$\left.\begin{aligned} \mathcal{E}_\varphi &= -\mathcal{H}_\vartheta = -\frac{ika^2}{2} E_{0x}\cos\gamma\frac{e^{ikR}}{R}, \\ \mathcal{E}_\vartheta &= \mathcal{H}_\varphi = -\frac{ika^2}{2} H_{0x}\cos\gamma\frac{e^{ikR}}{R}, \end{aligned}\right\} \qquad (1.7.8)$$

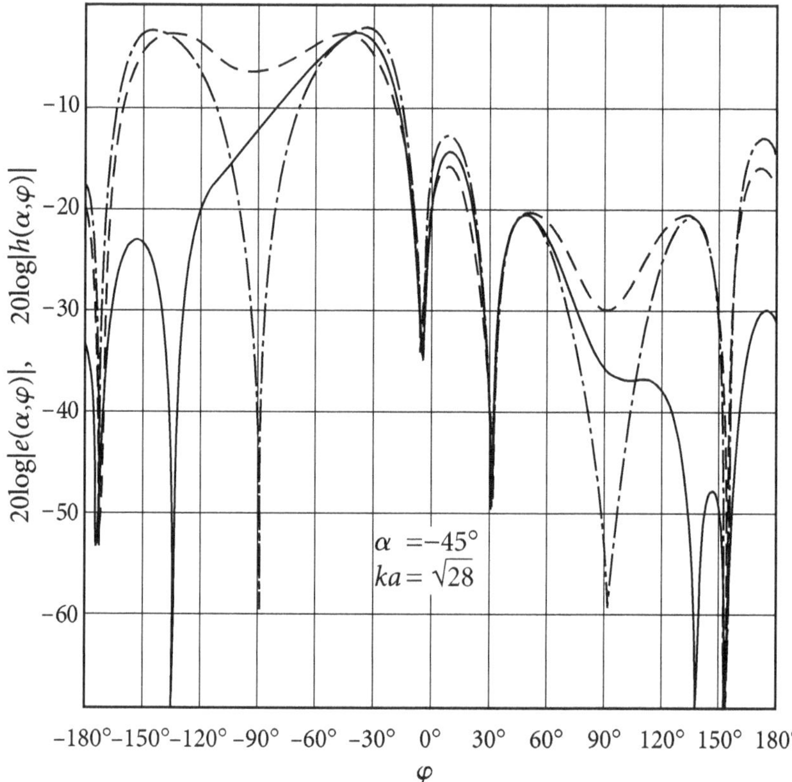

Figure 1.7.3: Field scattered by a strip. The solid line depicts the field scattered by a black strip. The dashed and dashed-dotted curves correspond to the field scattered by a perfectly conducting strip for different polarizations of the incident wave.

which agrees with the expression for the field diffracted by a perfectly conducting disk.

From Eqn. (1.7.6) it is easy to see that in the direction of specular reflection ($\vartheta = \pi - \gamma, \varphi = \pi/2$) and in the direction of the source ($\vartheta = \pi - \gamma$, $\varphi = -\pi/2$) the field scattered by the black disk is equal to zero. It can be shown that *in these directions the field scattered by a black plate of arbitrary shape is zero.*

In the plane of incidence ($\varphi = \pm\pi/2$) the scattered field is equal to

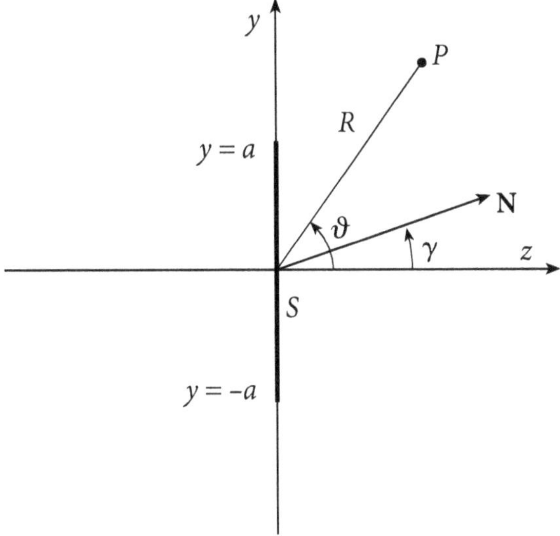

Figure 1.7.4: Cross section of a disk in the *yoz* plane. **N** depicts the direction of propagation of the incident wave. Reprinted with the permission of Kluwer Academic/Plenum Publishers.

$$
\left.
\begin{aligned}
\mathcal{E}_\varphi &= -\mathcal{H}_\vartheta = -\frac{ia}{2}E_{0x}(\cos\gamma + \cos\vartheta)\sin\varphi \frac{J_1(ka\mu)}{\mu}\frac{e^{ikR}}{R}, \\
\mathcal{E}_\vartheta &= \mathcal{H}_\varphi = -\frac{ia}{2}H_{0x}(\cos\gamma + \cos\vartheta)\sin\varphi \frac{J_1(ka\mu)}{\mu}\frac{e^{ikR}}{R},
\end{aligned}
\right\} \tag{1.7.9}
$$

and the bistatic radar cross section is equal to

$$
\sigma = \pi a^2 \left| (\cos\gamma + \cos\vartheta)\frac{J_1(ka\mu)}{\mu} \right|^2 \tag{1.7.10}
$$

and does not depend on the polarization of the incident wave.

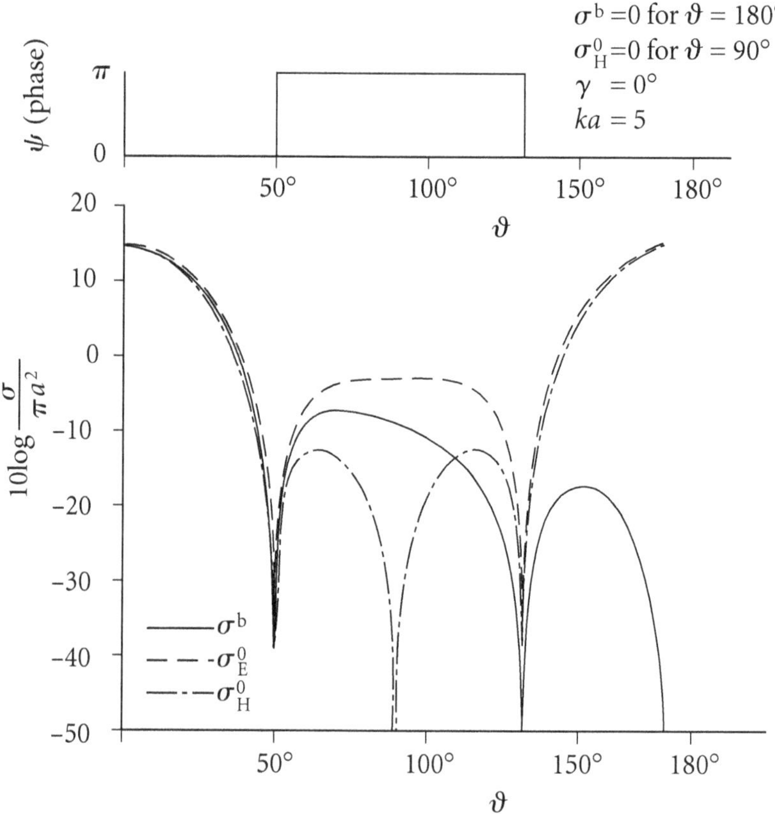

Figure 1.7.5: Scattering pattern produced by plane wave diffraction from a disk. for $\gamma = 0°$ and $ka = 5$.

This equation was used to numerically calculate the bistatic radar cross section of a circular plate for $ka = 5$. Calculations were performed for the scattered field in the incidence plane ($\varphi = \pi/2$). The incident angle γ was chosen to be equal to the values of 0° and 45°. The results of these calculations are presented in Figs. 1.7.5 and 1.7.6 using solid lines. These figures also display the phase ψ of the scattered field defined by

$$E_x = -i|E_x|e^{ikR}e^{i\psi}, \quad H_x = -i|H_x|e^{ikR}e^{i\psi}. \tag{1.7.11}$$

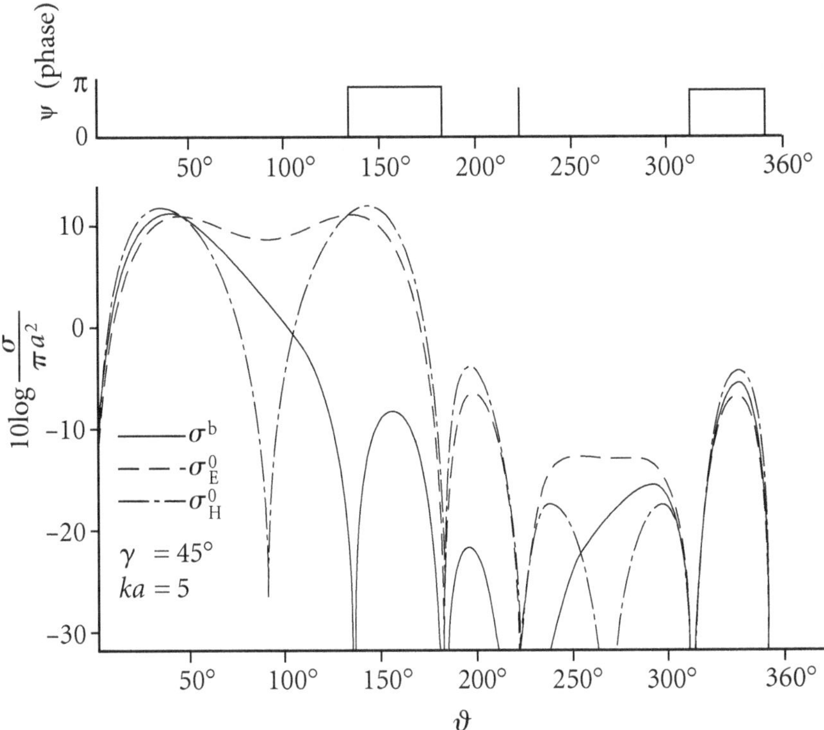

Figure 1.7.6: Scattering pattern produced by plane wave diffraction from a disk for $\gamma = 45°$ and $ka = 5$.

As can be seen from these figures, the phase is discontinuous (at π) as we traverse across each minimum. The narrow spike in the phase of Fig. 1.7.6 corresponds to a null that lies below the lower boundary of the graph.

For comparison, these results are presented with the bistatic radar cross section (within the physical optics approximation) σ_E^0 and σ_H^0 for a perfectly conducting disk. The functions σ_E^0 and σ_H^0 correspond to E-polarized ($H_{0x} = 0$) and H-polarized ($E_{0x} = 0$) incident waves, respectively. In directions other than the specularly reflected ray and the source, σ effectively duplicates the behavior of σ_E^0 and σ_H^0, closely resembling them near the shadow lobe. In some directions the black disk is more brilliant than the conducting disk.

Finally, note that in relation to the shadow contour theorem, the results obtained here have a broader meaning. The equations for the field scattered by a black strip also determine the field scattered by an infinitely long black cylinder of arbitrary cross section. The solution for a black disk is valid for an arbitrary black body with a shadow contour in the form of a circle. Here, specifically, are the cases of a black sphere and black body of revolution (for an incident plane wave propagating along the axis of symmetry).

§1.8 Physical Model of a Black Body

Above we researched the abstract model of a black body. Now, an example of a physical model of such a body will be studied. With this goal, let us examine the diffraction of the plane wave

$$E_x = -H_y = e^{-ikz} \tag{1.8.1}$$

from an isotropic inhomogeneous sphere with radius a and electromagnetic parameters ϵ and μ that are independent of ϑ and φ (Fig. 1.8.1). Using separation of variables, it is possible to show that the scattered far field ($r \gg ka^2$) is given by

$$\begin{aligned} E_\varphi = -H_\vartheta &= \frac{e^{ikr}}{kr}\sum_1^\infty \frac{2n+1}{n(n+1)}(-1)^n \times \\ &\times \left[-\frac{a_n}{\sin\vartheta}\frac{\partial P_n(\cos\vartheta)}{\partial\vartheta} + b_n \frac{\partial^2 P_n(\cos\vartheta)}{\partial\vartheta^2} \right]\sin\varphi, \\ E_\vartheta = H_\varphi &= \frac{e^{ikr}}{kr}\sum_1^\infty \frac{2n+1}{n(n+1)}(-1)^n \\ &\times \left[a_n \frac{\partial^2 P_n(\cos\vartheta)}{\partial\vartheta^2} - \frac{b_n}{\sin\vartheta}\frac{\partial P_n(\cos\vartheta)}{\partial\vartheta} \right]\cos\varphi, \\ E_r = H_r &= 0, \end{aligned} \tag{1.8.2}$$

where P_n is the Legendre polynomial of the first kind,

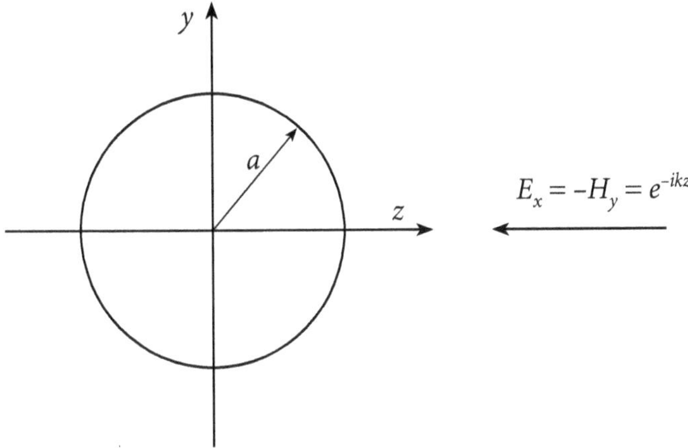

Figure 1.8.1: Cross section of a sphere in the *yoz* plane depicting a plane wave incident along the *z*-axis.

$$\left.\begin{aligned} a_n &= \frac{\psi_n(ka)u_n'(ka) - \epsilon(a)u_n(ka)\psi_n'(ka)}{\epsilon(a)u_n(ka)\zeta_n'(ka) - u_n'(ka)\zeta_n(ka)}, \\ b_n &= \frac{\psi_n(ka)v_n'(ka) - \mu(a)\psi_n'(ka)v_n(ka)}{\mu(a)v_n(ka)\zeta_n'(ka) - v_n'(ka)\zeta_n(ka)}, \end{aligned}\right\} \tag{1.8.3}$$

$$\psi_n(ka) = \sqrt{\frac{\pi ka}{2}} J_{n+1/2}(ka), \quad \zeta_n(ka) = \sqrt{\frac{\pi ka}{2}} H^{(1)}_{n+1/2}(ka), \tag{1.8.4}$$

and $J_{n+1/2}$ and $H^{(1)}_{n+1/2}$ are the Bessel function and Hankel function of the first kind, respectively. The prime represents the derivative with respect to the argument. The functions u_n and v_n are solutions to the differential equations

$$\epsilon \frac{\partial}{\partial \rho}\left[\frac{1}{\epsilon}\frac{\partial}{\partial \rho} u_n(\rho)\right] + \left[\epsilon\mu - \frac{n(n+1)}{\rho^2}\right] u_n(\rho) = 0, \quad \text{for } \rho = kr \tag{1.8.5}$$

$$\mu\frac{\partial}{\partial\rho}\left[\frac{1}{\mu}\frac{\partial}{\partial\rho}v_n(\rho)\right]+\left[\epsilon\mu-\frac{n(n+1)}{\rho^2}\right]v_n(\rho)=0, \quad \text{for } \rho=kr \qquad (1.8.5, \text{cont.})$$

and correspond to ψ_n when ϵ = const and μ = const.

Utilizing the properties of the Legendre polynomial

$$\left.\begin{aligned}\lim_{\vartheta\to 0}\frac{\partial^2 P_n(\cos\vartheta)}{\partial\vartheta^2}&=\lim_{\vartheta\to 0}\frac{1}{\sin\vartheta}\frac{\partial P_n(\cos\vartheta)}{\partial\vartheta}=-\frac{n(n+1)}{2},\\ \lim_{\vartheta\to\pi}\frac{\partial^2 P_n(\cos\vartheta)}{\partial\vartheta^2}&=-\lim_{\vartheta\to\pi}\frac{1}{\sin\vartheta}\frac{\partial P_n(\cos\vartheta)}{\partial\vartheta}=-(-1)^n\frac{n(n+1)}{2},\end{aligned}\right\} \qquad (1.8.6)$$

and Eqn. (1.8.2) we find that

$$E_x=H_y=-\frac{e^{ikr}}{kr}\sum_{n=1}^{\infty}(-1)^n\frac{2n+1}{2}(a_n-b_n) \quad \text{for} \quad \vartheta=0 \qquad (1.8.7)$$

and

$$E_x=-H_y=\frac{e^{ikr}}{kr}\sum_{n=1}^{\infty}\frac{2n+1}{2}(a_n+b_n) \quad \text{for} \quad \vartheta=\pi \qquad (1.8.8)$$

The value $\vartheta = 0$ corresponds to the direction to the source, and the value $\vartheta = \pi$ corresponds to the main shadow lobe of the diffraction pattern. Equation (1.8.7) shows that when $\epsilon(r) = \mu(r)$, the field scattered in the direction of the source is equal to zero. In other words, a sphere with parameters $\epsilon(r) = \mu(r)$ is a black sphere. The same result was established by a different method in [37]. It is evident that the magnitude of a wave reflected by any body can be significantly reduced if this body is surrounded by an absorbing sphere with slowly changing parameters $\epsilon(r) = \mu(r)$. The effect of such a decrease in amplitude is even for the case $\epsilon(r) \neq \mu(r)$, when $\mu(r) = 1$ and $\epsilon(r)$ varies sufficiently slowly [38].

§1.9 Observation by M. L. Levin

While examining this material, Dr. M. L. Levin made an interesting observation. He found that the field backscattered from *arbitrary bodies of revolution* with $\epsilon = \mu$ is equal to zero, when these bodies are excited by the wave incident along their axes of symmetry. He presented the following elegant proof based on the symmetry of the problem and duality of Maxwell's equations.

Let the incident plane wave $\mathbf{E}^0, \mathbf{H}^0$ propagate in free space toward a body of revolution along its axis of symmetry. It should be recalled that a Gaussian system of units is used. In this system, $\epsilon = \mu = 1$ for the free space and the moduli of electric and magnetic vectors of the plane wave are equal to each other ($E^0 = H^0$). The scattered far field will be denoted by $\mathbf{E}^S, \mathbf{H}^S$. Note that in this field, $E^S = H^S$. The scattering problem is illustrated in Fig. 1.9.1. We assume that inside the scattering object the quantities ϵ and μ are axisymmetrical and equal to each other ($\epsilon = \mu$).

Using the duality of Maxwell's equations, we make the following substitutions: $\mathbf{E} \to -\mathbf{H}, \mathbf{H} \to \mathbf{E}, \epsilon \to \mu, \mu \to \epsilon$. As a result, we obtain the problem shown in Fig. 1.9.2. Here, $E_1^0 = H^0, H_1^0 = -E^0, E_1^S = H^S, H_1^S = -E^S$.

We will now rotate E_1^0 and H_1^0 by 90° and take into account that inside the scattering object the quantities ϵ and μ are axisymmetrical. Due to this symmetry, we obtain the situation shown in Fig. 1.9.3. Here, $E_2^0 = E^0$, $H_2^0 = H^0, E_2^S = -E^S, H_2^S = -H^S$. However, because the incident wave and the

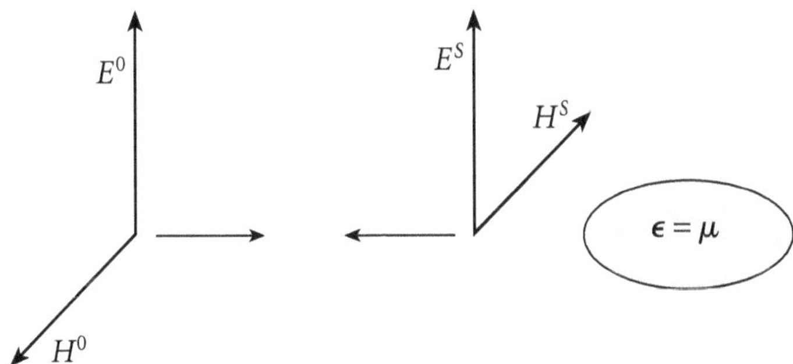

Figure 1.9.1: Field scattered by a plane wave incident upon a black body with $\epsilon = \mu$.

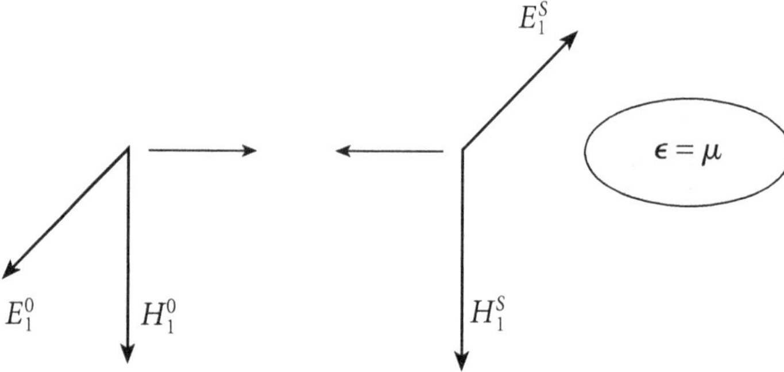

Figure 1.9.2: Field scattered by a plane wave incident upon a black body with $\epsilon = \mu$ (dual of Fig. 1.9.1).

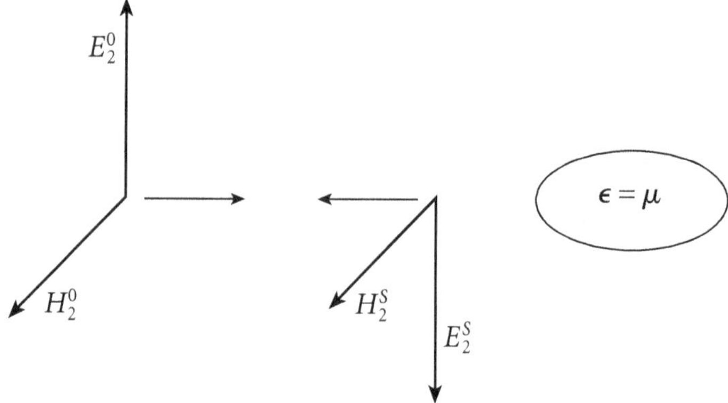

Figure 1.9.3: Field scattered by a plane wave incident upon a black body with $\epsilon = \mu$ (vector rotation of Fig. 1.9.2 by 90°).

scattering objects are the same as those in Fig. 1.9.1, we must have the same scattered field, i.e., $E_2^S = E^S$, $H_2^S = H^S$. Thus, we obtain two sets of opposite equalities: $E_2^S = -E^S$, $H_2^S = -H^S$, and $E_2^S = E^S$, $H_2^S = H^S$. They can be consistent only in the case when $E_2^S = 0$, and $H_2^S = 0$. This conclusion completes the proof of the above statement.

§1.10 Fundamental Properties of Scattering from Black Bodies

- With the aid of the vector analog of the Helmholtz theorems, we generalized the well known Kirchhoff-Kottler theory for volumetric black bodies.
- The field scattered by large black bodies can be interpreted as shadow radiation because it concentrates mainly in the vicinity of the shadow lobe. The well known phenomenon of forward scattering is actually the physical manifestation of this radiation.
- The shadow contour theorem was proved. This theorem shows that during the excitation of a black body by an arbitrary electromagnetic wave, the scattered field does not depend on its whole shape and is determined only by the shadow contour (the border between the illuminated and shadowed portions of the body).
- It was established that the total scattering cross-section for black bodies is half the total scattering cross-section of metallic bodies with identical shadow contours. This result is of fundamental importance. It shows that even with application of perfectly absorbing coatings on a metallic object, its total scattered power can be reduced solely by factor of two. This means that against bistatic radar, it is impossible to completely mask the scattering object by any radar absorbing materials.
- A uniform asymptotic expression, valid for an arbitrary observation angle, was developed for the field scattered by a black half-plane.
- The reflected plane wave vanishes for the case of a black half-plane. It was shown that the electric and magnetic currents excited on the surface of a black half-plane create separately reflected plane waves that cancel one another.
- The scattering pattern of a black disk was examined. In the specular direction and in the direction of the source, the black disk does not radiate. In the shadow region, the scattering pattern for a black disk is similar to the scattering pattern of a perfectly conducting disk.
- In accordance with shadow contour theorem, the equations obtained for the scattering from a black disk and black strip are also solutions to the scattering of a plane wave from an arbitrary black body possessing a shadow contour in the form of a circle or two parallel lines.

- An example of physical model of a black body is provided. It was rigorously shown that a sphere with parameters $\epsilon(r) = \mu(r)$ produces no scattered field in the direction of the source. As shown by M.L. Levin, this result is valid for arbitrary bodies of revolution illuminated by radar along their axis of symmetry.

- Shadow radiation cannot be eliminated by any radar absorbing materials. Therefore, it can be used, in particular, for the detection of stealth objects designed to evade detection by monostatic radar.

- Additional results related to the scattering from black bodies are presented in [25, 39–41,150].

- In the framework of the generalized Kirchhoff-Kottler theory presented in this chapter, it was shown (pp. 18–19 in [25]) that the backscattered field from an arbitrary black body is equal to zero.

- The shadow radiation concept is developed further in Section VI of [40]. The first-order and second-order types of the shadow radiation are introduced there. The first-order shadow radiation is concentrated near the shadow boundary behind an opaque scattering object. Actually, this is well known forward scattering. The second-order shadow radiation is the result of transformation of the first-order shadow radiation into edge waves, creeping waves, and surface diffracted rays. This transformation results from transverse diffusion of the wave field and its interaction with the surface of the scattering object.

- A new mathematical model for a black screen is introduced in [41]. Called the Riemann black screen, it is located on the boundary between two neighboring Riemann spaces and consists of two facets. Together, these facets are not transparent to the wave field. But, each separately is transparent. The well known classical models of black screens are interpreted as special sets of the Riemann screens.

2

Edge Diffraction at Convex Perfectly Conducting Bodies: Elements of the Physical Theory of Diffraction

§2.1 Uniform and Nonuniform Currents

Beginning with this chapter, diffraction problems for perfectly conducting bodies will be investigated. These problems can be rigorously formulated using the Maxwell's equations and boundary conditions. However, their solutions in the analytic form can be found only for geometrically simple bodies, such as infinitely long cylinders, spheres, disks, etc. Besides, these solutions can be effectively used for numerical calculations only then when the dimensions of the scattering objects are smaller than a wavelength. For large objects which are much greater than a wavelength, rigorous solutions usually loose their practical value. Numerical methods of solving boundary value problems for large complex objects are also ineffective. That is why significant attention in diffraction theory is given to approximate methods that can analyze scattering from large complex objects at high frequencies.

One of the methods that is applied to a wide class of problems is the physical optics (PO) approximation [196], which can be viewed as the generalization of Kirchhoff-Kottler theory for perfectly conducting bodies. According to this method, the current density induced by an incident wave on the illuminated portion of a body is

$$\mathbf{j}^0 = \frac{c}{2\pi}\mathbf{n}\times\mathbf{H}_0\,, \tag{2.1.1}$$

where c is the speed of light in a vacuum, $\mathbf{n}$ is the exterior surface normal, and $\mathbf{H}_0$ is the magnetic field of the incident wave. On the shadowed side of the body the surface current is equal to zero ($\mathbf{j} = 0$). Equation (2.1.1) implies that the current induced at each point of the illuminated surface is equal to the current that would be induced on an infinite plane tangent to the surface. The scattered field created by $\mathbf{j}^0$ is then determined using retarded vector potentials.

In this chapter and in Chapters 3, 4, and 7, we present some elements of the Physical Theory of Diffraction (PTD) in its original form [3–15, 21, 25]. This theory can be considered as a natural extension of the physical optics. The central and original idea of PTD is the separation of the surface current into the two components

$$\mathbf{j} = \mathbf{j}^0 + \mathbf{j}^1\,, \tag{2.1.2}$$

where $\mathbf{j}^1$ is the surface density of an additional current caused by the perturbation of the surface. By a perturbation we mean ***any deviation of the surface from an infinite plane*** (a smooth curve, a ridge, a boss, an aperture, etc.). If the body is convex and smooth and its dimension and radius of curvature are large in comparison with a wavelength, then this additional current is generally concentrated near the boundary between the illuminated and shadowed surfaces of the body. If this body possesses a sharp vertex or edge, then this additional current concentrates in their vicinity as well and radiates corner or edge waves.

Since the current excited by a plane wave on a perfectly conducting planar surface is distributed uniformly, (the absolute value of the surface current density is constant), then $\mathbf{j}^0$ can be referred to as the *uniform current*. The additional current $\mathbf{j}^1$ created by deformations in the surface of the body will be called the *nonuniform current*. The physical optics approximation takes into account only the uniform current. Therefore, in some cases, this method provides unsatisfactory results. A more accurate estimation of the scattered field requires the calculation of the radiation generated by the nonuniform current.

One should note that the uniform currents are often called the geometrical optics (or physical optics) currents and the nonuniform currents near sharp edges are often called ***fringe currents***.

It is pertinent to emphasize here that the PTD idea of the separation of the total current into uniform and nonuniform components has proved to be very effective, and it is widely used in the diffraction theory. For more information about its applications, one can see the references given in § 2.7. The modern form of PTD is presented in [61–66, 71–75].

In this chapter and in Chapter 3, the PTD is applied for the investigation of scattering from bodies of revolution with edges (sharp bends) on their surface. Linear dimensions of these bodies are assumed to be large compared to the wavelength. We proceed from the evident physical hypothesis that in a sufficiently small region near an edge the nonuniform current is approximately equal to that on a wedge. Having determined the edge wave being radiated by the nonuniform current, we can calculate the radar cross section for several bodies that are of practical interest.

Chapter 2 will examine the problem for bodies with an illuminated convex surface, and Chapter 3 will investigate the case when the illuminated region is concave. The fundamental results that are presented in Chapters 2 and 3 were previously published and reported in [8, 12, 13].

In § 2.2 an investigation of the field radiated by the nonuniform current excited on a wedge is conducted. These results are then utilized to calculate the edge waves scattered by a circumferential edge. Finally, §§ 2.4, 2.5, and 2.6 calculate the radar cross section of some specific bodies of revolution.

§2.2 Edge Waves Scattered by a Wedge

Let there be an electromagnetic plane wave

$$\begin{bmatrix} E_z \\ H_z \end{bmatrix} = \begin{bmatrix} E_{0z} \\ H_{0z} \end{bmatrix} e^{-ikr\cos(\varphi - \varphi_0)} \tag{2.2.1}$$

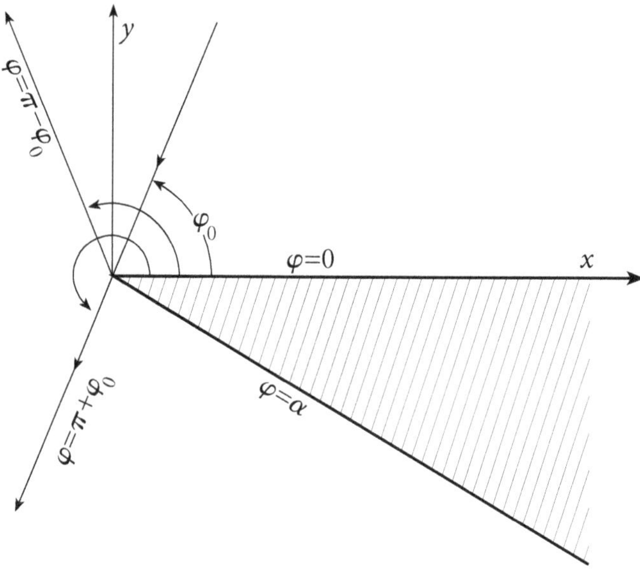

Figure 2.2.1: Wedge diffraction of a plane wave. The boundary of the reflected wave is given by $\varphi = \pi - \varphi_0$. The shadow boundary is given by $\varphi = \pi + \varphi_0$.

incident upon a conducting wedge located in free space. The wedge possesses an exterior angle α. All angles will be measured from the illuminated side of the wedge (Fig. 2.2.1).

This problem was first rigorously solved by Macdonald [42] and later by Sommerfeld [43]. Calculating the difference between the rigorous solution and the physical optics approximation, we can determine the field radiated by the nonuniform current. Omitting the mathematical steps found in reference [13], the final expressions are presented here. For $kr \gg 1$ the field is in the form of a cylindrical wave

$$E_z = -H_\varphi = E_{0z} f^1 \frac{e^{i(kr+\pi/4)}}{\sqrt{2\pi kr}}, \tag{2.2.2}$$

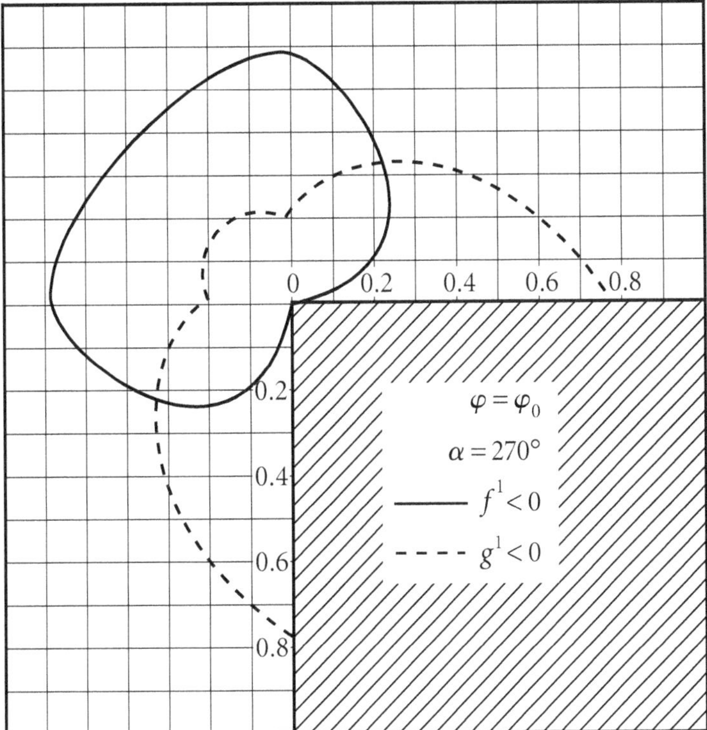

Figure 2.2.2: Directivity patterns of edge waves radiated by nonuniform (fringe) currents. The function f^1 (g^1) corresponds to the case of electric (magnetic) vector being parallel to the ridge of the wedge. Reprinted from [3] with the permission of *Zhurnal Tekhnicheskoi Fiziki*.

$$E_\varphi = H_z = H_{0z} g^1 \frac{e^{i(kr+\pi/4)}}{\sqrt{2\pi kr}}, \qquad (2.2.2, \text{cont.})$$

scattering from the tip of the wedge. Functions f^1 and g^1 depend only on φ, φ_0, and α and are

$$f^1 = f - f^0, \quad g^1 = g - g^0, \qquad (2.2.3)$$

where

$$\left.\begin{matrix} f(\varphi,\varphi_0,\alpha) \\ g(\varphi,\varphi_0,\alpha) \end{matrix}\right\} = \frac{1}{n}\sin\frac{\pi}{n}\times\left[\frac{1}{\cos\frac{\pi}{n}-\cos\frac{\varphi-\varphi_0}{n}} \mp \frac{1}{\cos\frac{\pi}{n}-\cos\frac{\varphi+\varphi_0}{n}}\right], \tag{2.2.4}$$

$$n=\frac{\alpha}{\pi}$$

and

$$f^0(\varphi,\varphi_0)=\frac{\sin\varphi_0}{\cos\varphi+\cos\varphi_0},\quad g^0(\varphi,\varphi_0)=\frac{-\sin\varphi}{\cos\varphi+\cos\varphi_0}, \tag{2.2.5}$$

if only one side of the wedge is illuminated (*i.e.*, $0\le\varphi_0\le\alpha-\pi$), and

$$\begin{aligned} f^0(\varphi,\varphi_0)&=\frac{\sin\varphi_0}{\cos\varphi+\cos\varphi_0}+\frac{\sin(\alpha-\varphi_0)}{\cos(\alpha-\varphi)+\cos(\alpha-\varphi_0)}, \\ g^0(\varphi,\varphi_0)&=-\frac{\sin\varphi}{\cos\varphi+\cos\varphi_0}-\frac{\sin(\alpha-\varphi)}{\cos(\alpha-\varphi)+\cos(\alpha-\varphi_0)}, \end{aligned} \tag{2.2.6}$$

if both sides of the wedge are illuminated ($\alpha-\pi<\varphi_0<\pi$). The functions f and g represent cylindrical waves radiated by the total current, i.e., the sum of the uniform and nonuniform currents, and the functions f^0 and g^0 are related to the cylindrical wave radiated only by the uniform current ($\mathbf{j}^0$).

Some of the properties of f^1 and g^1 should be noted. The function $f^1(\alpha,\varphi,\varphi_0)$ is continuous, whereas $g^1(\alpha,\varphi,\varphi_0)$ has a finite discontinuity at $\varphi_0=\alpha-\pi$. The reason for this discontinuity is that the uniform current is non-zero on the side along which the plane wave of Eqn. (2.2.1) propagates (for $\varphi_0=\alpha-\pi$). Under backscatter conditions, when the source and point of observation are co-located, ($\varphi=\varphi_0$), both f^1 and g^1 are continuous. The function g^1 has no discontinuity at $\varphi=\varphi_0=\alpha-\pi$, because the current element does not radiate in the longitudinal direction.

On the geometrical optics boundaries of the incident and reflected plane waves (i.e., for $\varphi=\pi+\varphi_0, \varphi=\pi-\varphi_0$, and $\varphi=2\alpha-\pi-\varphi_0$), the functions f, f^0 and g, g^0 approach infinity while f^1 and g^1 remain finite. The coo-responding relationships for f^1 and g^1 are given in Section 4 of [13] and are

$$\left.\begin{matrix} f^1 \\ g^1 \end{matrix}\right\} = \frac{\frac{1}{n}\sin\frac{\pi}{n}}{\cos\frac{\pi}{n} - \cos\frac{\varphi-\varphi_0}{n}} + \frac{1}{2}\cot\varphi_0 \pm \frac{1}{2n}\cot\frac{\pi}{n} \tag{2.2.7}$$

if $\varphi = \pi - \varphi_0$ and $\varphi_0 < \alpha - \pi$;

$$\left.\begin{matrix} f^1 \\ g^1 \end{matrix}\right\} = \frac{\frac{1}{n}\sin\frac{\pi}{n}}{\cos\frac{\pi}{n} - \cos\frac{\varphi-\varphi_0}{n}} + \frac{1}{2}\cot\varphi_0 \pm \frac{1}{2n}\cot\frac{\pi}{n} + \left\{\begin{matrix} \dfrac{-\sin(\alpha-\varphi_0)}{\cos(\alpha-\varphi)+\cos(\alpha-\varphi_0)} \\ \dfrac{\sin(\alpha-\varphi)}{\cos(\alpha-\varphi)+\cos(\alpha-\varphi_0)} \end{matrix}\right. \tag{2.2.8}$$

if $\varphi = \pi - \varphi_0$ and $\alpha - \pi < \varphi_0 < \pi$;

$$\left.\begin{matrix} f^1 \\ g^1 \end{matrix}\right\} = \mp \frac{\frac{1}{n}\sin\frac{\pi}{n}}{\cos\frac{\pi}{n} - \cos\frac{\varphi+\varphi_0}{n}} \pm \frac{1}{2}\cot\varphi_0 - \frac{1}{2n}\cot\frac{\pi}{n} \tag{2.2.9}$$

if $\varphi = \pi + \varphi_0$ and $\varphi_0 < \alpha - \pi$;

$$\left.\begin{matrix} f^1 \\ g^1 \end{matrix}\right\} = \frac{\frac{1}{n}\sin\frac{\pi}{n}}{\cos\frac{\pi}{n} - \cos\frac{\varphi-\varphi_0}{n}} + \frac{1}{2}\cot(\alpha-\varphi_0) \pm \frac{1}{2n}\cot\frac{\pi}{n} + \left\{\begin{matrix} \dfrac{-\sin\varphi_0}{\cos\varphi+\cos\varphi_0} \\ \dfrac{\sin\varphi}{\cos\varphi+\cos\varphi_0} \end{matrix}\right. \tag{2.2.10}$$

if $\varphi = 2\alpha - \pi - \varphi_0$ and $\alpha - \pi < \varphi_0 < \pi$.

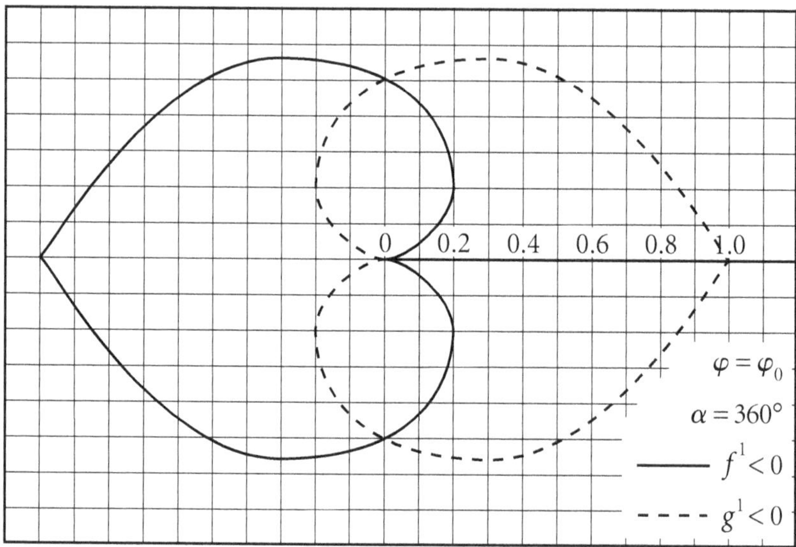

Figure 2.2.3: Directivity patterns of edge waves radiated by nonuniform (fringe) currents induced on a half-plane $\varphi = 0°$, $\varphi = 360°$. The function f^1 (g^1)corresponds to the case of electric (magnetic) vector being parallel to the edge of the half-plane. Reprinted from [3] with the permission of *Zhurnal Tekhnicheskoi Fiziki.*

An exception exists only for the case of grazing incidence ($\varphi_0 = \pi$ or $\varphi_0 = \alpha - \pi$), when the functions f^1 and g^1 are singular in the direction of the incident wave ($\varphi = 0$ or $\varphi = \alpha$) as is shown by Eqn. 4.04 in [13]. The reason for this singularity is that the field radiated by the current j^1 does not have a ray structure in the vicinity of the wedge faces, and therefore, cannot be represented in the form of a cylindrical wave there.

Plots of f^1 and g^1 (Figs. 2.2.2 and 2.2.3) provide a visual representation of the of edge waves radiated by nonuniform currents. In particular, they show that these waves significantly depend on the polarization of the incident wave.

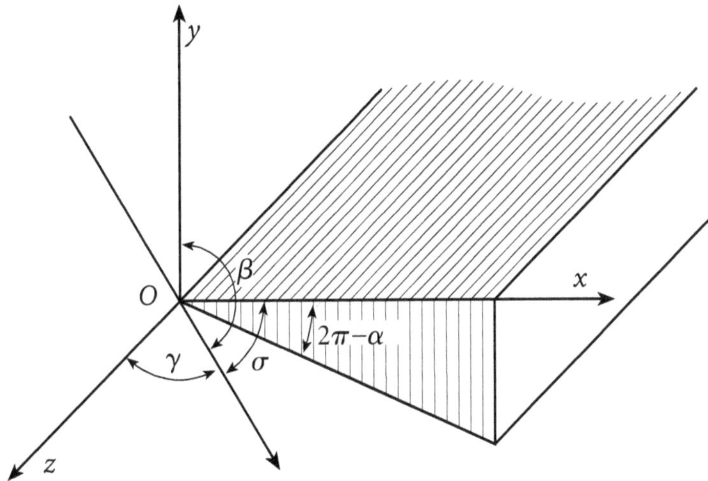

Figure 2.2.4: Oblique incidence of a plane wave illuminating a wedge. The angle γ is subtended between the direction of propagation and the z-axis.

When the incident wave is given by

$$\mathbf{E} = \mathbf{E}_0 e^{ik(x\cos\sigma + y\cos\beta + z\cos\gamma)} \tag{2.2.11}$$

forming an arbitrary angle γ ($0 < \gamma < \pi/2$) relative to the edge of the wedge subtending an angle α (Fig. 2.2.4), the field radiated by the nonuniform current is equal to (for $kr \gg 1$)

$$\left.\begin{aligned} E_z &= \left[H_{0z}\cos\gamma + E_{0z} f^1(\varphi,\varphi_0,\alpha)\right]\frac{e^{i(k_1 r+\pi/4)}}{\sqrt{2\pi k_1 r}}e^{ikz\cos\gamma}, \\ E_\varphi &= H_{0z} g^1(\varphi,\varphi_0,\alpha)\frac{e^{i(k_1 r+\pi/4)}}{\sqrt{2\pi k_1 r}}e^{ikz\cos\gamma}, \end{aligned}\right\} \quad k_1 = k\sin\gamma. \tag{2.2.12}$$

where the first term occurring in the brackets for the expression for E_z represents the polarization coupling in the physical optics approximation approach, i.e., in the field generated by the uniform current $\mathbf{j}^0$. The incident angle φ_0 is determined by the condition

$$e^{ik(x\cos\sigma+y\cos\beta)}=e^{-ik_1(x\cos\varphi_0+y\sin\varphi_0)}, \tag{2.2.13}$$

where

$$\tan\varphi_0=\frac{\cos\beta}{\cos\sigma}.$$

The remaining components of the field radiated by the nonuniform current are determined by using Maxwell's equations and are equal to

$$\left.\begin{aligned} E_r&=-\cot\gamma E_z, \quad H_r=-\cot\gamma H_z,\\ E_\varphi&=\frac{1}{\sin\gamma}H_z, \quad H_\varphi=-\frac{1}{\sin\gamma}E_z.\end{aligned}\right\} \tag{2.2.14}$$

The planes of equal phase are given by

$$r\sin\gamma+z\cos\gamma=\text{const} \tag{2.2.15}$$

and are conic surfaces with generatrices that form a subtended angle of $\pi/2+\gamma$ with the positive z-axis. Thus, the scattered waves radiated by the fringe currents under oblique incidence of a plane wave are conical waves diverging from the edge. The normals to the surfaces of constant phase are oriented at an angle γ with respect to the z-axis as is shown in Fig. 2.2.5. They can be interpreted as diffracted rays. These waves can be presented in a more visual form, using the components (see Fig. 2.2.5)

$$\left.\begin{aligned} E_\gamma&=E_r\cos\gamma-E_z\sin\gamma,\\ H_\gamma&=H_r\cos\gamma-H_z\sin\gamma.\end{aligned}\right\} \tag{2.2.16}$$

The final expression for the scattered field in the far zone is of the form

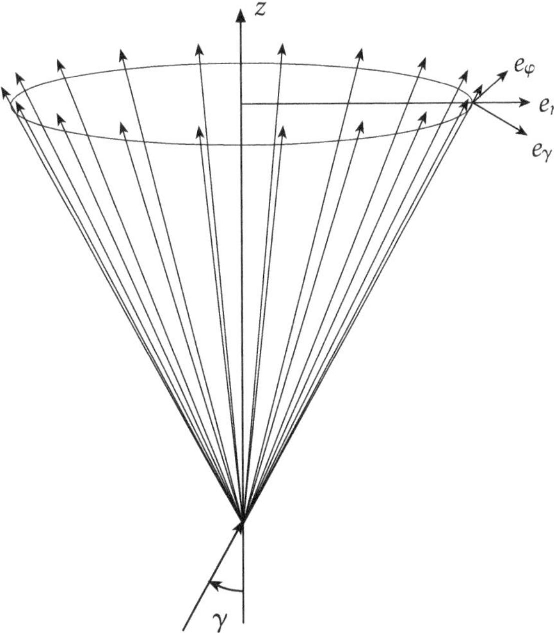

Figure 2.2.5: Cone of diffracted rays.

$$\left.\begin{aligned} E_\gamma &= H_\varphi = -\frac{1}{\sin\gamma} E_z, \\ H_\gamma &= -E_\varphi = -\frac{1}{\sin\gamma} H_z. \end{aligned}\right\} \tag{2.2.17}$$

Using these results, the following section will determine the edge waves scattered from a circumferential edge.

§2.3 Diffraction at a Circumferential Edge

An edge on a body of revolution is a circle. If the diameter of the circle is much greater than a wavelength, then the nonuniform current near the edge of any body of revolution can be considered to be identical to that

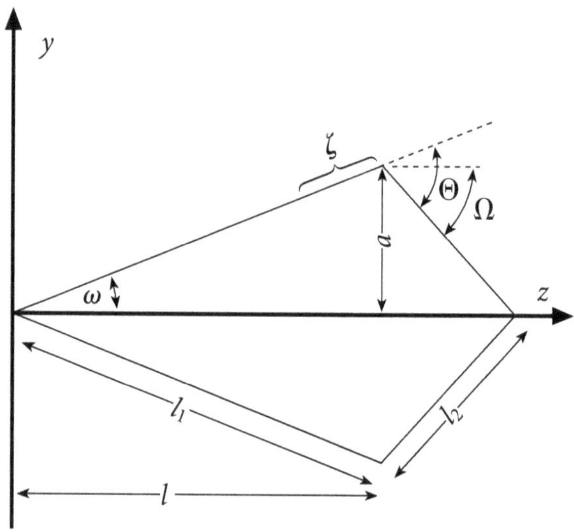

Figure 2.3.1: Cross section of a body of revolution in the *yoz* plane

near the edge of a corresponding tangential conic surface. The calculation of the field radiated by this equivalent current is the canonical problem investigated in the present section. Its solution will be used in the next sections to determine the radar cross section of certain bodies of revolution. This canonical problem is studied for the case when the incident wave propagates in the direction along the symmetry axis of the conic surface. The scattered field is calculated at this axis, which represents the focal line for edge waves.

Thus, let there be an electromagnetic plane wave incident upon a conical body on the positive *z*-axis (Fig. 2.3.1). From

$$\left.\begin{aligned} \mathbf{E} &= -\frac{1}{ik}\left(\nabla\nabla\cdot\mathbf{A}+k^2\mathbf{A}\right), \\ \mathbf{H} &= \nabla\times\mathbf{A}, \end{aligned}\right\} \tag{2.3.1}$$

the following expressions can be used to determine the scattered far field.

$$\left.\begin{aligned} E_x &= H_y = ikA_x\,, \\ E_y &= -H_x = ikA_y\,, \end{aligned}\right\} \text{ for } \vartheta = 0 \tag{2.3.2}$$

and

$$\left.\begin{aligned} E_x &= -H_y = ikA_x\,, \\ E_y &= H_x = ikA_y\,, \end{aligned}\right\} \text{ for } \vartheta = \pi, \tag{2.3.3}$$

where the vector potential is determined by

$$\begin{aligned} \mathbf{A} = \frac{1}{c}\frac{e^{ikr}}{r}\int_0^{2\pi}\Bigg[&\int_0^{l_1} \mathbf{j}_1(\zeta)e^{\pm ik\zeta\cos\omega}(a-\zeta\sin\omega)\,d\zeta + \\ &+\int_0^{l_2} \mathbf{j}_2(\zeta)e^{\mp ik\zeta\cos\Omega}(a-\zeta\sin\Omega)\,d\zeta\Bigg]d\psi\,. \end{aligned} \tag{2.3.4}$$

Here, r is the distance from the discontinuity to the point of observation, $\mathbf{j}_1(\zeta)$ is the surface current density flowing on the illuminated portion of the body ($z < l$), and $\mathbf{j}_2(\zeta)$ is the surface current density on the shadowed portion of the body ($z > l$). The upper sign in the exponential is for the case $\vartheta = 0$, and the lower sign is for $\vartheta = \pi$. Since the nonuniform current is principally concentrated near the edge, its corresponding vector potential is

$$\mathbf{A} = \frac{a}{c}\frac{e^{ikr}}{r}\int_0^{2\pi}\left[\int_0^{\infty} \mathbf{j}_1^1(\zeta)e^{\pm ik\zeta\cos\omega}d\zeta + \int_0^{\infty} \mathbf{j}_2^1(\zeta)e^{\mp ik\zeta\cos\Omega}d\zeta\right]d\psi\,. \tag{2.3.5}$$

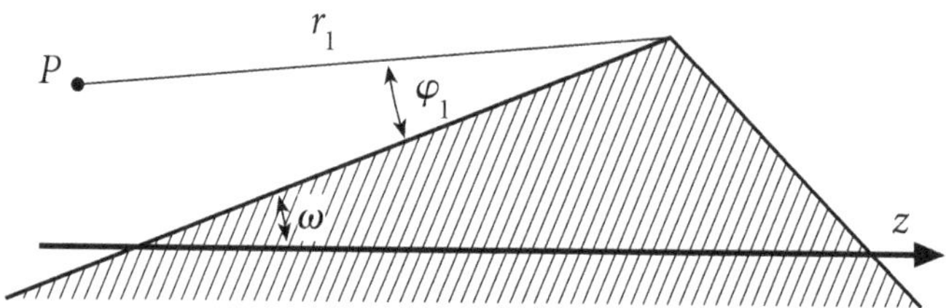

Figure 2.3.2: Dihedral corresponding to the ridge of a conical surface.

It is evident that for $ka \gg 1$ the nonuniform current near the discontinuity of a conical surface will be approximately the same as that of a wedge (Fig. 2.3.2). In a local polar coordinate system r_1, φ_1, z_1, the far field radiated by the nonuniform currents on such a wedge is

$$\left.\begin{aligned} E_{z_1}(\psi) &= -H_{\varphi_1}(\psi) = ikA_{z_1}(\psi), \\ H_{z_1}(\psi) &= E_{\varphi_1}(\psi) = ikA_{\varphi_1}(\psi), \end{aligned}\right\} \tag{2.3.6}$$

where

$$\mathbf{A} = \frac{1}{c}\sqrt{\frac{2\pi}{kr_1}}\left[\int_0^\infty \mathbf{j}_1^1(\zeta)e^{\pm ik\zeta\cos\omega}d\zeta + \int_0^\infty \mathbf{j}_2^1(\zeta)e^{\mp ik\zeta\cos\Omega}d\zeta\right]e^{i(kr_1+\pi/4)}. \tag{2.3.7}$$

Here the upper sign in the exponential is for $\varphi_1 = \pi + \omega_1$ and the lower sign for $\varphi_1 = \omega$. On the other hand, it was shown in § 2.2 that the field is equal to

$$\left.\begin{aligned} E_{z_1}(\psi) &= E_{0z_1}(\psi)f^1\frac{e^{i(kr_1+\pi/4)}}{\sqrt{2\pi kr_1}}, \\ H_{z_1}(\psi) &= H_{0z_1}(\psi)g^1\frac{e^{i(kr_1+\pi/4)}}{\sqrt{2\pi kr_1}}, \end{aligned}\right\} \tag{2.3.8}$$

where $E_{0z}(\psi)$ and $H_{0z}(\psi)$ are the amplitudes of the incident wave on the edge of the wedge and f^1 and g^1 are the angular functions characterizing the scattering pattern.

Let us introduce the definition

$$\mathbf{J} = \int_0^\infty \mathbf{j}_1^1(\zeta) e^{\pm ik\zeta\cos\omega} d\zeta + \int_0^\infty \mathbf{j}_2^1(\zeta) e^{\mp ik\zeta\cos\Omega} d\zeta . \tag{2.3.9}$$

Setting Eqn. (2.3.6) equal to Eqn. (2.3.8), we find that

$$J_{z_1} = \frac{cE_{0z_1}(\psi)}{ik2\pi} f^1, \quad J_{\varphi_1} = \frac{cH_{0z_1}(\psi)}{ik2\pi} g^1 . \tag{2.3.10}$$

The components J_{z_1} and J_{φ_1} are mutually perpendicular. They are parallel to the x-y plane when $\vartheta = 0$ and $\vartheta = \pi$ (Fig. 2.3.3). The different orientation of the unit vector e_{φ_1} for $\vartheta = 0$ and $\vartheta = \pi$ is connected with the fact that the angle φ_1 is referenced from the illuminated side of the wedge. In the original x, y, z system of coordinates $\mathbf{J}$ possesses components equal to

$$\left.\begin{aligned} J_x &= J_{z_1}\sin\psi - J_{\varphi_1}\cos\psi , \\ J_y &= -J_{z_1}\cos\psi - J_{\varphi_1}\sin\psi \end{aligned}\right\} \quad \text{for} \quad \vartheta = 0, \tag{2.3.11}$$

and

$$\left.\begin{aligned} J_x &= J_{z_1}\sin\psi + J_{\varphi_1}\cos\psi , \\ J_y &= -J_{z_1}\cos\psi + J_{\varphi_1}\sin\psi \end{aligned}\right\} \quad \text{for} \quad \vartheta = \pi . \tag{2.3.12}$$

Substituting Eqn. (2.3.10) into Eqns. (2.3.11) and (2.3.12) we obtain

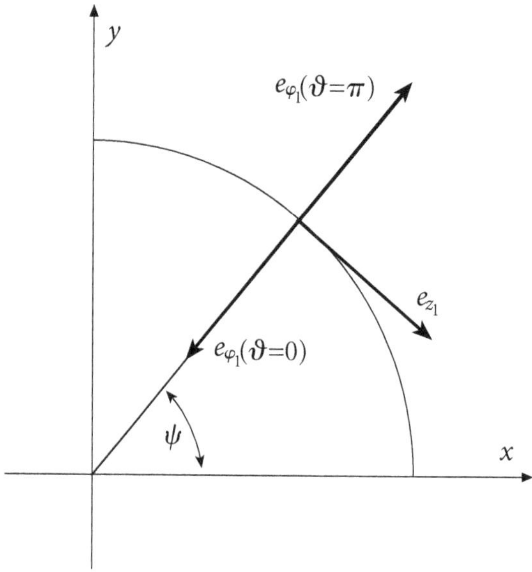

Figure 2.3.3: Relative orientation of e_{φ_1} and e_{z_1} for $\vartheta = 0$ and $\vartheta = \pi$.

$$\left.\begin{aligned} J_x &= \frac{c}{ik2\pi}\left[f^1 E_{0z_1}(\psi)\sin\psi - g^1 H_{0z_1}(\psi)\cos\psi\right], \\ J_y &= -\frac{c}{ik2\pi}\left[f^1 E_{0z_1}(\psi)\cos\psi + g^1 H_{0z_1}(\psi)\sin\psi\right] \end{aligned}\right\} \text{ for } \vartheta = 0. \tag{2.3.13}$$

and

$$\left.\begin{aligned} J_x &= \frac{c}{ik2\pi}\left[f^1 E_{0z_1}(\psi)\sin\psi + g^1 H_{0z_1}(\psi)\cos\psi\right], \\ J_y &= -\frac{c}{ik2\pi}\left[f^1 E_{0z_1}(\psi)\cos\psi - g^1 H_{0z_1}(\psi)\sin\psi\right] \end{aligned}\right\} \text{ for } \vartheta = \pi. \tag{2.3.14}$$

Identifying the current near the discontinuity of a conical surface with the current on a wedge, we determine the components of the vector potential are

$$\left.\begin{aligned} A_x &= \frac{a}{ik2\pi}\frac{e^{ikr}}{r}\int_0^{2\pi}\left[f^1 E_{0z_1}(\psi)\sin\psi - g^1 H_{0z_1}(\psi)\cos\psi\right]d\psi\,, \\ A_y &= -\frac{a}{ik2\pi}\frac{e^{ikr}}{r}\int_0^{2\pi}\left[f^1 E_{0z_1}(\psi)\cos\psi + g^1 H_{0z_1}(\psi)\sin\psi\right]d\psi \end{aligned}\right\}\text{for } \vartheta = 0 \quad (2.3.15)$$

and

$$\left.\begin{aligned} A_x &= \frac{a}{ik2\pi}\frac{e^{ikr}}{r}\int_0^{2\pi}\left[f^1 E_{0z_1}(\psi)\sin\psi + g^1 H_{0z_1}(\psi)\cos\psi\right]d\psi\,, \\ A_y &= -\frac{a}{ik2\pi}\frac{e^{ikr}}{r}\int_0^{2\pi}\left[f^1 E_{0z_1}(\psi)\cos\psi - g^1 H_{0z_1}(\psi)\sin\psi\right]d\psi \end{aligned}\right\}\text{for } \vartheta = \pi. \quad (2.3.16)$$

Furthermore, let there be a plane wave polarized such that $\mathbf{E}_0 \parallel ox$. Then,

$$E_{0z_1}(\psi) = E_{0x}\sin\psi\,, \quad H_{0z_1}(\psi) = -E_{0x}\cos\psi\,. \qquad (2.3.17)$$

Taking into account these relationships and substituting Eqns. (2.3.15) and (2.3.16) into Eqns. (2.3.2) and (2.3.3), we find that the field scattered by a circumferential edge on a conical surface is

$$E_x = H_y = \frac{aE_{0x}}{2}\left(f^1 + g^1\right)\frac{e^{ikr}}{r}, \quad E_y = H_x = 0 \quad \text{for} \quad \vartheta = 0 \qquad (2.3.18)$$

and

$$E_x = -H_y = \frac{aE_{0x}}{2}\left(f^1 - g^1\right)\frac{e^{ikr}}{r}, \quad E_y = H_x = 0 \quad \text{for} \quad \vartheta = \pi\,. \qquad (2.3.19)$$

Equation (2.3.18) is valid for $0 < \omega < \pi/2$ and $\omega < \Theta < \pi$, and Eqn. (2.3.19) is valid for $0 \le \omega \le \pi/2$ and $\omega \le \Theta \le \pi$. In the case of a disk ($\omega = \pi/2$, $\Theta = \pi$) the field radiated by the fringe currents is equal (within the given approximation) to zero on the z-axis, since $f^1 = -g^1 = -1/2$ for $\vartheta = 0$ and $f^1 = g^1 = -1/2$ for $\vartheta = \pi$.

The theory developed in this section allows the prediction of the radar cross section for bodies of revolution with edges. Below, we will determine the expressions for some bodies of practical interest.

§2.4 Cones

Let there be a perfectly conducting cone (Fig. 2.3.1) illuminated by an electromagnetic plane wave given by

$$E_x = H_y = E_{0x} e^{ikz}, \quad H_x = 0. \tag{2.4.1}$$

The uniform current excited on the surface of the cone possesses the components

$$j_x^0 = \frac{c}{2\pi} E_{0x} \sin\omega e^{ikz}, \quad j_y^0 = 0, \quad j_z^0 = \frac{c}{2\pi} E_{0x} \cos\omega \cos\psi e^{ikz}. \tag{2.4.2}$$

In the $\vartheta = \pi$ direction (for $R \gg ka^2$, $R \gg kl^2$) this current creates the scattered field

$$E_x = -H_y = -E_{0x} \frac{i}{4k} \tan^2\omega \frac{e^{ikR}}{R} + E_{0x} \left(\frac{i}{4k} \tan^2\omega + \frac{a}{2} \tan\omega \right) \frac{e^{ikR}}{R} e^{i2kl}, \tag{2.4.3}$$
$$E_y = H_x = 0.$$

It is pertinent to remember that the field radiated by the uniform current is often called the physical optics (PO) approximation for the scattered field. Thus, Eqn. (2.4.3) represents the PO approximation. Here, the first term describes a spherical wave emanating from the tip of the cone, and the remaining terms represent a spherical wave propagating from the circumferential edge.

According to Eqn. (2.3.19), the nonuniform current near the circumferential edge creates a spherical wave equal to

$$E_x = -H_y = -\frac{aE_{0x}}{2}\left(\tan\omega + \frac{\frac{2}{n}\sin\frac{\pi}{n}}{\cos\frac{\pi}{n} - \cos\frac{2\omega}{n}}\right)\frac{e^{ikR}}{R}e^{i2kl}, \tag{2.4.4}$$

$$E_y = H_x = 0,$$

where

$$n = 1 + \frac{\omega + \Omega}{\pi}. \tag{2.4.5}$$

The asymptotic analysis of the exact solution for the diffraction from a semi-infinite cone [44, 45] shows that the radiation by the nonuniform currents excited near the cone's vertex can be neglected in the $\vartheta = \pi$ direction. Therefore, summing Eqns. (2.4.3) and (2.4.4), we obtain the following expression for the scattered field:

$$E_x = -H_y = -E_{0x}\left[\frac{i}{2}\tan^2\omega\left(1 - e^{i2kl}\right) + ka\frac{\frac{2}{n}\sin\frac{\pi}{n}}{\cos\frac{\pi}{n} - \cos\frac{2\omega}{n}}e^{i2kl}\right]\frac{e^{ikR}}{2kR}, \tag{2.4.6}$$

$$E_y = H_x = 0.$$

One significant property of this expression should be noted. The expression for the spherical wave from the base of a cone not only contains a term depending on f and g — the last term in the square brackets in Eqn. (2.4.6) — but also contains an additional term, $-i\tan^2\omega e^{i2kl}/2$, that does not depend on f and g and is determined only by the uniform current. In other words, the resulting spherical wave from the base of the cone is not exclusively determined by f and g which characterize the total edge wave diffracted from the edge of a wedge. This interesting phenomenon cannot be revealed within the approximations made in [46, 47, 48], which is why they are not applicable for cones of large flare angle $\omega \approx \pi/2$.

According to Eqn. (2.4.6), the radar cross section is determined by

$$\sigma = \pi a^2 |\Sigma|^2 \tag{2.4.7}$$

where Σ is related to the scattered field by

$$E_x = -H_y = -\frac{aE_{0x}}{2}\frac{e^{ikR}}{R}\Sigma, \tag{2.4.8}$$

and is equal to

$$\Sigma = \frac{1}{ka}\tan^2\omega \sin kl\, e^{ikl} + \frac{\frac{2}{n}\sin\frac{\pi}{n}}{\cos\frac{\pi}{n} - \cos\frac{2\omega}{n}} e^{i2kl}. \tag{2.4.9}$$

An analogous expression using the physical optics approximation can be written in accordance with Eqn. (2.4.3) as

$$\Sigma^0 = \frac{1}{ka}\tan^2\omega \sin kl\, e^{ikl} - \tan\omega e^{i2kl}. \tag{2.4.10}$$

When transforming a cone into a disk ($\omega \to \pi/2, l \to 0$) Eqns. (2.4.9) and (2.4.10) are transformed respectively into

$$\left.\begin{aligned} \Sigma &= -ika - \frac{1}{n}\cot\frac{\pi}{n}, \\ \Sigma^0 &= -ika. \end{aligned}\right\} \tag{2.4.11}$$

Equations (2.4.9) and (2.4.10) can further show that for large values of ka ($ka \gg \tan^2\omega$), Σ and Σ^0 are equal to

$$\left.\begin{aligned}\Sigma &= \frac{\dfrac{2}{n}\sin\dfrac{\pi}{n}}{\cos\dfrac{\pi}{n}-\cos\dfrac{2\omega}{n}}e^{i2kl}, \\ \Sigma^0 &= \tan\omega e^{i2kl}.\end{aligned}\right\} \tag{2.4.12}$$

Thus, even at high frequencies ($ka \gg \tan^2\omega$ and $R \gg kl^2$), Eqn. (2.4.9) differs from the physical optics approximation significantly, since

$$\sigma = \pi a^2 \left| \frac{\dfrac{2}{n}\sin\dfrac{\pi}{n}}{\cos\dfrac{\pi}{n}-\cos\dfrac{2\omega}{n}} \right|^2, \tag{2.4.13}$$

and

$$\sigma^0 = \pi a^2 \tan^2\omega. \tag{2.4.14}$$

Thus,

$$\sigma = \frac{\sigma^0}{\tan^2\omega} \left| \frac{\dfrac{2}{n}\sin\dfrac{\pi}{n}}{\cos\dfrac{\pi}{n}-\cos\dfrac{2\omega}{n}} \right|^2, \tag{2.4.15}$$

i.e., at sufficiently high frequencies (or for a sufficiently large cone), σ is proportional to σ^0, and the coefficient of proportionality does not depend on the dimensions of the cone but only on its shape.

This result is illustrated by curves that show the radar cross section of a cone ($\omega = 10°25'$, $\lambda = 2$ cm, $\Omega = 90°$) as a function of its length (Fig. 2.4.1). Whereas our expression (solid line) is in agreement with experimental measurements (triangles), the physical optics approximation (dashed line) provides values 10–15 dB below experimental data. Figure 2.4.2 presents curves for the radar cross section of a cone ($ka = 2.75\pi$, $\Omega = 90°$) as ω is varied

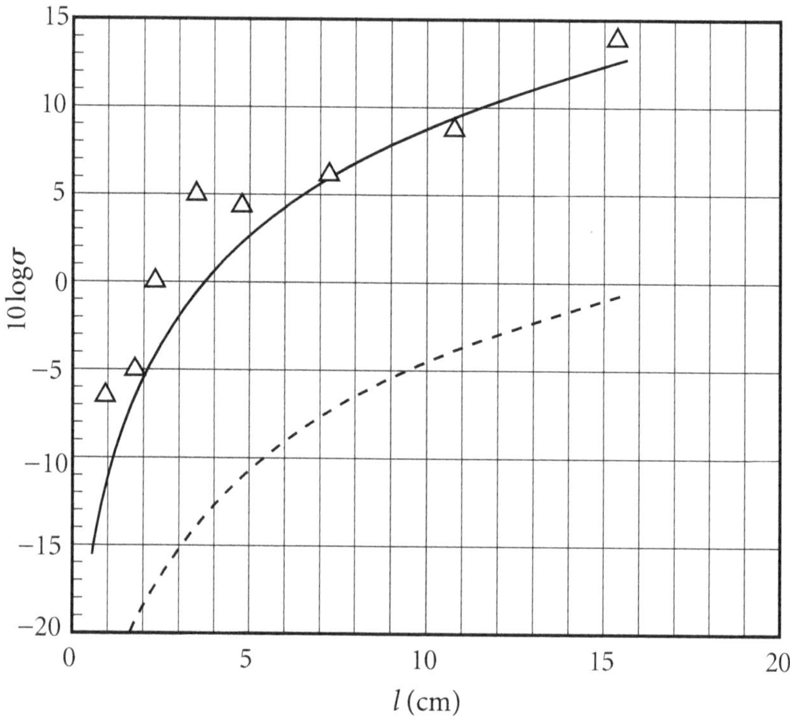

Figure 2.4.1: Behavior of the RCS of a cone as a function of its length. The solid line corresponds to the RCS (σ) according to Eqns. (2.4.7) and (2.4.9) which includes the nonuniform current near the edge. The dashed line (σ^0) corresponds to the physical optics approximation. Reprinted from [12] with the permission of *Radiotekhnika i elektronika.*

between 0° and 90° (disk). The difference between our curve and the physical optics approximation approaches almost 30 dB for $\omega = 2°$.

As opposed to the result given by Eqn. (2.4.10) using the physical optics approximation, Eqn. (2.4.9) also estimates the role of the form of the shadowed portion of the body and shows that the more the body is funnel-shaped ($\Omega \approx \pi - \omega$) the larger reflected field will be. Thus, for example, when $\omega = 10°$, $kl = 10\pi$ ($k = \pi$), and $\Omega = 170°$, the amplitude of the wave

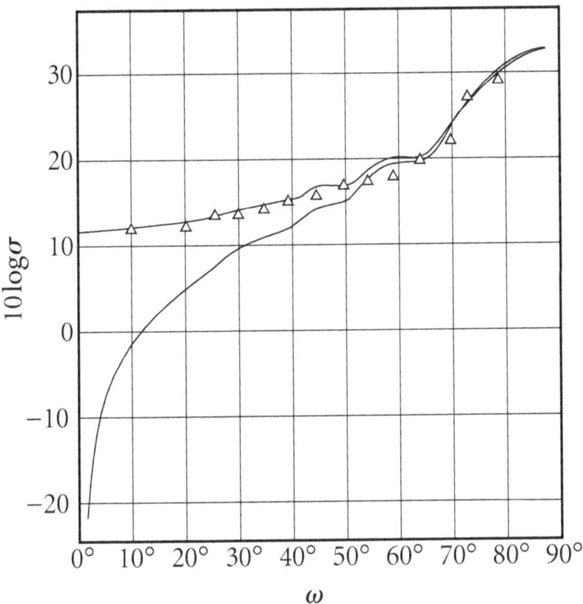

Figure 2.4.2: Dependence of the RCS of a finite cone on the flare angle (radius of the base remaining constant). The lower curve corresponds to the physical optics approximation σ^0. Reprinted from [12] with the permission of *Radiotekhnika i elektronika.*

reflected by the cone is 15 dB greater than the value provided by the physical optics approximation (Fig. 2.4.3).

Equation (2.4.13) is equivalent to an expression previously introduced in articles [46, 47, 48]. However, it is applicable only for narrow cone angles, whereas our Eqn. (2.4.9) is valid for cones of any flare angle, ω $(0 \leq \omega \leq \pi/2)$.

This method of calculation can be generalized for asymmetrically incident plane waves. However, in this case, generally speaking, it is necessary to pay attention to the nonuniform current occurring at the cone vertex.

The next section will calculate the radar cross section of a body of revolution formed by rotating the generatrix presented in Fig. 2.4.4 around the z-axis. Integrating the uniform current, it is easy to show that the field scattered from the surface of a truncated cone $(l_1 \leq z \leq l_1 + l_2)$in the $\vartheta = \pi$ direction (Fig. 2.4.5), is given by

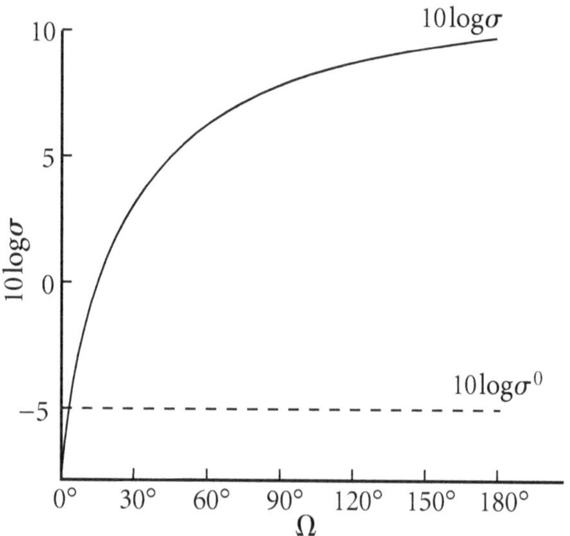

Figure 2.4.3: Dependence of the RCS of a finite cone on form of the shadowed region. Reprinted from [12] with the permission of *Radiotekhnika i elektronika.*

$$E_x = -H_y = E_{0x}\left[-\left(\frac{i}{4k}\tan^2\omega_1 + \frac{a_1}{2}\tan\omega_1\right)e^{i2kl_1} + \right.$$
$$\left. + \left(\frac{i}{4k}\tan^2\omega_1 + \frac{a_2}{2}\tan\omega_1\right)\right]e^{i2k(l_1+l_2)}\frac{e^{ikR}}{R}. \tag{2.4.16}$$

Combining this expression with Eqn. (2.4.3), where the values of l and a are replaced by l_1 and a_1, we can determine the field generated by the uniform current flowing on the illuminated surface of the body

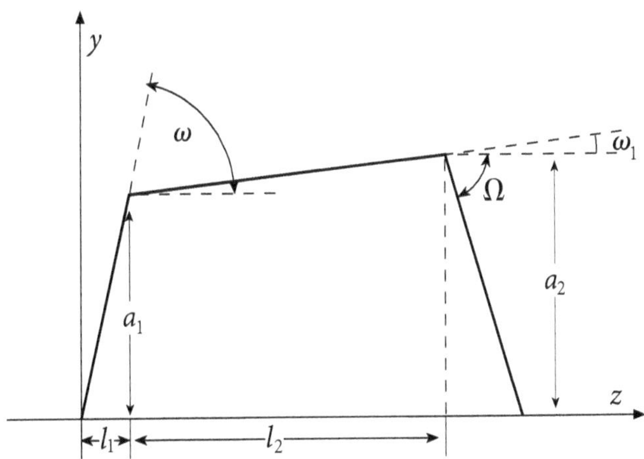

Figure 2.4.4: Generatrix of surface of revolution.

$$E_x = -H_y = -\frac{a_1 E_{0x}}{2}\left\{\frac{1}{ka_1}\tan^2\omega \sin kl_1 e^{ikl_1} - \tan\omega e^{i2kl_1} + \right. \tag{2.4.17}$$
$$\left. + \left[\frac{1}{ka_1}\tan^2\omega_1 \sin kl_2 e^{ikl_2} + \left(1 - \frac{a_2}{a_1}e^{i2kl_2}\right)\tan\omega_1\right]e^{i2kl_1}\right\}\frac{e^{ikR}}{R}.$$

According to § 2.3, the field radiated by the nonuniform current is

$$E_x = -H_y = -\frac{a_1 E_{0x}}{2}\left[\left(\frac{\frac{2}{n_1}\sin\frac{\pi}{n_1}}{\cos\frac{\pi}{n_1} - \cos\frac{2\omega}{n_1}} + \tan\omega - \tan\omega_1\right)e^{i2kl_1} + \right. \tag{2.4.18}$$
$$\left. + \frac{a_2}{a_1}\left(\frac{\frac{2}{n_2}\sin\frac{\pi}{n_2}}{\cos\frac{\pi}{n_2} - \cos\frac{2\omega_1}{n_2}} + \tan\omega_1\right)e^{i2k(l_1+l_2)}\right]\frac{e^{ikR}}{R},$$

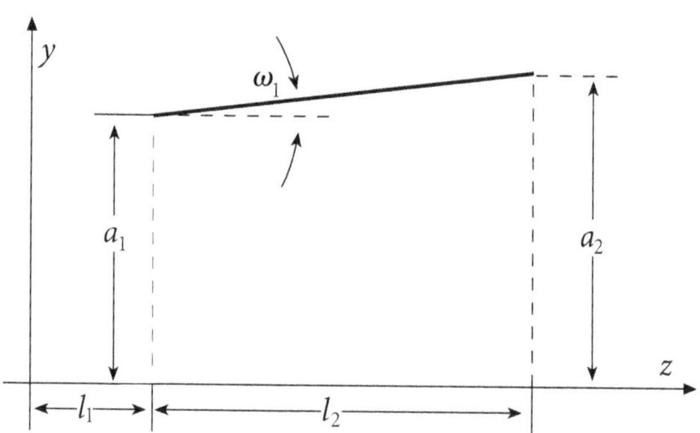

Figure 2.4.5: Generatrix of a conical surface.

where

$$n_1 = 1 + \frac{\omega - \omega_1}{\pi}, \quad n_2 = 1 + \frac{\omega_1 + \Omega}{\pi}. \tag{2.4.19}$$

Summing Eqns. (2.4.17) and (2.4.18) obtains a more accurate expression for the field scattered in the $\vartheta = \pi$ direction:

$$E_x = -H_y = -\frac{a_1 E_{0x}}{2} \left[\frac{1}{ka_1} \tan^2\omega \sin kl_1 e^{ikl_1} + \frac{\frac{2}{n_1}\sin\frac{\pi}{n_1}}{\cos\frac{\pi}{n_1} - \cos\frac{2\omega}{n_1}} e^{i2kl_1} + \right.$$
$$\left. + \left(\frac{1}{ka_1} \tan^2\omega_1 \sin kl_2 e^{ikl_2} + \frac{a_2}{a_1} \frac{\frac{2}{n_2}\sin\frac{\pi}{n_2}}{\cos\frac{\pi}{n_2} - \cos\frac{2\omega_1}{n_2}} e^{i2kl_2} \right) e^{i2kl_1} \right] \frac{e^{ikR}}{R}. \tag{2.4.20}$$

It follows that the radar cross section is equal to

$$\sigma = \pi a_1^2 \left| \frac{1}{ka_1} \tan^2\omega \sin kl_1 e^{ikl_1} + \frac{\frac{2}{n_1}\sin\frac{\pi}{n_1}}{\cos\frac{\pi}{n_1} - \cos\frac{2\omega}{n_1}} e^{i2kl_1} + \right.$$
$$\left. + \left(\frac{1}{ka_1} \tan^2\omega_1 \sin kl_2 e^{ikl_2} + \frac{a_2}{a_1} \frac{\frac{2}{n_2}\sin\frac{\pi}{n_2}}{\cos\frac{\pi}{n_2} - \cos\frac{2\omega_1}{n_2}} e^{i2kl_2} \right) e^{i2kl_1} \right|^2 . \tag{2.4.21}$$

The corresponding value according to the physical optics approximation is

$$\sigma^0 = \pi a_1^2 \left| \frac{1}{ka_1} \tan^2\omega \sin kl_1 e^{ikl_1} - \tan\omega e^{i2kl_1} + \right.$$
$$\left. + \left[\frac{1}{ka_1} \tan^2\omega_1 \sin kl_2 e^{ikl_2} + \left(1 - \frac{a_2}{a_1} e^{i2kl_2} \right) \tan\omega_1 \right] e^{i2kl_1} \right|^2 . \tag{2.4.22}$$

If the conical surface is transformed into a disk ($\omega \to \pi/2$, $l_1 \to 0$), Eqns. (2.4.21) and (2.4.22) become

$$\sigma = \pi a_1^2 \left| -ika_1 - \frac{1}{n_1} \cot\frac{\pi}{n_1} + \right.$$
$$\left. + \frac{1}{ka_1} \tan^2\omega_1 \sin kl_2 e^{ikl_2} + \frac{a_2}{a_1} \frac{\frac{2}{n_2}\sin\frac{\pi}{n_2}}{\cos\frac{\pi}{n_2} - \cos\frac{2\omega_1}{n_2}} e^{i2kl_2} \right|^2 , \tag{2.4.23}$$

$$\sigma^0 = \pi a_1^2 \left| -ika_1 + \frac{1}{ka_1} \tan^2\omega_1 \sin kl_2 e^{ikl_2} + \left(1 - \frac{a_2}{a_1} e^{i2kl_2}\right) \tan\omega_1 \right|^2 . \quad (2.4.24)$$

Setting $\omega_1 = 0$, the values for the radar cross section of a finite cylinder are

$$\sigma = \pi a_1^2 \left| -ika_1 - \frac{1}{n_1} \cot\frac{\pi}{n_1} + \frac{\frac{2}{n_2}\sin\frac{\pi}{n_2}}{\cos\frac{\pi}{n_2} - 1} e^{i2kl_2} \right|^2 , \quad (2.4.25)$$

$$\sigma^0 = \pi a_1^2 (ka_1)^2 , \quad (2.4.26)$$

where

$$n_1 = \frac{3}{2}, \quad n_2 = 1 + \frac{\Omega}{\pi} . \quad (2.4.27)$$

Equation (2.4.25) is more accurate than Eqn. (15.06) of [13], where the value of the field in the $\vartheta = \pi$ direction was determined using the physical optics approximation.

§2.5 Paraboloids of Revolution

Let us calculate the radar cross section of a paraboloid of revolution $r^2 = 2pz$ (Fig. 2.5.1) illuminated by a plane wave described by Eqn. (2.4.1). The uniform current excited on the surface of the paraboloid is given by

$$j_x^0 = \frac{c}{2\pi} E_{0x} \sin\alpha e^{ikz} , \quad j_y^0 = 0 , \quad j_z^0 = \frac{c}{2\pi} E_{0x} \cos\alpha \cos\psi e^{ikz} . \quad (2.5.1)$$

Integrating the current, it is not difficult to show that in the $\vartheta = \pi$ direction the radiated field is

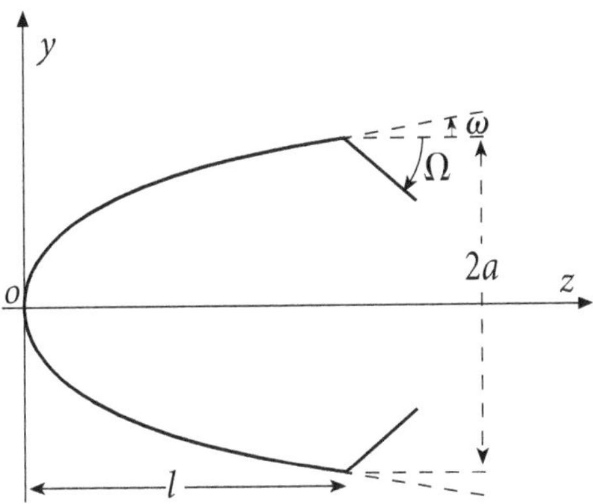

Figure 2.5.1: Cross section of a body of revolution in the *yoz* plane.

$$E_x = -H_y = -E_{0x}\frac{a}{2}\left(1-e^{i2kl}\right)\tan\omega\frac{e^{ikR}}{R}, \quad E_y = H_x = 0. \tag{2.5.2}$$

Here, a is the radius of the paraboloid base, $l = a^2/(2p) = (a/2)\cot\omega$ is its length, α is the angle between the z-axis and the tangent to the surface of the paraboloid ($r^2 = 2pz$). At $z = l, \alpha = \omega$ ($\tan\omega = p/a$).

The scattered field from the nonuniform current created by the sharp circumferential bend on the surface of the paraboloid can be determined in the $\vartheta = \pi$ direction by Eqn. (2.4.4). The field from the nonuniform current caused by the surface curvature of the paraboloid is equal to zero [49]. Therefore, by summing Eqns. (2.5.2) and (2.4.4) we determine that the equation for the scattered field is

$$E_x = -H_y = -\frac{aE_{0x}}{2}\left(\tan\omega + \frac{\frac{2}{n}\sin\frac{\pi}{n}}{\cos\frac{\pi}{n} - \cos\frac{2\omega}{n}}e^{i2kl}\right)\frac{e^{ikR}}{R}, \tag{2.5.3}$$

$$E_y = H_x = 0,$$

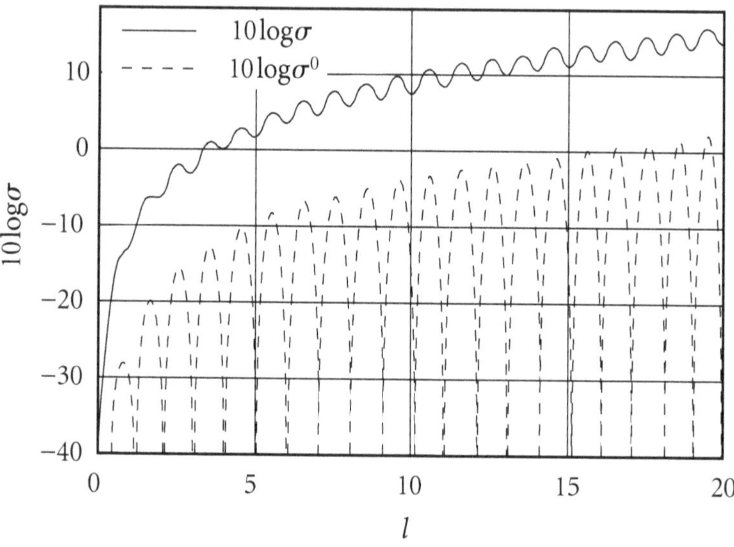

Figure 2.5.2: Dependence of the RCS of a finite paraboloid on its length for a constant value of ω ($\tan\omega = 0.1$). Diameter of the base is changed. Reprinted from [12] with the permission of *Radiotekhnika i elektronika.*

where

$$n = 1 + \frac{\omega + \Omega}{\pi}.$$

It follows that the radar cross section of the paraboloid will be equal to

$$\sigma = \pi a^2 \left| \tan\omega + \frac{\frac{2}{n}\sin\frac{\pi}{n}}{\cos\frac{\pi}{n} - \cos\frac{2\omega}{n}} e^{i2kl} \right|^2. \tag{2.5.4}$$

If the paraboloid is transformed into a disk ($\omega \to \pi/2, l \to 0, \Omega = \text{const}$), the expression for the radar cross section becomes

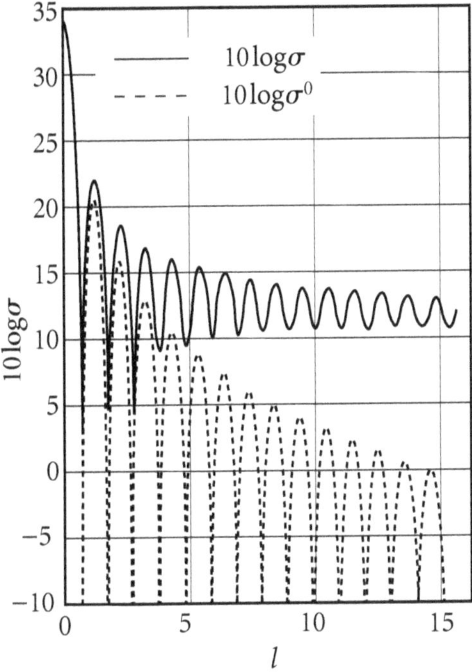

Figure 2.5.3: Dependence of the RCS of a finite paraboloid on its length with a constant diameter base. Reprinted from [12] with the permission of *Radiotekhnika i elektronika.*

$$\sigma = \pi a^2 \left| ika + \frac{1}{n}\cot\frac{\pi}{n} \right|^2 . \tag{2.5.5}$$

Comparing Eqn. (2.5.4) with

$$\sigma^0 = \pi a^2 \tan^2\omega \left| 1 - e^{i2kl} \right|^2 , \tag{2.5.6}$$

which gives the radar cross section according to the physical optics approximation, we see that they differ from one another significantly. First of all,

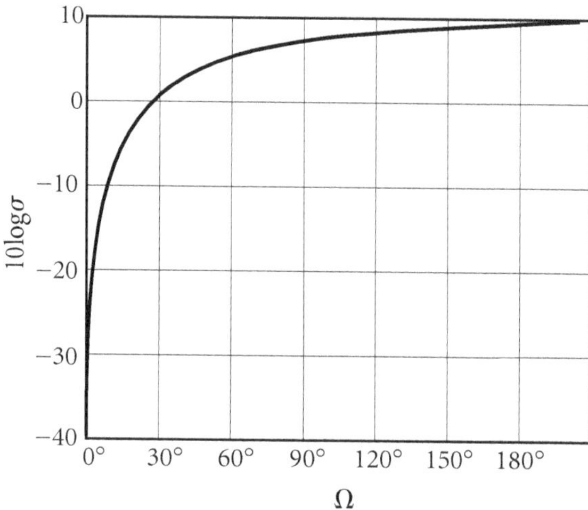

Figure 2.5.4: Dependence of RCS of a finite paraboloid on the form of the shadow region. Reprinted from [12] with the permission of Radiotekhnika i elektronika.

our attention will focus upon the oscillatory character of σ^0: the reflected field is equal to zero if the length of the paraboloid is equal to a multiple of half-wavelengths ($l = n\lambda/2, n = 1, 2, 3, \ldots$) and reaches a maximum value if $l = (n+1/2)\lambda/2, n = 1, 2, 3, \ldots$.

Figure 2.5.2 shows the radar cross section of paraboloids with $\Omega = 90°$, $\tan\omega = 0.1$ ($k = \pi$). Although Eqn. (2.5.4) maintains the oscillatory character of the RCS, the amplitude of the oscillation is approximately 2 dB, and the maximum value of σ exceeds σ^0 by almost 13 dB. An even stronger discrepancy between our results and physical optics occurs when the paraboloid is transformed into a disk (Fig. 2.5.3, $ka = 3\pi$, $\mathrm{k} = \pi$, $\omega \to 90°$, $\Omega \to 90°$).

As is the case for the cone, the shadow region exerts a significant influence on the value of the reflected field. For instance, Fig. 2.5.4 shows that for a paraboloid with $ka = 2\pi$, $kl = 10\pi$, $\tan\omega = 0.1$ ($k = \pi$), and increasing Ω ($\omega < \Omega < \pi - \omega$), the value of the reflected field is as much as 44 dB above the values predicted by physical optics.

§2.6 Spherical Surfaces

Let there be a perfectly conducting sphere (with radius ρ and centered on the z-axis at $z = \rho$) illuminated by a plane wave given by Eqn. (2.4.1). The uniform part of the surface current will be

$$j_x^0 = -\frac{c}{2\pi} E_{0x} \cos\theta e^{ikz} , \quad j_y^0 = 0, \quad j_z^0 = \frac{c}{2\pi} E_{0x} \sin\theta \cos\psi e^{ikz} . \tag{2.6.1}$$

Such a current flowing on a spherical sector (Fig. 2.6.1) creates a field in the $\vartheta = \pi$ direction equal to

$$E_x = -H_y = E_{0x}\left[-\frac{a}{2\cos\omega} + \frac{i}{4k} + \left(\frac{a}{2}\tan\omega - \frac{i}{4k}\right)e^{i2kl}\right]\frac{e^{ikR}}{R}, \tag{2.6.2}$$

where

$$l = \rho(1 - \sin\omega), \tag{2.6.3}$$

and a is the radius of the sphere cross section at $z = l$. For $ka \gg 1$, Eqn. (2.6.2) becomes

$$E_x = -H_y = -\frac{aE_{0x}}{2}\left(\sec\omega - \tan\omega e^{i2kl}\right)\frac{e^{ikR}}{R}. \tag{2.6.4}$$

Setting $\omega = 0$,

$$E_x = -H_y = -\frac{aE_{0x}}{2}\frac{e^{ikR}}{R}, \tag{2.6.5}$$

which gives us the value of the field scattered by a hemisphere. The corresponding value for the radar cross section is

$$\sigma^0 = \pi a^2 . \tag{2.6.6}$$

Now the field radiated by the nonuniform current flowing near the circumferential edge formed with the base will be determined. The nonuniform

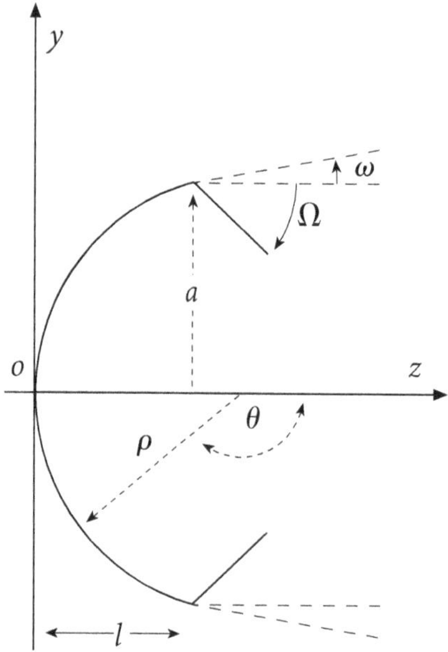

Figure 2.6.1: Spherical segment with a conical base.

current caused by the curvature of the smooth part of the surface is small and can be neglected provided $ka \gg 1$. The nonuniform current excited by the circumferential edge produces a field in the $\vartheta = \pi$ direction that is given by Eqn. (2.4.4). Summing this with the field given by Eqn. (2.6.4), we find that the expression for the scattered field is

$$E_x = -H_y = -\frac{aE_{0x}}{2}\left(\frac{1}{\cos\omega} + \frac{\frac{2}{n}\sin\frac{\pi}{n}}{\cos\frac{\pi}{n} - \cos\frac{2\omega}{n}} e^{i2kl}\right)\frac{e^{ikR}}{R}. \tag{2.6.7}$$

Subsequently, the radar cross section of a spherical segment is equal to

$$\sigma = \pi a^2 \left| \frac{1}{\cos\omega} + \frac{\frac{2}{n}\sin\frac{\pi}{n}}{\cos\frac{\pi}{n} - \cos\frac{2\omega}{n}} e^{i2kl} \right|^2 , \tag{2.6.8}$$

$$n = 1 + \frac{\omega + \Omega}{\pi} .$$

According to the physical optics approximation, the radar cross section resulting from the field described by Eqn. (2.6.4) is equal to

$$\sigma^0 = \pi a^2 \left| \frac{1}{\cos\omega} - \tan\omega e^{i2kl} \right|^2 . \tag{2.6.9}$$

If the spherical region of the body is transformed into a flat surface ($\omega \to \pi/2, l \to 0, \Omega = \text{const}$), Eqns. (2.6.8) and (2.6.9) become, respectively,

$$\sigma = \pi a^2 \left| ika + \frac{1}{n}\cot\frac{\pi}{n} \right|^2 , \tag{2.6.10}$$

$$\sigma^0 = \pi a^2 (ka)^2 .$$

From Eqns. (2.6.8) and (2.6.9) it follows that the radar cross section of a spherical segment is an oscillatory function of its length. The period of oscillation is equal to $\lambda/2$. These equations were used to calculate the numerical results seen in Fig. 2.6.2 and demonstrate that for small angular discontinuities ($\Omega = 15°$) the radiation by nonuniform current can be neglected. Figure 2.6.3 presents a graph of the radar cross section of a spherical segment and a finite cone (dashed-dotted curve) possessing the same diameter and type of base.

The results obtained in this chapter indicate that the value of the reflected field significantly depends on the form of the shadowed portion of the body and increases with increasing angular discontinuity. However, since the nonuniform current is generally concentrated near the discontinuity, the portion of the shadowed surface that is separated from the dis-

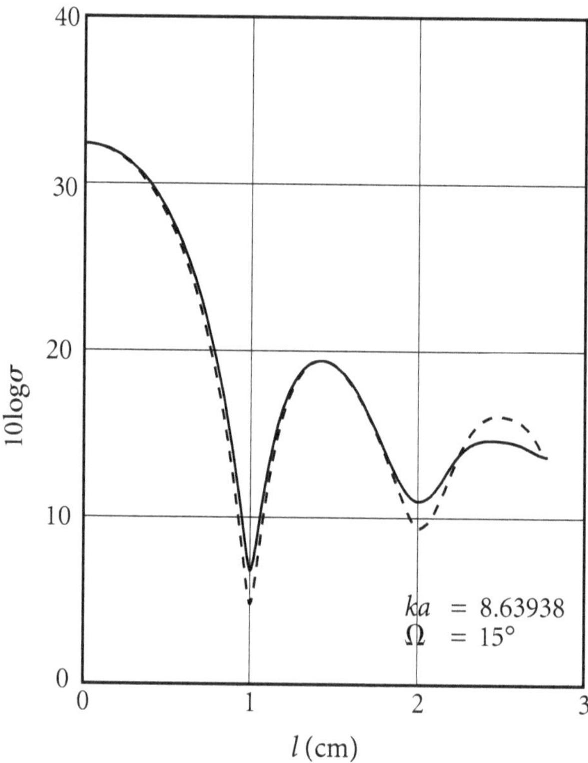

Figure 2.6.2: The RCS of a spherical segment of varying length but constant diameter base. The function σ (solid line) is calculated using Eqn. (2.6.8), which includes the nonuniform current near the edge. The physical optics approximation value σ^0 (dashed line) is calculated according to Eqn. (2.6.9). Reprinted from [13] with the permission of *Radio i Svyaz.*

continuity by a few wavelengths does not appear to affect the value of the reflected field appreciably and can be arbitrary in shape.

It is interesting that even at large (in comparison with a wavelength) body dimensions, our expressions are in good agreement with experimental results but differ significantly from the expressions provided by the physical optics approximation. Also, the physical optics approximation, despite the popular opinion regarding its reliability in such cases, diverges dramatically from experimental results.

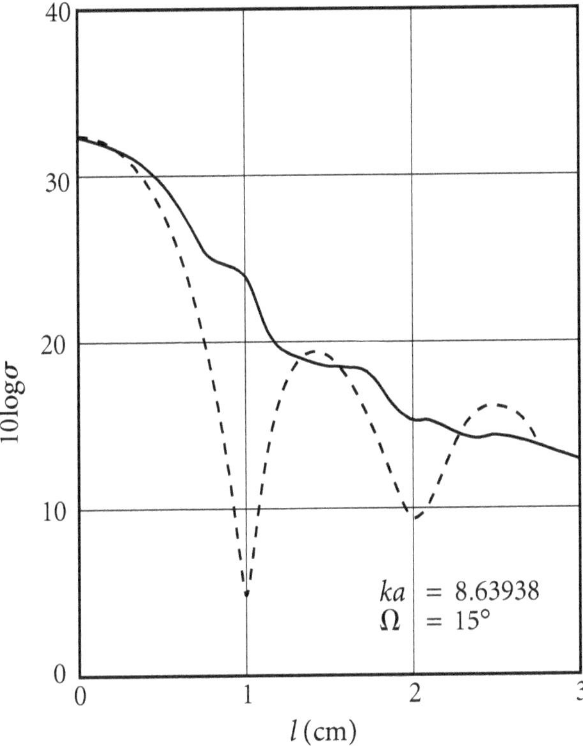

Figure 2.6.3: The RCS of a spherical segment (solid line) and a finite cone (dashed line), possessing identical bases. Reproduced from [13] with the permission of *Radio i Svyaz.*

One should note that in parallel with experimental verification of this theory, we also performed theoretical comparisons using a disk (see, [6,13]). Our approximate solution of this problem was compared with the rigorous solution obtained by M. G. Belkina [50]. It was found that for $ka = 5$ (a is the radius of the disk) the approximate scattering pattern is in satisfactory agreement with the exact calculations.

§2.7 Additional Comments

- Equation (2.2.12) is the result of a correction of Eqn. (5.10) in [13]. The additional term $H_{0z}\cos\gamma$ is included for the E_z component. It is caused by the polarization coupling in the physical optics approach, i.e., in the field generated by the uniform current $\mathbf{j}^0$. This correction was first published in [51] and later discussed in [52] and [53]. The physics behind this phenomenon is explained in [197].
- The investigation of edge diffracted waves has a long history. Just as was demonstrated in [54], Newton was the first who made visual observations of diffracted rays, although his explanation of this phenomenon was incorrect.
- Kalashnikov was the first to introduce the term "diffracted rays" and carried out the first objective proof of their existence by recording them on a photographic plate [55].
- The next important step was taken by Rubinowicz [56]. He discovered theoretically that the wave field arising from edge diffraction at an aperture has a ray structure. Particularly, he showed that every stationary point on the edge creates a cone of diffracted rays that satisfy Fermat's principle. The angle between each diffracted ray and the tangent to the edge equals the angle between the incident ray and the tangent (Fig. 2.2.5). This cone concept was used later by Keller [57, 58]. Now, this cone is well known as a diffraction cone. Its existence was also proved experimentally by Senior and Uslenghi [59].
- In [60] an extension of Rubinowicz theory [56] was suggested which, in turn, led directly to Keller's Geometrical Theory of Diffraction (GTD).
- The modern Physical Theory of Diffraction (PTD) based on the concept of elementary edge waves was developed in [61–65, 195]. The directivity pattern of these waves can be interpreted as the incremental diffraction coefficient [66–70] or as the equivalent edge current [53, 71–74]. From the mathematical point of view, these quantities are determined, in particular, by Eqns. (2.3.9)–(2.3.14).
- Fast convergent integrals convenient for numerical calculation of the nonuniform current on wedge faces are derived in [75]. Possible generaliza-

tions of the definition of the nonuniform current which can be useful in the solution of various diffraction problems are also presented in this article.

- Further, applications and developments of the PTD concept demonstrated in this chapter are presented in [151–191].

- In [63, 64, 195] it was shown that GTD follows from PTD when the stationary phase technique is applied to the PTD integrals for the scattered field. In other words, GTD actually represents a ray asymptotic form of PTD. However, in contrast to GTD, PTD is more general and is applicable to all diffraction regions (shadow boundaries, foci, and caustics) where the diffracted field does not have a ray structure.

- In this chapter we considered scattering from bodies of revolution when the incident wave propagates along their axes of symmetry. The scattering problem for an oblique incidence can be easily investigated using the modern PTD concept of elementary edge waves [61–65, 195]. Alternative techniques to solving this problem are the Uniform Theory of Diffraction [76, 77] and the Incremental Theory of Diffraction [69, 70]. Notice also the asymptotic theory for plane screens [192] similar to PTD and the method of matched asymptotic expansions [193, 194] that leads to the PTD ray asymptotics.

- Article [196] presents a new form of the PO field as a sum of the reflected field and the shadow radiation.

3

Edge Diffraction at Concave Surfaces: Extension of the Physical Theory of Diffraction

This chapter will study edge waves scattered from an edge of a concave body. To do this, § 3.1 will examine the plane wave excitation of a wedge-shaped horn. The main objective here is to determine the field radiated by the non-uniform/fringe current. The results of this section are used in subsequent sections to calculate the radar cross section of a concave bodies of revolution.

§3.1 Field Inside a Wedge-Shaped Horn

The problem here is to determine the field excited inside the wedge-shaped region $0 \le \varphi \le \alpha$ $(\alpha \le \pi)$ by the plane waves

$$E_z = E_{0z} e^{-ikr\cos(\varphi-\varphi_0)}, \quad H_z = 0 \tag{3.1.1}$$

and

$$H_z = H_{0z} e^{-ikr\cos(\varphi-\varphi_0)}, \quad E_z = 0. \tag{3.1.2}$$

The source of these waves is a filament of electric or magnetic current located at infinity. The boundary of the region is considered to be perfectly conducting. Then, according to Eqns. (1.25) and (1.26) of reference [13], the field excited in the wedge-shaped region is

$$\left.\begin{aligned} E_z &= E_{0z}\left[u(r,\varphi-\varphi_0)-u(r,\varphi+\varphi_0)\right], \\ H_z &= H_{0z}\left[u(r,\varphi-\varphi_0)+u(r,\varphi+\varphi_0)\right], \end{aligned}\right\} \tag{3.1.3}$$

where

$$u(r,\psi)=\frac{1}{2\alpha}\int\limits_C \frac{e^{-ikr\cos\beta}}{1-e^{i\pi(\beta+\psi)/\alpha}}d\beta, \tag{3.1.4}$$

and the contour C is shown in Fig. 3.1.1. The integrand in Eqn. (3.1.4) possesses poles at the points

$$\beta_m = 2\alpha m-\psi, \quad m=0,\pm1,\pm2,\ldots \tag{3.1.5}$$

Evaluating the integral along the contour D and taking the poles of Eqn. (3.1.5) into consideration, we obtain

$$u(r,\varphi-\varphi_0)\pm u(r,\varphi+\varphi_0)=v(r,\varphi-\varphi_0)\pm v(r,\varphi+\varphi_0)+A^{\pm}-B^{\pm}, \tag{3.1.6}$$

where

$$v(r,\psi)=\frac{1}{2\alpha}\int\limits_D \frac{e^{-ikr\cos\beta}}{1-e^{i\pi(\beta+\psi)/\alpha}}d\beta, \tag{3.1.7}$$

and

$$A^{\pm}=\sum_{0}^{N;M'}\delta_m'^{+}e^{-ikr\cos\left(\Phi_m^{+}-\varphi\right)}\pm\sum_{0}^{M';N-1}\delta_m^{+}e^{-ikr\cos\left(\Phi_m^{+}+\varphi\right)}, \tag{3.1.8}$$

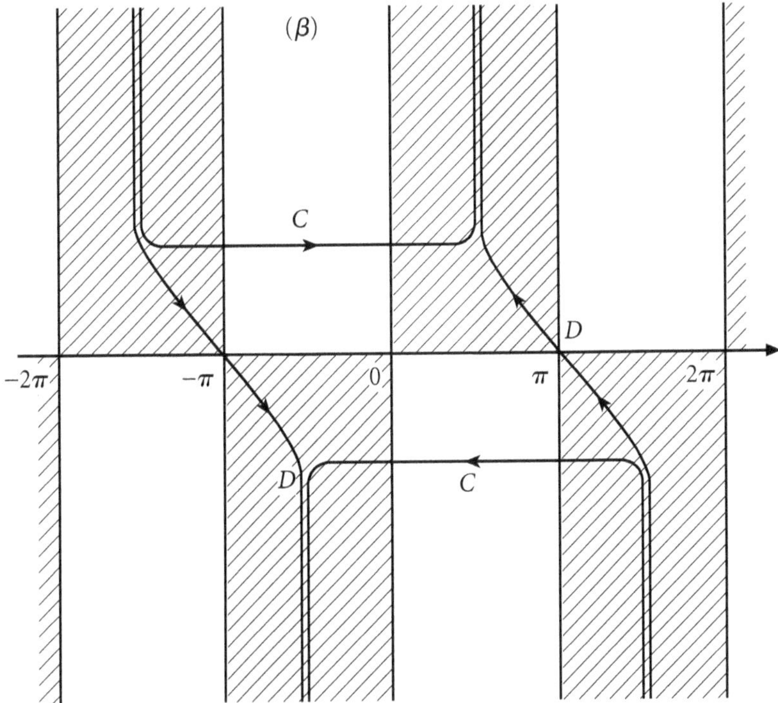

Figure 3.1.1: Contour of integration of Eqn. (3.1.4).

$$B^{\pm} = \sum_{1}^{M;N'-1} \delta_m^- e^{-ikr\cos\left(\Phi_m^- + \varphi\right)} \pm \sum_{1}^{N';M} \delta_m'^- e^{-ikr\cos\left(\Phi_m^- - \varphi\right)} . \tag{3.1.9}$$

Also,

$$\Phi_m^{\pm} = 2\alpha m \pm \varphi_0 , \tag{3.1.10}$$

$$\delta_m^{\pm} = \begin{cases} 1 & \text{for} \quad 0 < \varphi < \pi - \Phi_m^{\pm} \\ 0 & \text{for} \quad \pi - \Phi_m^{\pm} < \varphi < \pi \end{cases} , \tag{3.1.11}$$

$$\delta'^{\pm}_{m} = \begin{cases} 1 & \text{for} \quad \Phi^{\pm}_{m} - \pi < \varphi < \pi \\ 0 & \text{for} \quad 0 < \varphi < \Phi^{\pm}_{m} - \pi \end{cases}. \tag{3.1.12}$$

In addition, M, M', N, N' are whole numbers that satisfy the conditions

$$\frac{\pi-\varphi_0}{2\alpha} < N < \frac{\pi-\varphi_0}{2\alpha}+\frac{1}{2}, \quad \frac{\pi-\varphi_0}{2\alpha}-\frac{1}{2} < M' < \frac{\pi-\varphi_0}{2\alpha}, \tag{3.1.13}$$

$$\frac{\pi+\varphi_0}{2\alpha}-\frac{1}{2} < M < \frac{\pi+\varphi_0}{2\alpha}, \quad \frac{\pi+\varphi_0}{2\alpha} < N' < \frac{\pi+\varphi_0}{2\alpha}+\frac{1}{2}. \tag{3.1.14}$$

The physical meaning of this solution will now be examined.

The function $A^{\pm}$ is composed of the sum of the incident plane wave $\exp[-ikr\cos(\varphi-\varphi_0)]$and the plane waves arising as a result of the multi-bounce reflection of the wave $\pm\exp[-ikr\cos(\varphi+\varphi_0)]$ from the boundary of the horn. The wave $\pm\exp[-ikr\cos(\varphi+\varphi_0)]$ arises due to the reflection of the incident wave $\pm\exp[-ikr\cos(\varphi-\varphi_0)]$ from the $\varphi=0$ boundary. This superposition of waves will be called the A-series. It is terminated by either a wave with the number M', reflected from the $\varphi=0$ boundary (Fig. 3.1.2*a*), or by a wave with the number N, reflected from the $\varphi=\alpha$ boundary (Fig. 3.1.2*b*).

The function $B^{\pm}+\exp[-ikr\cos(\varphi-\varphi_0)]$ is the sum of the incident wave and the waves arising as a result of the multibounce reflection of the wave $\pm\exp[-ikr\cos(2\alpha-\varphi-\varphi_0)]$. The wave $\pm\exp[-ikr\cos(2\alpha-\varphi-\varphi_0)]$ arises due to a reflection of the incident wave from the $\varphi=\alpha$ boundary. This superposition of waves will be called the B-series. It is concluded by a wave with the number M, reflected from the $\varphi=0$ boundary (Fig. 3.1.3*a*) or a wave with the number N' reflected from the $\varphi=\alpha$ (Fig. 3.1.3*b*).

The function $v(r,\psi)$ can easily be presented in the form

$$v(r,\psi)=\frac{i}{2a}\sin\frac{\pi^2}{\alpha}\int_{0^-+i\infty}^{0^+-i\infty}\frac{e^{ikr\cos\zeta}\,d\zeta}{\cos\frac{\pi^2}{\alpha}-\cos\frac{\pi(\psi+\zeta)}{\alpha}}. \tag{3.1.15}$$

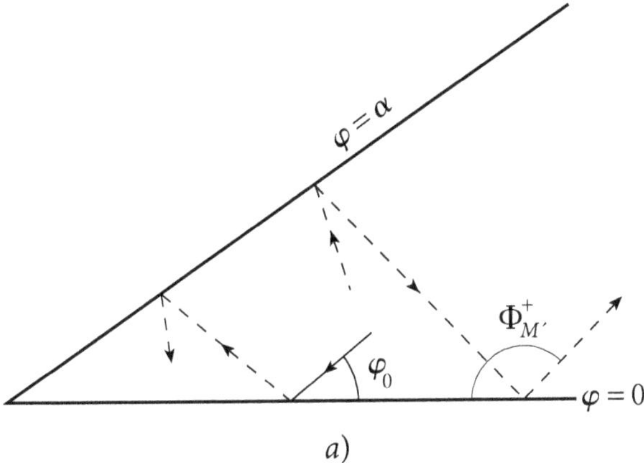

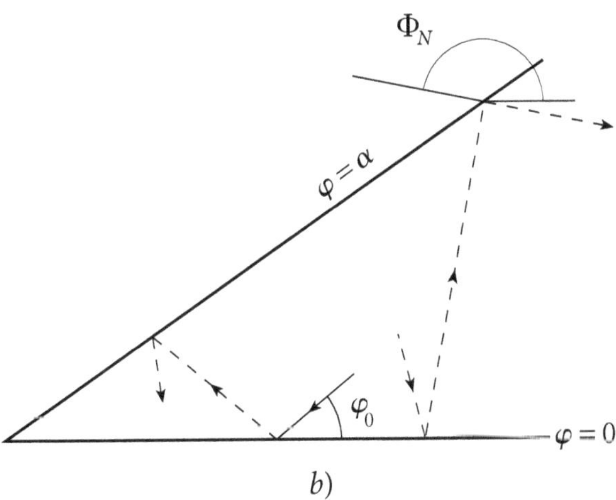

Figure 3.1.2: A-series of plane waves. In case *a*) the series ends with the M´ wave ($\Phi_{M'} = 2\alpha M' + \varphi_0$) with a total number of waves equal to 2M´+2. In case *b*) the series ends with the *N* wave ($\Phi_N = 2\alpha N + \varphi_0$) with a total number of waves equal to $2N+1$.

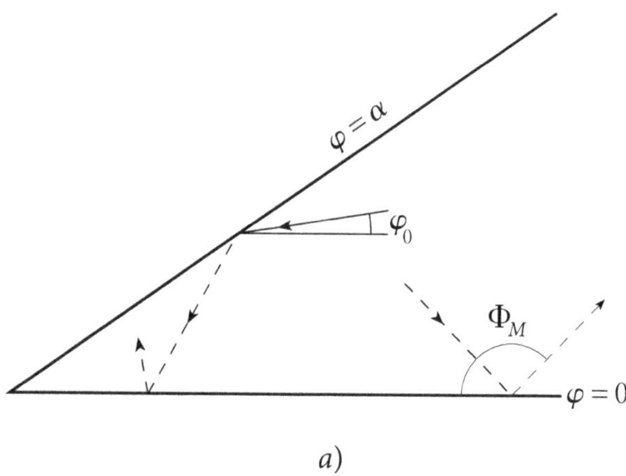

a)

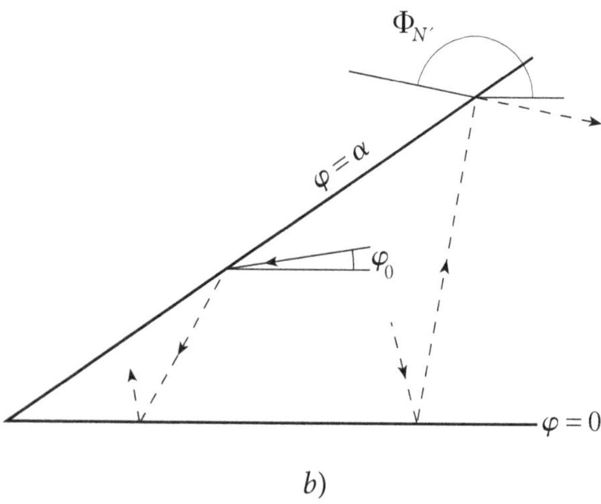

b)

Figure 3.1.3: B-series of plane waves. In case *a)* the series ends with the M wave ($\Phi_M = 2\alpha M - \varphi_0$) with a total number of waves equal to 2M+1. In case *b)* the series ends with the N' wave ($\Phi_{N'} = 2\alpha N' - \varphi_0$) with a total number of waves equal to $2N'$.

It follows that the due to the flare angle $\alpha = \pi/m$ ($m = 1, 2, 3\ldots$), the function $v(r, \psi)$ equals zero and the field inside the horn is only a superposition of plane waves. According to Eqn. (3.1.6), the waves incident upon the $\varphi = 0$ boundary are

$$\begin{bmatrix} E_z \\ H_z \end{bmatrix} = \begin{bmatrix} E_{0z} \\ H_{0z} \end{bmatrix} e^{-ikr\cos\left(\Phi_m^+ - \varphi\right)}, \quad m = 0,1,2,\ldots,M',N-1 \tag{3.1.16}$$

and

$$\begin{bmatrix} E_z \\ H_z \end{bmatrix} = \begin{bmatrix} -E_{0z} \\ H_{0z} \end{bmatrix} e^{-ikr\cos\left(\Phi_m^- - \varphi\right)}, \quad m = 1,2,\ldots,M,N'-1. \tag{3.1.17}$$

The $\varphi = \alpha$ boundary is excited, in a similar fashion, by

$$\begin{bmatrix} E_z \\ H_z \end{bmatrix} = \begin{bmatrix} -E_{0z} \\ H_{0z} \end{bmatrix} e^{-ikr\cos\left(\Phi_m^+ + \varphi\right)}, \quad m = 0,1,2,\ldots,N-1,M'-1 \tag{3.1.18}$$

and

$$\begin{bmatrix} E_z \\ H_z \end{bmatrix} = \begin{bmatrix} E_{0z} \\ H_{0z} \end{bmatrix} e^{-ikr\cos\left(\Phi_m^- + \varphi\right)}, \quad m = 0,1,2,\ldots,M-1,N'-1. \tag{3.1.19}$$

At these boundaries, these waves induce a surface current that creates the scattered field. Utilizing the method of references [3, 13], it is possible to show that the physical optics approximation of this field is equal to

$$\begin{aligned} \frac{E_z}{E_{0z}} &= A^- + B^- + \sum_{m=0}^{M',N-1} v_1\left(\varphi, \Phi_m^+\right) - \sum_{m=1}^{M,N'-1} v_1\left(\varphi, \Phi_m^-\right) - \\ &- \sum_{m=1}^{M',N} v_1\left(\alpha - \varphi, \Phi_m^+ - \alpha\right) + \sum_{m=1}^{M,N'} v_1\left(\alpha - \varphi, \Phi_m^- - \alpha\right), \end{aligned} \tag{3.1.20}$$

$$\frac{H_z}{H_{oz}} = A^+ + B^+ + \sum_{m=0}^{M',N-1} v_2\left(\varphi,\Phi_m^+\right) + \sum_{m=1}^{M,N'-1} v_2\left(\varphi,\Phi_m^-\right) + \\ + \sum_{m=1}^{M',N} v_2\left(\alpha-\varphi,\Phi_m^+-\alpha\right) + \sum_{m=1}^{M,N'} v_2\left(\alpha-\varphi,\Phi_m^- -\alpha\right), \tag{3.1.21}$$

where

$$v_1(\varphi,\phi) = i\frac{\sin\phi}{2\pi}\int_{0^- + i\infty}^{0^+ - i\infty} \frac{e^{ikr\cos\zeta}\,d\zeta}{\cos\phi+\cos(\zeta+\varphi)}, \tag{3.1.22}$$

$$v_2(\varphi,\phi) = -\frac{i}{2\pi}\int_{0^- + i\infty}^{0^+ - i\infty} \frac{\sin(\zeta+\varphi)e^{ikr\cos\zeta}\,d\zeta}{\cos\phi+\cos(\zeta+\varphi)}. \tag{3.1.23}$$

The functions $\nu_{1,2}(\varphi,\,\Phi_m^\pm)$ determine the field associated with the $\varphi = 0$ boundary due its excitation by the $\exp[-ikr\cos(\Phi_m^\pm - \varphi)]$ wave, and the function $\nu_{1,2}(\alpha-\varphi,\,\Phi_m^\pm-\alpha) = \nu_{1,2}(\alpha-\varphi,\,\alpha+\Phi_{m-1}^\pm)$ is the field associated with the $\varphi = \alpha$ boundary due to its excitation by the $\exp[-ikr\cos(\Phi_{m-1}^\pm+\varphi)]$ wave. The +/– superscripts associated with Φ_m^+ (Φ_m^-) indicate that the field being described by these variables is produced by an *A*-series (*B*-series) wave.

A few words are necessary regarding terminology. We refer to expressions described by Eqns. (3.1.20) and (3.1.21) as the physical optics (PO) approximation. The difference between the exact expression for the scattered field given by Eqn. (3.1.3) and PO approximation describes the field generated by the nonuniform or fringe currents. These notions are natural generalizations of analogous notions introduced in Chapter 2 for the analysis of convex surfaces.

The scattered field created by the nonuniform current is determined by the expressions

$$
\begin{aligned}
\frac{E_z}{E_{0z}} &= v(r,\varphi-\varphi_0) - v(r,\varphi+\varphi_0) - \sum_{m=0}^{M',N-1} v_1(\varphi,\Phi_m^+) + \sum_{m=1}^{M,N'-1} v_1(\varphi,\Phi_m^-) + \\
&+ \sum_{m=1}^{M',N} v_1(\alpha-\varphi,\Phi_m^+-\alpha) - \sum_{m=1}^{M,N'} v_1(\alpha-\varphi,\Phi_m^--\alpha)
\end{aligned}
\tag{3.1.24}
$$

$$
\begin{aligned}
\frac{H_z}{H_{0z}} &= v(r,\varphi-\varphi_0) + v(r,\varphi+\varphi_0) - \sum_{m=0}^{M',N-1} v_2(\varphi,\Phi_m^+) - \sum_{m=1}^{M,N'-1} v_2(\varphi,\Phi_m^-) - \\
&- \sum_{m=1}^{M',N} v_2(\alpha-\varphi,\Phi_m^+-\alpha) - \sum_{m=1}^{M,N'} v_2(\alpha-\varphi,\Phi_m^--\alpha).
\end{aligned}
\tag{3.1.25}
$$

Using the method of steepest decent [78], it is not difficult to find that the first term of the asymptotic expansion for Eqns. (3.1.24) and (3.1.25) for $kr \gg 1$ is

$$
\left.
\begin{aligned}
E_z &= E_{0z}\hat{f}^1(\varphi,\varphi_0,\alpha)\frac{e^{i(kr+\pi/4)}}{\sqrt{2\pi kr}}, \\
H_z &= H_{0z}\hat{g}^1(\varphi,\varphi_0,\alpha)\frac{e^{i(kr+\pi/4)}}{\sqrt{2\pi kr}},
\end{aligned}
\right\}
\tag{3.1.26}
$$

where

$$
\begin{aligned}
\hat{f}^1 &= \mathrm{f}(\varphi,\varphi_0) - \sum_{m=0}^{M',N-1} f^0(\varphi,\Phi_m^+) + \sum_{m=1}^{M,N'-1} f^0(\varphi,\Phi_m^-) + \\
&+ \sum_{m=1}^{M',N} f^0(\alpha-\varphi,\Phi_m^+-\alpha) - \sum_{m=1}^{M,N'} f^0(\alpha-\varphi,\Phi_m^--\alpha),
\end{aligned}
\tag{3.1.27}
$$

$$\hat{g}^1 = g(\varphi,\varphi_0) - \sum_{m=0}^{M',N-1} g^0(\varphi,\Phi_m^+) - \sum_{m=1}^{M,N'-1} g^0(\varphi,\Phi_m^-) - \\ - \sum_{m=1}^{M',N} g^0(\alpha-\varphi,\Phi_m^+ - \alpha) - \sum_{m=1}^{M,N'} g^0(\alpha-\varphi,\Phi_m^- - \alpha). \tag{3.1.28}$$

Here, the functions f and g are determined by Eqn. (2.2.4), and the functions f^0 and g^0 are determined by Eqn. (2.2.5). It should be noted that M', N, and M, N' are whole numbers defined by Eqns. (3.1.13) and (3.1.14). For each specific choice of values for φ, φ_0, and α, it is possible to satisfy only one number from each of the pairs M', N, and M, N'. For this reason, the upper bound of the summation index is shown as a pair of possible numbers.

Note that $f^0(\varphi, \Phi_{M'}^+)$, $f^0(\varphi, \Phi_M^-)$, $g^0(\varphi, \Phi_{M'}^+)$, and $g^0(\varphi, \Phi_M^-)$ approach infinity when $\varphi = \pi - \Phi_{M'}^+$ and $\varphi = \pi - \Phi_M^-$. In addition $f^0(\alpha-\varphi, \Phi_N^+-\alpha)$, $f^0(\alpha-\varphi, \Phi_{N'}^- -\alpha)$, $g^0(\alpha-\varphi, \Phi_N^+-\alpha)$, $g^0(\alpha-\varphi, \Phi_{N'}^- -\alpha)$ approach infinity when $\varphi = \Phi_N^+ - \pi$ and $\varphi = \Phi_{N'}^- - \pi$. However, these singularities are compensated by a corresponding singularity in the functions f and g such that $\hat{f}^1$ and $\hat{g}^1$ appear continuous and finite. Thus,

$$\left.\begin{aligned} -f^0(\varphi,\Phi_{M'}^+) - \frac{\frac{1}{n}\sin\frac{\pi}{n}}{\cos\frac{\pi}{n} - \cos\frac{\varphi+\varphi_0}{n}} &= \\ = \frac{1}{2}\cot\Phi_{M'}^+ + \frac{1}{2n}\cot\frac{\pi}{n} & \\ -g^0(\varphi,\Phi_{M'}^+) + \frac{\frac{1}{n}\sin\frac{\pi}{n}}{\cos\frac{\pi}{n} - \cos\frac{\varphi+\varphi_0}{n}} &= \\ = \frac{1}{2}\cot\Phi_{M'}^+ - \frac{1}{2n}\cot\frac{\pi}{n} & \end{aligned}\right\} \text{for } \varphi = \pi - \Phi_{M'}^+ \tag{3.1.29}$$

$$\left.\begin{aligned} & f^0(\varphi,\Phi_M^-)+\frac{\frac{1}{n}\sin\frac{\pi}{n}}{\cos\frac{\pi}{n}-\cos\frac{\varphi-\varphi_0}{n}}= \\ & \qquad =-\frac{1}{2}\cot\Phi_M^- -\frac{1}{2n}\cot\frac{\pi}{n} \\ & -g^0(\varphi,\Phi_M^-)+\frac{\frac{1}{n}\sin\frac{\pi}{n}}{\cos\frac{\pi}{n}-\cos\frac{\varphi-\varphi_0}{n}}= \\ & \qquad =\frac{1}{2}\cot\Phi_M^- -\frac{1}{2n}\cot\frac{\pi}{n} \end{aligned}\right\}\text{for } \varphi=\pi-\Phi_M^- \tag{3.1.30}$$

$$\left.\begin{aligned} & -f^0(\alpha-\varphi,\Phi_{N'}^- -\alpha)-\frac{\frac{1}{n}\sin\frac{\pi}{n}}{\cos\frac{\pi}{n}-\cos\frac{\varphi+\varphi_0}{n}}= \\ & \qquad =\frac{1}{2}\cot\left(\Phi_{N'}^- -\alpha\right)+\frac{1}{2n}\cot\frac{\pi}{n} \\ & -g^0(\alpha-\varphi,\Phi_{N'}^- -\alpha)+\frac{\frac{1}{n}\sin\frac{\pi}{n}}{\cos\frac{\pi}{n}-\cos\frac{\varphi+\varphi_0}{n}}= \\ & \qquad =\frac{1}{2}\cot\left(\Phi_{N'}^- -\alpha\right)-\frac{1}{2n}\cot\frac{\pi}{n} \end{aligned}\right\}\text{for } \varphi=\Phi_{N'}^- -\pi \tag{3.1.31}$$

$$\left.\begin{aligned} &f^0(\alpha-\varphi,\Phi_N^+-\alpha)+\frac{\frac{1}{n}\sin\frac{\pi}{n}}{\cos\frac{\pi}{n}-\cos\frac{\varphi-\varphi_0}{n}}= \\ &\qquad=-\frac{1}{2}\cot\left(\Phi_N^+-\alpha\right)-\frac{1}{2n}\cot\frac{\pi}{n} \\ &-g^0(\alpha-\varphi,\Phi_N^+-\alpha)+\frac{\frac{1}{n}\sin\frac{\pi}{n}}{\cos\frac{\pi}{n}-\cos\frac{\varphi-\varphi_0}{n}}= \\ &\qquad=\frac{1}{2}\cot\left(\Phi_N^+-\alpha\right)-\frac{1}{2n}\cot\frac{\pi}{n} \end{aligned}\right\}\text{for } \varphi=\Phi_N^+-\pi. \quad (3.1.32)$$

Equations (3.1.29) – (3.1.32) remove the indeterminacy arising in functions $\hat{f}^1$ and $\hat{g}^1$.

§3.2 Diffraction at a Circumferential Edge of a Concave Surface of Revolution

These results allow us to calculate edge waves created by a sharp concave bend on a surface of revolution. When the diameter of the edge is greater than the wavelength, it is immediately evident that the edge wave will be approximately the same as the edge wave created at the intersection of two corresponding conic surfaces (Fig. 3.2.1). Our goal is to determine the field radiated by the nonuniform current induced near this edge by the incident plane wave

$$E_x=E_{0x}e^{ikz}, H_x=0. \quad (3.2.1)$$

Using the method outlined in § 2.3, it is not difficult to show that this field in the far zone ($R\gg ka^2, R\gg kl^2$) in the $\vartheta=\pi$ direction may be expressed as

$$E_x = -H_y = \frac{a_1 E_{0x}}{2}\left(\hat{f}^1 - \hat{g}^1\right)\frac{e^{ikR}}{R}e^{2ikl_1}, \quad E_y = H_x = 0, \tag{3.2.2}$$

where a_1 is the radius of the first edge at $z = l_1$,

$$\begin{aligned}\hat{f}^1 &= f(\pi-\omega_2,\pi-\omega_2) - f^0(\pi-\omega_2,\pi-\omega_2) - \\ &\quad - f^0(\omega_1,\omega_1) + \epsilon f^0(\pi-\omega_2,\pi-\omega_2+2\omega_1),\end{aligned} \tag{3.2.3}$$

$$\begin{aligned}\hat{g}^1 &= g(\pi-\omega_2,\pi-\omega_2) - g^0(\pi-\omega_2,\pi-\omega_2) - \\ &\quad - g^0(\omega_1,\omega_1) - \epsilon g^0(\pi-\omega_2,\pi-\omega_2+2\omega_1),\end{aligned} \tag{3.2.4}$$

and

$$\epsilon = \begin{cases} 1 & \text{for} \quad 2\omega_1 < \omega_2 \\ 0 & \text{for} \quad \omega_1 < \omega_2 < 2\omega_1 \end{cases}. \tag{3.2.5}$$

In addition,

$$\left.\begin{matrix} f(\pi-\omega_2,\pi-\omega_2) \\ g(\pi-\omega_2,\pi-\omega_2) \end{matrix}\right\} = \frac{\sin\frac{\pi}{n_1}}{n_1}\left(\frac{1}{\cos\frac{\pi}{n_1}-1} \mp \frac{1}{\cos\frac{\pi}{n_1}-\cos\frac{2\pi-2\omega_2}{n_1}}\right), \tag{3.2.6}$$

$$n_1 = 1 - \frac{\omega_2-\omega_1}{\pi},$$

$$f^0(\pi-\omega_2,\pi-\omega_2) = -g^0(\pi-\omega_2,\pi-\omega_2) = -\frac{1}{2}\tan\omega_2, \tag{3.2.7}$$

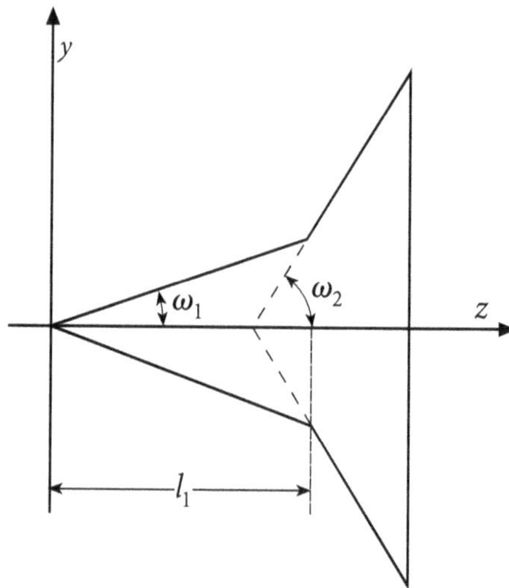

Figure 3.2.1: Cross section of a body of revolution in the *yoz* plane.

$$\left.\begin{aligned} f^0(\pi-\omega_2,\pi-\omega_2+2\omega_1) &= -\frac{\sin(\omega_2-2\omega_1)}{\cos\omega_2+\cos(\omega_2-2\omega_1)}, \\ g^0(\pi-\omega_2,\pi-\omega_2+2\omega_1) &= \frac{\sin\omega_2}{\cos\omega_2+\cos(\omega_2-2\omega_1)}, \end{aligned}\right\} \tag{3.2.8}$$

$$f^0(\omega_1,\omega_1) = -g^0(\omega_1,\omega_1) = \frac{1}{2}\tan\omega_1\,, \tag{3.2.9}$$

and

$$f^0(\pi-\omega_2,\pi-\omega_2+2\omega_1) + g^0(\pi-\omega_2,\pi-\omega_2+2\omega_1) = \tan\omega_1\,. \tag{3.2.10}$$

Equation (3.2.2) can easily be written as

$$E_x = -H_y = -\frac{a_1 E_{0x}}{2} \frac{e^{ikR}}{R} \sum_1^1, \tag{3.2.11}$$

where

$$\sum_1^1 e^{-i2kl_1} = \hat{g}^1 - \hat{f}^1 = \frac{\frac{2}{n_1}\sin\frac{\pi}{n_1}}{\cos\frac{\pi}{n_1} - \cos\frac{2\pi - 2\omega_2}{n_1}} - \tan\omega_2 + \begin{cases} \tan\omega_1 & \text{for} \quad \omega_1 < \omega_2 < 2\omega_1 \\ 0 & \text{for} \quad 2\omega_1 < \omega_2 \end{cases}. \tag{3.2.12}$$

The interval of ω_2, interesting from a practical point of view, is equal to

$$\omega_1 \leqslant \omega_2 \leqslant \frac{\pi}{2}. \tag{3.2.13}$$

It should be noted that when $\omega_2 = \pi/2$, Eqn. (3.2.12) is equal to

$$\hat{g}^1 - \hat{f}^1 = -\frac{1}{n_1}\cot\frac{\pi}{n_1}. \tag{3.2.14}$$

Section 3.4 will apply these results to calculate the radar cross section of realistic bodies possessing a finite cone structure. To this end, it is beneficial if we first use the physical optics approximation to study waves scattered from a semi-infinite cone.

§3.3 Field of a Reflected Conical Wave

Let there be an electromagnetic wave given by Eqn. (3.2.1) incident upon a perfectly conducting semi-infinite cone, the cross section of which in the meridianal plane is shown in Fig. 3.3.1. Recall that Maxwell's equations are equivalent to

$$\nabla^2\mathbf{E}+k^2\mathbf{E}=0\,,\quad \nabla\cdot\mathbf{E}=0 \tag{3.3.1}$$

and

$$\nabla^2\mathbf{H}+k^2\mathbf{H}=0\,,\quad \nabla\cdot\mathbf{H}=0\,. \tag{3.3.2}$$

A solution to **E** and **H** will be sought in the form

$$\mathbf{E}=\mathcal{E}e^{ikS},\quad \mathbf{H}=\mathcal{H}e^{ikS}, \tag{3.3.3}$$

where

$$\left.\begin{aligned}\mathcal{E}&=\mathcal{E}_0+\frac{1}{ik}\mathcal{E}_1+\ldots,\\ \mathcal{H}&=\mathcal{H}_0+\frac{1}{ik}\mathcal{H}_1+\ldots.\end{aligned}\right\} \tag{3.3.4}$$

Substituting these expressions into Eqns. (3.3.1) and (3.3.2) and setting first-order terms in k equal to zero, we obtain the well known geometrical optics relationships

$$(\nabla S)^2=1\,, \tag{3.3.5}$$

$$\mathcal{E}_0\cdot\nabla S=\mathcal{H}_0\cdot\nabla S=0\,, \tag{3.3.6}$$

$$\mathcal{H}_0=-\mathcal{E}_0\times\nabla S,\quad \mathcal{E}_0=\mathcal{H}_0\times\nabla S, \tag{3.3.7}$$

$$\left.\begin{aligned}\mathcal{E}_0\nabla^2S+2(\nabla S\cdot\nabla)\mathcal{E}_0&=0,\\ \mathcal{H}_0\nabla^2S+2(\nabla S\cdot\nabla)\mathcal{H}_0&=0.\end{aligned}\right\} \tag{3.3.8}$$

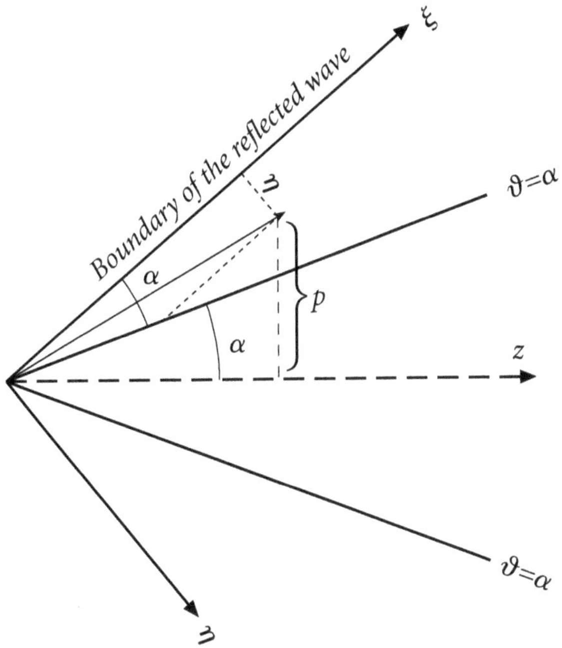

Figure 3.3.1: Semi-infinite cone with flare angle, 2α.

We are interested in the field in the region of the reflected conical wave. Therefore, it is convenient to introduce ray coordinates ξ, η, and ζ (see, Fig. 3.3.1) with Lamé coefficients

$$h_\xi = h_\eta = 1, h_\zeta = p = \xi \sin 2\alpha - \eta \cos 2\alpha . \tag{3.3.9}$$

In these coordinates, the equation for the surface of the cone is

$$\eta = \xi \tan \alpha . \tag{3.3.10}$$

In addition,

$$S=\xi,\quad \nabla S=\mathbf{e}_\xi,\quad \nabla^2 S=\frac{\sin 2\alpha}{\xi\sin 2\alpha-\eta\cos 2\alpha}. \tag{3.3.11}$$

Translating Eqn. (3.3.8) to ray coordinates and enforcing the boundary conditions on the surface of the perfectly conducting cone, we obtain

$$E_\xi=H_\xi=0,\quad E_\eta=H_\zeta=-H_\varphi,\quad H_\eta=-E_\zeta=E_\varphi, \tag{3.3.12}$$

where

$$\left.\begin{aligned} E_\varphi&=-E_{0\varphi}\sqrt{\frac{\eta}{\xi\sin 2\alpha-\eta\cos 2\alpha}}\,e^{ik\xi},\\ H_\varphi&=H_{0\varphi}\sqrt{\frac{\eta}{\xi\sin 2\alpha-\eta\cos 2\alpha}}\,e^{ik\xi},\end{aligned}\right\} \tag{3.3.13}$$

$$E_{0\varphi}=-E_{0x}\sin\varphi,\quad H_{0\varphi}=E_{0x}\cos\varphi. \tag{3.3.14}$$

In spherical coordinates, the reflected field of Eqn. (3.3.13) may be written as

$$\left.\begin{aligned} E_\varphi&=-E_{0\varphi}\sqrt{\frac{\sin(2\alpha-\vartheta)}{\sin\vartheta}}\,e^{ikr\cos(2\alpha-\vartheta)},\\ H_\varphi&=H_{0\varphi}\sqrt{\frac{\sin(2\alpha-\vartheta)}{\sin\vartheta}}\,e^{ikr\cos(2\alpha-\vartheta)},\end{aligned}\right\} \tag{3.3.15}$$

It is interesting to note that for conical waves the amplitude front ($\vartheta=\text{const}$) does not coincide with the phase front ($\xi=\text{const}$).

§3.4 Radar Cross Section of a Conical Body

In this section the previous results are applied to calculate the radar cross section of bodies of revolution shown in Fig. 3.4.1. The incident plane wave is given by Eqn. (3.2.1). The scattered field will be found in the direction of the source ($\vartheta = \pi$).

According to Eqn. (2.4.17), the physical optics component of the scattered field is equal to

$$E_x = -H_y = -\frac{a_1 E_{0x}}{2}\Sigma_0 \frac{\mathrm{e}^{ikR}}{R}, \tag{3.4.1}$$

where

$$\begin{aligned}\Sigma_0 = {} & \frac{1}{ka_1}\tan^2\omega_1 \sin kl_1 e^{ikl_1} - \tan\omega_1 e^{2ikl_1} + \\ & + \left[\frac{1}{ka_1}\tan^2\omega_2 \sin kl_2 e^{ikl_2} + \left(1 - \frac{a_2}{a_1}e^{2ikl_2}\right)\tan\omega_2\right]e^{2ikl_1}.\end{aligned} \tag{3.4.2}$$

The value of the radar cross section will be

$$\sigma_0 = \pi a_1^2 \left|\Sigma_0\right|^2. \tag{3.4.3}$$

The nonuniform current, induced by the incident plane wave near the edge at $z = l_1 + l_2, x^2 + y^2 = a_2^2$, radiates a field in the $\vartheta = \pi$ direction given by

$$E_x = -H_y = -\frac{a_1 E_{0x}}{2}\frac{e^{ikR}}{R}\Sigma_0^1, \tag{3.4.4}$$

where

$$\Sigma_0^1 = \frac{a_2}{a_1}\left(\frac{\frac{2}{n_2}\sin\frac{\pi}{n_2}}{\cos\frac{\pi}{n_2} - \cos\frac{2\omega_2}{n_2}} + \tan\omega_2\right)e^{2ik(l_1+l_2)}, \tag{3.4.5}$$

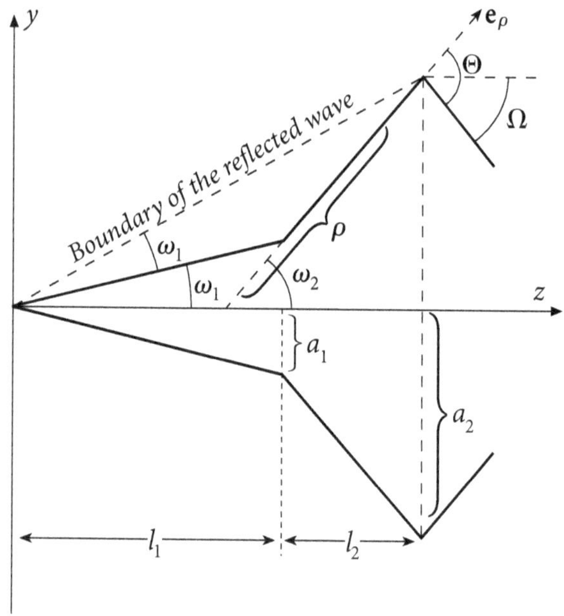

Figure 3.4.1: Cross section of a conical body in the *yoz* plane.

and

$$n_2 = 1 + \frac{\omega_2 + \Omega}{\pi}.$$

This equation follows directly from Eqn. (2.4.18) if we eliminate the terms with phase $\exp[2ikl_1]$ and replace ω_1 with ω_2. Summing Eqns. (3.4.1) and (3.4.4), we obtain an approximate expression for the scattered field

$$E_x = -H_y = -\frac{a_1 E_{0x}}{2} \frac{e^{ikR}}{R} \Sigma_1 , \tag{3.4.6}$$

where

$$\Sigma_1 = \frac{1}{ka_1}\tan^2\omega_1 \sin kl_1 e^{ikl_1} + \frac{1}{ka_1}\tan^2\omega_2 \sin kl_2 e^{ik(l_2+2l_1)} +$$

$$+(\tan\omega_2 - \tan\omega_1)e^{2ikl_1} + \frac{a_2}{a_1}\frac{\frac{2}{n_2}\sin\frac{\pi}{n_2}}{\cos\frac{\pi}{n_2} - \cos\frac{2\omega_2}{n_2}} e^{2ik(l_1+l_2)}. \qquad (3.4.7)$$

Incorporating Eqn. (3.2.11), which describes the field radiated by the non-uniform current induced in the vicinity of the edge at $z = l_1, x^2 + y^2 = a_1^2$, we obtain a more accurate expression

$$E_x = -H_y = -\frac{a_1 E_{0x}}{2}\frac{e^{ikR}}{R}\left(\Sigma_1 + \Sigma_1^1\right), \qquad (3.4.8)$$

where Σ_1^1 is determined according to Eqn. (3.2.12).

Equations (3.4.4), (3.4.5), (3.4.8) represent the fields caused only by the diffraction of the incident wave. However, due to the condition that $\omega_1 < \omega_2 < 2\omega_1$, the surface of the body bound by $l_1 \le z \le l_1 + l_2$ is illuminated not only by the plane wave, but also by the conical wave scattered by the front section of the body ($0 \le z \le l_1$). This conical wave is given by Eqn. (3.3.15). Using the notations shown in Fig. 3.4.1, the tangential components of this wave on the body surface, $l_1 < z < l_1 + l_2$, can be represented by:

$$\begin{bmatrix} H_\varphi \\ H_\rho \end{bmatrix} = \begin{bmatrix} \cos\varphi \\ -\sin\varphi\sin(\omega_2 - 2\omega_1) \end{bmatrix} E_{0x} Q e^{ik\xi}, \qquad (3.4.9)$$

where

$$Q = \left[\frac{2a_1\sin(\omega_2 - \omega_1)\cos\omega_1 - \rho\sin\omega_2\sin(\omega_2 - 2\omega_1)}{\rho\sin^2\omega_2}\right]^{1/2}, \qquad (3.4.10)$$

$$\xi = l_1 - a_1\frac{\cos(\omega_2 - 2\omega_1)}{\sin\omega_2} + \rho\cos(\omega_2 - 2\omega_1). \qquad (3.4.11)$$

This field excites a surface current, the density of which according to geometrical optics is

$$\left.\begin{aligned} j_x &= \frac{c}{2\pi} E_{0x} \left[\cos^2 \varphi \sin\omega_2 - \sin^2 \varphi \sin(\omega_2 - 2\omega_1) \right] Q e^{ik\xi} , \\ j_y &= \frac{c}{2\pi} E_{0x} \sin 2\varphi \sin(\omega_2 - \omega_1) \cos\omega_1 Q e^{ik\xi} . \end{aligned}\right\} \tag{3.4.12}$$

The field radiated by this current in the $\vartheta = \pi$ direction is equal to

$$E_x = -H_y = -\frac{a_1 E_{0x}}{2} \frac{e^{ikR}}{R} \Sigma_2^0 , \tag{3.4.13}$$

where

$$\Sigma_2^0 = \frac{2}{ika_1} \cos(\omega_2 - \omega_1) \sin\omega_1 \mathrm{I} e^{ika_1\phi} , \tag{3.4.14}$$

$$\mathrm{I} = \int\limits_{k\rho_1}^{k\rho_2^0} \sqrt{ka_1 \tilde{\lambda} t - \tilde{\mu} t^2} \, e^{i\nu t} dt . \tag{3.4.15}$$

Here,

$$\tilde{\lambda} = \frac{2\sin(\omega_2 - \omega_1)\cos\omega_1}{\sin^2 \omega_2} , \quad \tilde{\mu} = \frac{\sin(\omega_2 - 2\omega_1)}{\sin\omega_2} , \tag{3.4.16}$$

$$\nu = 2\cos(\omega_2 - \omega_1)\cos\omega_1 , \quad \phi = 2\frac{\cos^2 \omega_1}{\sin\omega_1} \frac{\sin(\omega_2 - \omega_1)}{\sin\omega_2} , \tag{3.4.17}$$

$$\rho_1 = \frac{a_1}{\sin\omega_2} , \tag{3.4.18}$$

$$\rho_2^0 = \begin{cases} \dfrac{a_2}{\sin\omega_2} & \text{for} \quad a_2 \cot 2\omega_1 \leqslant l_1 + l_2 \\ 2a_1 \dfrac{\sin(\omega_2 - \omega_1)\cos\omega_1}{\sin(\omega_2 - 2\omega_1)\sin\omega_2} & \text{for} \quad a_2 \cot 2\omega_1 \geqslant l_1 + l_2 \end{cases} . \qquad (3.4.18, \text{cont.})$$

It is also understood that $2\omega_1 < \omega_2 < \pi/2$. The integral of Eqn. (3.4.15) is not contained in any known table of integrals. Therefore, numerical methods must be used.

The reflected conical wave given by Eqn. (3.4.9) generates on the second section of the body both the uniform and nonuniform currents. The field radiated by the uniform currents has already been determined. It is determined by Eqn. (3.4.13). The main component of the nonuniform currents is the fringe current in the vicinity of the edge $z = l_1+l_2$, $x^2+y^2 = a_2^2$. The problem is to find the field radiated by the fringe current. Calculation this field requires some simplifying assumptions. Assume that field of the conical wave (near each sufficiently small element of the edge) may be approximated as a plane wave propagating in the $\vartheta = 2\omega_1$ direction. The amplitude and phase of the equivalent plane wave at the edge is chosen to be equal to the amplitude and phase of the corresponding conical wave. Further, let us assume that the nonuniform current induced near the circular discontinuity will be the same as that induced near the edge of a corresponding wedge. As usual, such an assumption may be considered valid provided $ka \gg 1$.

The field radiated in the $\vartheta = \pi$ direction by this fringe current can be evaluated using Eqn. (2.3.16), taking into account that

$$E_{0z_1} = -E_{0x} \sin\psi \cdot Q_0 e^{ik\xi_0}, \quad H_{0z_1} = -E_{0x} \cos\psi \cdot Q_0 e^{ik\xi_0}, \qquad (3.4.19)$$

$$f^1 = f(\varphi_1, \varphi_1^0) - f^0(\varphi_1, \varphi_1^0), \quad g^1 = g(\varphi_1, \varphi_1^0) - g^0(\varphi_1, \varphi_1^0), \qquad (3.4.20)$$

where

$$Q_0 = \left[\frac{2ka_1 \sin(\omega_2 - \omega_1)\cos\omega_1 - ka_2 \sin\omega_2 \sin(\omega_2 - 2\omega_1)}{ka_2 \sin\omega_2} \right]^{1/2}, \qquad (3.4.21)$$

$$\xi_0 = l_1 + (a_2 - a_1)\frac{\cos(\omega_2 - 2\omega_1)}{\sin\omega_2}, \tag{3.4.22}$$

$$\varphi_1 = \omega_2, \quad \varphi_1^0 = \omega_2 - 2\omega_1, \quad n_2 = 1 + \frac{\omega_2 + \Omega}{\pi}. \tag{3.4.23}$$

Here, n_2 is entered into Eqn. (2.2.4) for the functions f and g. As a result, the components of the vector potential of the scattered field are

$$\begin{aligned} A_x &= -\frac{a_2 E_{0x}}{2ik}\left(f^1 + g^1\right)Q_0 e^{ik\xi_0}\frac{e^{ikR}}{R}e^{ik(l_1+l_2)}, \\ A_y &= 0. \end{aligned} \tag{3.4.24}$$

This leads to

$$E_x = -H_y = -\frac{a_1 E_{0x}}{2}\frac{e^{ikR}}{R}\Sigma_2^1, \tag{3.4.25}$$

where

$$\Sigma_2^1 = \frac{a_2}{a_1}\left(f^1 + g^1\right)Q_0 e^{ik(\xi_0 + l_1 + l_2)}, \tag{3.4.26}$$

and

$$f^1 + g^1 = \frac{\frac{2}{n_2}\sin\frac{\pi}{n_2}}{\cos\frac{\pi}{n_2} - \cos\frac{2\omega_1}{n_2}} + \tan\omega_1. \tag{3.4.27}$$

We will note that

$$Q_0 = 0 \quad \text{for} \quad l_1 + l_2 = a_2\cot 2\omega_1, \tag{3.4.28}$$

Therefore, when $l_1 + l_2 < a_2\tan 2\omega_1$, we assume that $\Sigma_2^1 = 0$.

As a result, the field in the $\vartheta = \pi$ direction may be written (with the inclusion of all of the aforementioned effects) as

$$E_x = -H_y = -\frac{a_1 E_{0x}}{2} \frac{e^{ikR}}{R} \Sigma, \tag{3.4.29}$$

where

$$\Sigma = \Sigma_0 + \Sigma_0^1 + \Sigma_1^1 + \Sigma_2^0 + \Sigma_2^1 . \tag{3.4.30}$$

The radar cross section is determined by Σ and is equal to

$$\sigma = \pi a_1^2 |\Sigma|^2 . \tag{3.4.31}$$

One should observe that this equation holds true for the following inequalities

$$\left.\begin{array}{l} ka_1 \gg 1, \quad ka_2 \gg 1, \quad 0 < \omega_1 < \dfrac{\pi}{2}, \\ 2\omega_1 < \omega_2 \leqslant \dfrac{\pi}{2}, \quad \omega_2 < \theta \leqslant \pi, \quad 0 < \Omega \leqslant \pi - \omega_2 . \end{array}\right\} \tag{3.4.32}$$

§3.5 Numerical Calculation of Radar Cross Section

The equations derived in § 3.4, were evaluated numerically to calculate and construct the following curves in Figs. 3.5.1 and 3.5.2:

Curve1: $10\log|\Sigma_0|^2$

Curve2: $10\log|\Sigma_0 + \Sigma_0^1|^2$

Curve3: $10\log|\Sigma_0 + \Sigma_0^1 + \Sigma_1^1|^2$

Curve4: $10\log|\Sigma_0 + \Sigma_0^1 + \Sigma_1^1 + \Sigma_2^1|^2$

Curve5: $10\log|\Sigma_0 + \Sigma_0^1 + \Sigma_1^1 + \Sigma_2^1 + \Sigma_2^0|^2$

Curve6: $10\log|\Sigma_0 + \Sigma_0^1 + \Sigma_1^1 + \Sigma_2^0|^2$

The calculations were conducted for the following values of parameters: $ka_1 = 2\pi, ka_2 = 5\pi, kl_1 = 15\pi, \tan\omega_1 = 2/15, \Omega = \pi/2, \omega_1 < \omega_2 < \pi/2.$

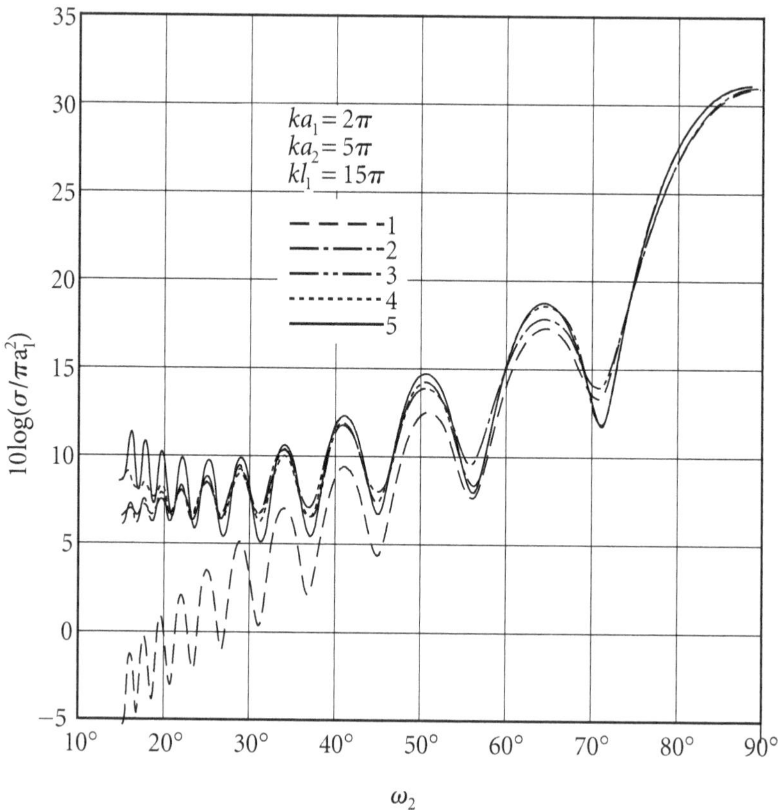

Figure 3.5.1: Dependence of the RCS of a conical body on its form.

Let us clarify the meaning of these curves. Curve 1 corresponds to the physical optics approximation under the condition that the surface currents are excited only by the incident plane wave given by Eqn. (3.2.1). Curve 2 includes the edge effect at $z = l_1 + l_2, x^2+y^2 = a_2^2$ created by the incident plane wave. The difference between the curves 1 and 2 is 3 dB for $\omega_2 = 45°$ which increases to 11 dB for $\omega_2 \approx 2\omega_1$.

Curve 3 includes the edge effect introduced by both edges, but under the condition that the edge at $z = l_1 + l_2, x^2+y^2 = a_2^2$ is illuminated only by a plane wave. By comparing Curves 2 and 3, it is possible to estimate the effect

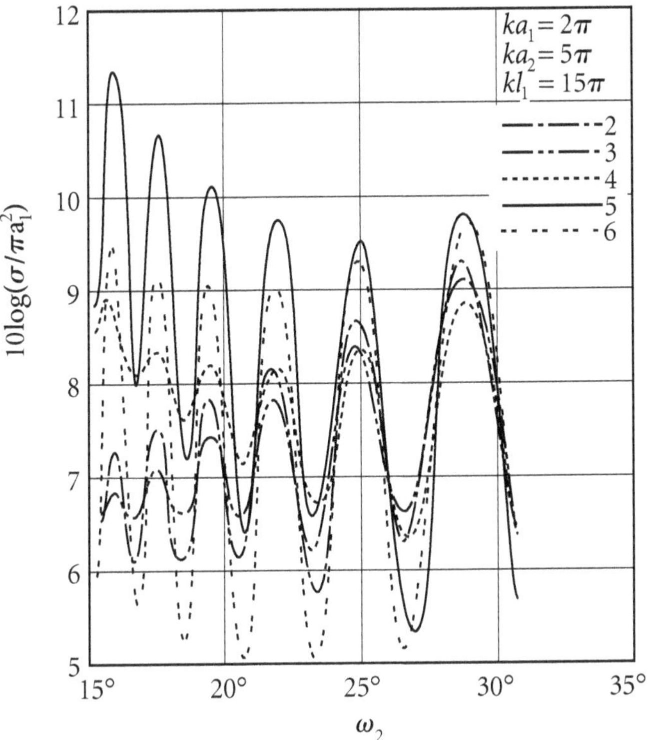

Figure 3.5.2: The RCS of a conical body.

of the nonuniform current arising at $z = l_1, x^2+y^2 = a_1^2$ due to the concave edge. The difference between these curves does not exceed 1 dB for $\omega_2 \leq 50°$ and attains the greatest values (~2 dB) at minima where $\omega_2 \approx 56°, 70°$.

Curve 4 differs from Curve 3 in that it includes the influence of the nonuniform current excited by the conical wave near the edge at $z = l_1+l_2$, $x^2+y^2 = a_2^2$. When $\omega_2 > 41°$ both curves flow together because the conical wave no longer illuminates this discontinuity. In other words, when $\omega_2 > 41°$ this edge is located in the shadow region with respect to the incident conical wave.

Lastly, Curve 5 shows the total scattered field. In contrast, to Curve 4, these data incorporate (within the physical optics approximation!) the diffraction of the reflected conical wave from the second section of the body $(z > l_1)$. This curve characterizes, therefore, the diffraction interaction of the two conical surfaces which form this body. Curve 5 needs to be compared with Curve 2 to estimate the effect of the diffraction interaction of different elements of a convex body. It can be seen that the maximum deviation of Curve 5 from Curve 2 is 5 dB. It is interesting to mention that the greatest differences between Curve 5 and Curve 2 occur for $\omega_2 < 25°$ at the maxima and for $\omega_2 > 25°$ at the minima.

To better present the relative influence of the uniform and nonuniform surface currents excited on the body $(l_1 < z < l_1 + l_2)$ by the reflected conical wave, a sixth curve is included in Fig. 3.5.2.

§3.6 Additional Comments

- The theory presented in this chapter was developed by the author in 1964–1965. The asymptotic expressions in § 3.1 for the edge waves inside a wedge-shaped horn were also derived in 1982 by Tran and Kim [79].
- Some results for scattering from concave dihedrals and trihedrals are presented in [18, 70, 99, 111, 151–153] of reference [40].
- Recent numerical results for the corner wave scattered from a semi-infinite perfectly conducting cones are presented in [80, 81]. They were obtained by exact solutions of relevant boundary value problems for Maxwell's equations.
- Paper [198] contains investigation of the corner waves diffracted at semi-infinite impedance cones.
- A review on asymptotic investigation of the edge waves diffracted at structures with concave and convex faces is presented in [199].

4

Measurement of Radiation from Diffraction / Nonuniform Currents

Chapters 2 and 3 theoretically researched the field radiated by the nonuniform part of the current. This chapter presents a method of measuring this field and examine the depolarization phenomenon of the reflected field.

The method of measuring the field radiated by the nonuniform current was first proposed for bodies of revolution in an article by E. N. Maizel's and the author [7]. Subsequently, it is shown that this method is universal in nature and is suitable for any metallic object [11].

§4.1 Backscattering of Waves with Circular Polarization

Let free space contain a perfectly conducting body of arbitrary shape, an element of which is displayed in Fig. 4.1.1. A system of coordinates is chosen such that its origin lies near the body and the source Q is positioned in the $x = 0$ plane. If the distance between the body and the source is much greater than the dimension of the body, then the incident wave can be considered to be locally plane. Let it be defined by

$$E_x = E_{0x} e^{ik(y\sin\gamma + z\cos\gamma)}, \quad E_y = 0. \tag{4.1.1}$$

Here, γ is the angle between the direction of propagation $\mathbf{N}$ and the z-axis.

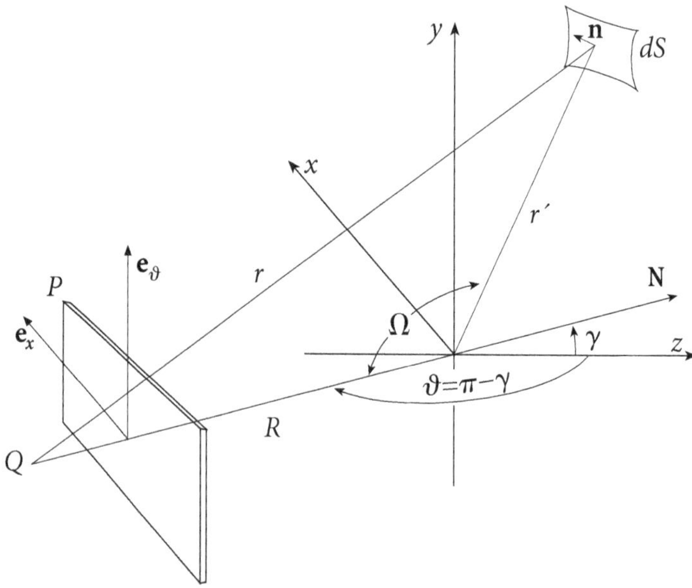

Figure 4.1.1: Diffraction from an arbitrary body. **N** is the direction of propagation of the incident wave, dS is the differential element of the body surface, Q is the source, and P is the polarizer transforming a linearly polarized wave into a circularly polarized wave. Reprinted from [11] with the permission of *Radiotekhnika i elektronika.*

Let us place in front of the source and parallel to the wavefront, a polarizer, P, which transforms linearly polarized radiation into circularly polarized radiation. After passing through the polarizer let there be a field with an electric vector E_n that lags 90° out of phase with the field with electric vector E_τ (Fig. 4.1.2). In this case, the polarizer creates right-hand circular polarization. A simple example of such a polarizer is a system of metallic strips, parallel to E_τ. As a result, the field of the incident wave at the coordinate origin will be equal to

$$E_x = \frac{e^{i\frac{\pi}{4}}}{\sqrt{2}} E_{0x}, \quad H_x = \frac{e^{-i\frac{\pi}{4}}}{\sqrt{2}} E_{0x}. \tag{4.1.2}$$

The field scattered by the body may be described in the far field as

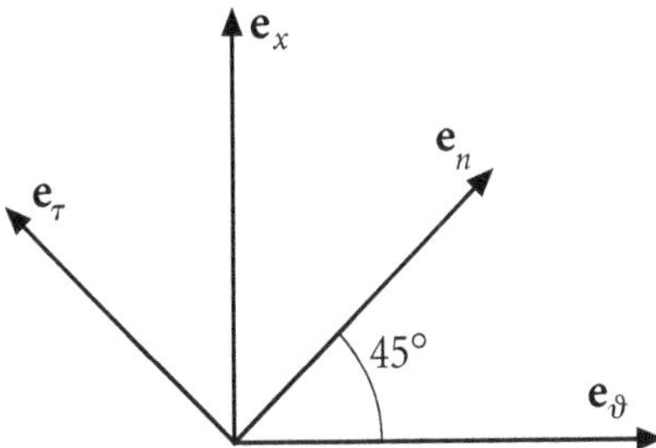

Figure 4.1.2: Components of the electric field after passage through the polarizer.

$$\left.\begin{aligned} E_\varphi = -H_\vartheta &= \frac{iaE_{0x}}{2}\frac{e^{i\frac{\pi}{4}}}{\sqrt{2}}\overline{\Sigma}(\vartheta,\varphi)\frac{e^{ikR}}{R}, \\ E_\vartheta = H_\varphi &= \frac{iaE_{0x}}{2}\frac{e^{-i\frac{\pi}{4}}}{\sqrt{2}}\Sigma(\vartheta,\varphi)\frac{e^{ikR}}{R}, \end{aligned}\right\} \tag{4.1.3}$$

where a is a length characterizing the dimension of the body and $\overline{\Sigma}(\vartheta,\varphi)$ and $\Sigma(\vartheta,\varphi)$ are scattering patterns. In the general case, the field given by Eqn. (4.1.3) is an elliptically polarized wave. In the direction of the source ($\vartheta = \pi - \gamma, \varphi = -\pi/2$), this wave transmits through the polarizer and generates the field

$$\left.\begin{aligned} E_x = -H_\vartheta &= \frac{iaE_{0x}}{2}\Sigma_+(\vartheta,\varphi)\frac{e^{ikR}}{R}e^{i\frac{\pi}{2}}, \\ E_\vartheta = H_x &= \frac{iaE_{0x}}{2}\Sigma_-(\vartheta,\varphi)\frac{e^{ikR}}{R}, \end{aligned}\right\} \tag{4.1.4}$$

where

$$\Sigma_\pm = \frac{1}{2}\left(\Sigma \pm \overline{\Sigma}\right). \tag{4.1.5}$$

If the source radiates a field with the other polarization ($\mathbf{H}_0 \perp yoz$), then the wave reflected from the body and transmitted through the polarizer is described at Q by analogous relationships given by

$$\left.\begin{aligned} H_x &= E_\vartheta = \frac{iaH_{0x}}{2}\Sigma_+(\vartheta,\varphi)\frac{e^{ikR}}{R}e^{i\frac{\pi}{2}}, \\ H_\vartheta &= -E_x = \frac{iaH_{0x}}{2}\Sigma_-(\vartheta,\varphi)\frac{e^{ikR}}{R}. \end{aligned}\right\} \tag{4.1.6}$$

Using the physical optics approximation, we will now examine the diffraction of a linearly polarized plane wave from this body. According to definition, the uniform current excited on the surface of the body by a plane wave is equal to

$$\left.\begin{aligned} j_x^0 &= -\frac{c}{2\pi}E_{0x}\left(n_y \sin\gamma + n_z \cos\gamma\right)e^{i\psi}, \\ j_y^0 &= \frac{c}{2\pi}E_{0x}n_x \sin\gamma e^{i\psi}, \\ j_z^0 &= \frac{c}{2\pi}E_{0x}n_x \cos\gamma e^{i\psi}, \end{aligned}\right\} \tag{4.1.7}$$

for an E-polarized wave ($\mathbf{E}_0 \perp yoz$) and

$$\left.\begin{aligned} j_x^0 &= 0, \\ j_y^0 &= \frac{c}{2\pi}H_{0x}n_z e^{i\psi}, \\ j_z^0 &= -\frac{c}{2\pi}H_{0x}n_y e^{i\psi}, \end{aligned}\right\} \tag{4.1.8}$$

for an H-polarized wave ($\mathbf{H}_0 \perp yoz$). Here, E_{ox} and H_{ox} are the amplitudes of the electric and magnetic fields of the incident wave, respectively, for E- and H-polarization, $\psi = k(y'\sin\gamma + z'\cos\gamma)$ is the phase of the incident wave at the point (x', y', z') on the surface of the body, and n_x, n_y, n_z, are the components of the surface normal.

Using these expressions for the current and vector potentials we can determine the scattered field

$$E_\varphi = -H_\vartheta = ikA_\varphi, \quad E_\vartheta = H_\varphi = ikA_\vartheta . \tag{4.1.9}$$

For E-polarization it is equal to

$$\begin{aligned} E_\varphi = -H_\vartheta &= \frac{ikE_{0x}}{2\pi}\frac{e^{ikR}}{R}\int\big[n_x \sin\gamma\cos\varphi + \\ &\quad + (n_y \sin\gamma + n_z \cos\gamma)\sin\varphi\big]e^{i\phi}dS, \\ E_\vartheta = H_\varphi &= \frac{ikE_{0x}}{2\pi}\frac{e^{ikR}}{R}\int\big[n_x(\sin\gamma\cos\vartheta\sin\varphi - \cos\gamma\sin\vartheta) - \\ &\quad - (n_y \sin\gamma + n_z \cos\gamma)\cos\varphi\cos\vartheta\big]e^{i\phi}dS, \end{aligned} \tag{4.1.10}$$

and for H-polarization

$$\begin{aligned} E_\varphi = -H_\vartheta &= \frac{ik}{2\pi}H_{0x}\cos\varphi\frac{e^{ikR}}{R}\int n_z e^{i\phi}dS, \\ E_\vartheta = H_\varphi &= \frac{ik}{2\pi}H_{0x}\frac{e^{ikR}}{R}\int(n_y \sin\vartheta + n_z \sin\varphi\cos\vartheta)e^{i\phi}dS. \end{aligned} \tag{4.1.11}$$

Here, R, ϑ, φ, are spherical coordinates at the point of observation, and $\phi = \psi - kr'\cos\Omega$. The integrals are evaluated over the illuminated region of the body. For the monostatic case, when the observation point and source point coincide ($\vartheta = \pi - \gamma$, $\varphi = -\pi/2$), Eqns. (4.1.10) and (4.1.11) respectively show that

$$E_x = -H_\vartheta = -\frac{ik}{2\pi} E_{0x} \frac{e^{ikR}}{R} \int (n_y \sin\gamma + n_z \cos\gamma) e^{i\phi} dS, \tag{4.1.12}$$

$$E_\vartheta = H_x = 0$$

and

$$E_\vartheta = H_x = \frac{ik}{2\pi} H_{0x} \frac{e^{ikR}}{R} \int (n_y \sin\gamma + n_z \cos\gamma) e^{i\phi} dS, \tag{4.1.13}$$

$$E_x = H_\vartheta = 0.$$

Further assuming that the amplitudes of the incident wave are given by Eqn. (4.1.2), we can write Eqns. (4.1.12) and (4.1.13) as

$$\left.\begin{aligned} E_x &= -H_\vartheta = \frac{iaE_{0x}}{2} \frac{e^{i\frac{\pi}{4}}}{\sqrt{2}} \frac{e^{ikR}}{R} \overline{\Sigma}^0, \\ E_\vartheta &= H_x = \frac{iaE_{0x}}{2} \frac{e^{-i\frac{\pi}{4}}}{\sqrt{2}} \frac{e^{ikR}}{R} \Sigma^0, \end{aligned}\right\} \tag{4.1.14}$$

where

$$\Sigma^0 = -\overline{\Sigma}^0 = \frac{k}{\pi a} \int (n_y \sin\gamma + n_z \cos\gamma) e^{i\phi} dS. \tag{4.1.15}$$

Let us introduce the angular functions of the scattered field given by Eqn. (4.1.3) as

$$\overline{\Sigma} = \overline{\Sigma}^0 + \overline{\Sigma}^1, \quad \Sigma = \Sigma^0 + \Sigma^1, \tag{4.1.16}$$

where the functions $\Sigma^0, \overline{\Sigma}^0$ and $\Sigma^1, \overline{\Sigma}^1$ are related to the fields radiated by the *uniform* and *nonuniform* currents, respectively. Substituting these expressions into Eqns. (4.1.4) and (4.1.6) and utilizing Eqn. (4.1.15), we determine the scattered field passing through the polarizer to the receiver at Q. In the case of E-polarization it is equal to

$$\left.\begin{aligned} E_x = -H_\vartheta &= \frac{iaE_{0x}}{4}\left(\Sigma^1 + \overline{\Sigma}^1\right)\frac{e^{ikR}}{R}e^{i\frac{\pi}{2}}, \\ E_\vartheta = H_x &= \frac{iaE_{0x}}{4}\left(2\Sigma^0 + \Sigma^1 - \overline{\Sigma}^1\right)\frac{e^{ikR}}{R}, \end{aligned}\right\} \qquad (4.1.17)$$

and for the case of H-polarization we have

$$\left.\begin{aligned} H_x = E_\vartheta &= \frac{iaH_{0x}}{4}\left(\Sigma^1 + \overline{\Sigma}^1\right)\frac{e^{ikR}}{R}e^{i\frac{\pi}{2}}, \\ H_\vartheta = -E_x &= \frac{iaH_{0x}}{4}\left(2\Sigma^0 + \Sigma^1 - \overline{\Sigma}^1\right)\frac{e^{ikR}}{R}. \end{aligned}\right\} \qquad (4.1.18)$$

The physical meaning of these results can be summarized as follows. The field scattered by the body at Q is the sum of two waves polarized in mutually perpendicular directions. The reflected wave, polarized in the same manner as the incident radiation from source, is described by the function $\Sigma_+ = (\Sigma^1 + \overline{\Sigma}^1)/2$ and is related only to the nonuniform current. The reflected wave with perpendicular polarization is described by the function $\Sigma_- = (2\Sigma_0 + \Sigma^1 - \overline{\Sigma}^1)/2$ and is a field radiated by both currents. Note that in general Σ^1 and $\overline{\Sigma}^1$ are not equal and therefore do not cancel in the expression for Σ_-. In other words, field radiated by the uniform current is not separable from the scattered field.

Thus, this method separates the scattered component resulting from the distortion of the surface of any perfectly conducting body of finite dimension (bend, fracture, point, protuberance, aperture, etc.) from the total scattered field. It should be noted that for the case of scattering of an electromagnetic wave by a group of scatterers, the separated component of the field results from not only surface perturbations, but also from mutual diffraction occurring between the scatterers. This last case is very important because it is particularly connected to the study of the diffraction from groups of scatterers, which is presently of significant practical interest.

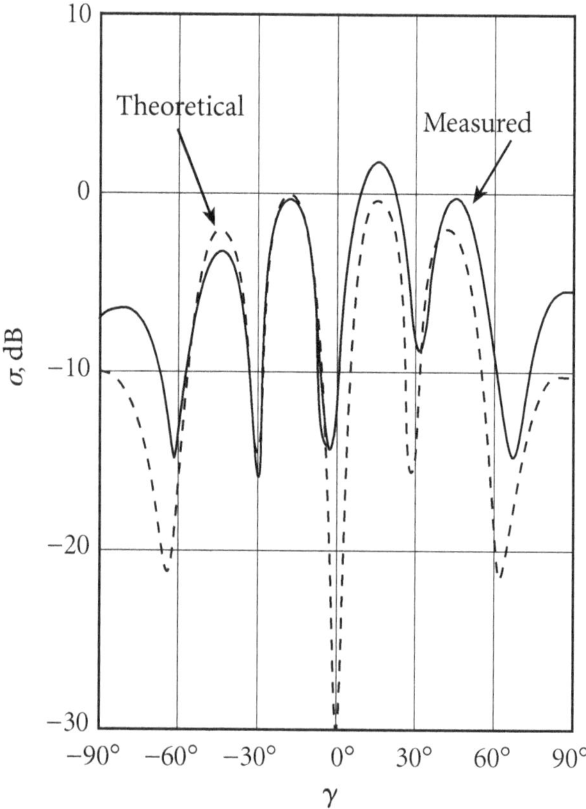

Figure 4.1.3: Backscattering from a disk. The field radiated only by the nonuniform/fringe current. Reprinted from [7] with the permission of *Radiotekhnika i elektronika.*

It is necessary, however, to keep in mind that the described separation of the scattered field cannot be realized in an arbitrary direction of observation, but only in those directions for which $\Sigma^0 = -\overline{\Sigma}^0$, as is the case, for instance, in the direction of the source.

Figure 4.1.3 presents results of the measurements and calculations of radar cross section

$$\sigma^{+} = \pi a^2 \left|\Sigma_{+}\right|^2 = \frac{1}{4}\pi a^2 \left|\Sigma^1 + \overline{\Sigma}^1\right|^2, \tag{4.1.19}$$

which is created by the nonuniform part of the current excited on a disk by a plane wave. The diameter of the disk is equal to $2a = 5\lambda/\pi$. Calculations were conducted taking into account second-order diffraction using approximate expressions for Σ and $\overline{\Sigma}$ derived in reference [6,13]. Since the preparation of a very thin planar disk is difficult, measurements were conducted using an obtuse cone, closely resembling a disk with a height equal to approximately one-tenth its diameter.

As is shown in Fig. 4.1.3, the theoretical and experimental curves are sufficiently close to one another. Some differences between them, especially in the range where γ approaches 90°, can be explained by the conical form of the target, as well as the approximate nature of the derived expressions. The value $\gamma = 90°$ corresponds to the direction along the surface of the disk, and the value $\gamma = 0°$ is the direction normal to the disk.

§4.2 Depolarization of Backscattering

The concept of the uniform and nonuniform surface currents can explain the physical nature of the depolarization phenomena of reflected signals. The scattering of an electromagnetic wave from an arbitrary metallic body will now be reexamined. The relative position of the source Q, the element of the surface of the illuminated body, and the system of coordinates are shown in Fig. 4.1.1. Once again, the source Q is located in the yoz plane and radiates a linearly polarized wave. However, now the polarizer P presented in Fig. 4.1.1 is absent.

Let α be defined as the angle between the yoz plane and the orientation of the electric field, $\mathbf{E}_0$, of the incident wave, as is shown in Fig. 4.2.1. The field of this wave is of the form

$$\left.\begin{aligned} E_x &= H_\gamma = E_{0x} e^{ik(y\sin\gamma + z\cos\gamma)}, \\ H_x &= -E_\gamma = H_{0x} e^{ik(y\sin\gamma + z\cos\gamma)}, \end{aligned}\right\} \tag{4.2.1}$$

where

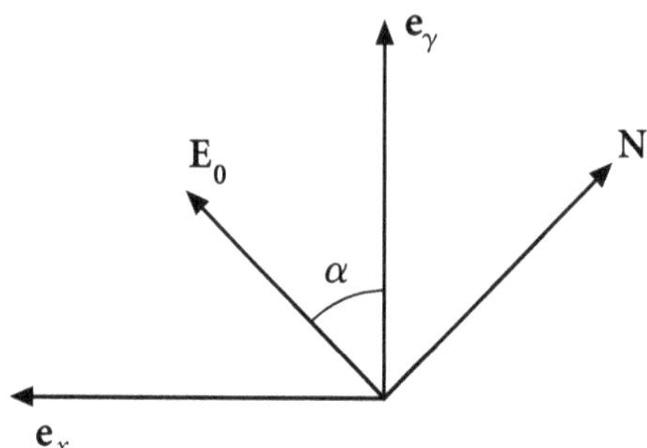

Figure 4.2.1: Orientation of the electric field with respect to the *yoz*-plane.

$$E_{0x} = E_0 \sin\alpha, \quad H_{0x} = -E_0 \cos\alpha, \quad \frac{E_{0x}}{E_{o\gamma}} = \tan\alpha. \tag{4.2.2}$$

The field scattered by the body is expressed in the far field as

$$\left.\begin{aligned} E_\varphi &= -H_\vartheta = \frac{ia}{2}\left[E_{0x}\overline{\Sigma}_1(\gamma,\vartheta,\varphi) + H_{0x}\Sigma_1(\gamma,\vartheta,\varphi)\right]\frac{e^{ikR}}{R}, \\ E_\vartheta &= H_\varphi = \frac{ia}{2}\left[E_{0x}\overline{\Sigma}_2(\gamma,\vartheta,\varphi) + H_{0x}\Sigma_2(\gamma,\vartheta,\varphi)\right]\frac{e^{ikR}}{R}. \end{aligned}\right\} \tag{4.2.3}$$

Here, a is a length characterizing the dimension of the body, R, ϑ, φ, are the spherical coordinates at the point of observation, $\overline{\Sigma}_{1,2}(\gamma,\vartheta,\varphi)$ and $\Sigma_{1,2}(\gamma,\vartheta,\varphi)$ are partial scattering patterns.

Clearly, the polarization of the scattered field, that is the orientation of its electric field in space, depends on the complex relationship between the directions to the source and observation points. In the direction of the source it is possible for the polarization of the scattered field to differ from the polarization of the source. This phenomenon is called the ***depolarization*** of the scattered field.

The reason for depolarization is easy to establish if the scattered field is viewed as radiation from uniform and nonuniform components of the surface current. In fact, in accordance with § 4.1 the uniform current radiates the field

$$\left.\begin{aligned} E_x = H_\vartheta &= \frac{iaE_{0x}}{2}\frac{e^{ikR}}{R}\overline{\Sigma}^0, \\ E_\vartheta = H_\varphi &= \frac{iaH_{0x}}{2}\frac{e^{ikR}}{R}\Sigma^0, \end{aligned}\right\} \tag{4.2.4}$$

in the direction of the source ($\vartheta = \pi - \gamma$, $\varphi = -\pi/2$). The functions Σ^0 and $\overline{\Sigma}^0$ satisfy the condition that $\Sigma^0 = -\overline{\Sigma}^0$, and are given by Eqn. (4.1.15). From Eqn. (4.2.4) we quickly arrive at the equality

$$\frac{E_x}{E_\vartheta} = \tan\alpha, \tag{4.2.5}$$

under the condition that $H_{0x} = -E_{0\vartheta}$. This means that within the physical optics approximation, the reflected wave does not undergo depolarization. It follows that depolarization of the reflected wave is caused only by the nonuniform current, or as we have in other words expressed, by the deviation of every surface element from a tangential plane.

We will introduce an expression for the value of δ, the angle subtended between the electric vector of the reflected wave and the electric vector of the field radiated by the source. To do this, the functions $\overline{\Sigma}_{1,2}$ and $\Sigma_{1,2}$ are written in the form

$$\overline{\Sigma}_{1,2} = \overline{\Sigma}^0_{1,2} + \overline{\Sigma}^1_{1,2}, \quad \Sigma_{1,2} = \Sigma^0_{1,2} + \Sigma^1_{1,2}, \tag{4.2.6}$$

where the terms $\overline{\Sigma}^0_{1,2}, \Sigma^0_{1,2}$ and $\overline{\Sigma}^1_{1,2}, \Sigma^1_{1,2}$ correspond to the fields radiated by the uniform and nonuniform currents, respectively. By making a comparison of Eqns. (4.2.4) and (4.2.3) we find that

$$\overline{\Sigma}^0_1 = \overline{\Sigma}^0, \quad \Sigma^0_1 = 0, \quad \overline{\Sigma}^0_2 = 0, \quad \Sigma^0_2 = -\overline{\Sigma}^0. \tag{4.2.7}$$

Therefore, the field scattered in the direction of the source ($\vartheta = \pi - \gamma$, $\varphi = -\pi/2$) is equal to

$$\left.\begin{aligned} E_x &= -H_\vartheta = \frac{ia}{2}\left[E_{0x}\left(\overline{\Sigma}^0 + \overline{\Sigma}_1^1\right) + H_{ox}\Sigma_1^1\right]\frac{e^{ikR}}{R}, \\ E_\vartheta &= H_x = \frac{ia}{2}\left[E_{0x}\overline{\Sigma}_2^1 - H_{0x}\left(\overline{\Sigma}^0 - \Sigma_2^1\right)\right]\frac{e^{ikR}}{R}. \end{aligned}\right\} \qquad (4.2.8)$$

The electric vector of this field forms the angle β with the *yoz* plane and is determined by the relationship

$$\tan\beta = \frac{|E_x|}{|E_\vartheta|} = \frac{\left|\overline{\Sigma}^0 + \overline{\Sigma}_1^1 - \Sigma_1^1 \cot\alpha\right|}{\left|\overline{\Sigma}^0 - \overline{\Sigma}_2^1 + \Sigma_2^1 \tan\alpha\right|}\tan\alpha \qquad (4.2.9)$$

under the condition $H_{0x} = -E_{0\vartheta}$. As a result, the desired angle, δ, which characterizes the value of depolarization, is equal to

$$\delta = \alpha - \beta. \qquad (4.2.10)$$

Though Eqn. (4.2.9) contains the terms related to the uniform currents, it also reveals the nature of depolarization. Indeed, if the nonuniform current were absent, the functions $\Sigma_{1,2}^1$ and $\overline{\Sigma}_{1,2}^1$, would be equal to zero and the angle β would be equal to α. In other words, the polarization of the scattered field would be exactly the same as that of the incident wave. This result demonstrates again that depolarization of backscattering is totally determined only by the nonuniform currents.

Specific results on calculating the depolarization of waves reflected from some bodies can be found, for instance, in references [82–84]. In particular, reference [83] shows that the influence of depolarization on the radar cross section of convex bodies can be neglected when $ka > 4$.

§4.3 Fundamental Results

- Chapters 2–4 presented edge waves created by the nonuniform currents induced on perfectly conducting bodies with edges. Asymptotic expressions obtained here were utilized to calculate the radar cross section of some bodies of interest. Numerical and experimental results are in satisfactory agreement.

- These asymptotic expressions do not reduce to and significantly differ from those related to the physical optics even for large objects compared to the wavelength.
- This is a current based theory. It is valid in diffraction regions such as foci and caustics where the ray based techniques are not applicable.
- The physical nature of the depolarization in backscattering was established. This phenomenon is produced only by the nonuniform current which is caused by the curvature of the object surface and its discontinuities (edges, corners).
- A technique was proposed for the separation and measurement of the backscattered field radiated only by the nonuniform current. It can be used to estimate the individual properties of body shape, and therefore, is of interest for the solution of inverse problems in diffraction theory.
- This theory studied only first-order edge waves. They provide a main contribution to the field scattered by large three-dimensional objects. Higher-order edge waves are significant components in the field scattered by objects with pronounced resonant properties. These waves are studied in the next sections.

5

Analysis of Wedge Diffraction Using the Parabolic Equation Method

§5.1 Parabolic Equation

The study of wave field properties at high frequencies is connected with the investigation of asymptotic solutions to the related boundary value problems. The method that was the first applied for this purpose, and which may be thought of as classical, is described as follows. It consists of approximating an existing exact solution when value of $1/(ka)$ is small, where k is the wave number and a is the characteristic dimension of the problem. The idea of this approximation is to transform rigorous series that are poorly converging at high frequencies into fast converging asymptotic series. The principal shortcoming of this method is its limited applicability: only relatively few problems possess rigorous solutions. In addition, we must mention that obtaining asymptotic representations even for rigorous analytic solutions is not a simple task. For instance, the exact analytical solution for the field diffracted at a semi-infinite circular cone has been well known for a long time. It is described by a series of spherical Legendre and Bessel functions. However, its proper transformation into fast convergent asymptotic series still proves to be a difficult problem.

A fundamentally different, more effective, and more direct method is one proposed by Leontovich in 1944 [85]. He proposed making an approximation directly into the field equations themselves. Utilizing this type of approximation he reduced the boundary value problem for the Helmholtz equation to one of parabolic form. Therefore, it was shown that by using this method, the necessary asymptotic solution can be simply and rapidly obtained. This method has proved even more universally applicable, having been used to obtain approximate solutions to many problems that possess no known rigorous solution. This method, known as the parabolic equation method, has been used with considerable success. Here the work of Leontovich, Fock, Mal'uzhinets, and Vainshtein [86–91] must be mentioned. This approach was also successfully used in [92–94].

However, a majority of diffraction theory problems examined using parabolic equations analyze objects possessing smooth boundaries. There are few problems connected with objects possessing discontinuous boundaries. In 1946, Mal'uzhinets showed that Sommerfeld's asymptotic expression for edge waves can be found by the parabolic equation method [91]. The importance of using parabolic equations to solve the diffraction from a disk is mentioned by Schensted [95]. Vainshtein obtained interesting results by solving the propagation of waves over piecewise inhomogeneous ground [96]. Popov accomplished a numerical solution of the wedge diffraction problem [97].

Chapter 5 will provide the asymptotic solution to the wedge diffraction problem. Chapter 6 will modify the parabolic equation method to analyze current waves on thin conducting cylinders.

The fundamental results of Chapters 5 and 6 were published in [16, 17, 19].

§5.2 Formulation of the Problem

Let there be a plane wave incident on a wedge, the boundaries of which are $\varphi = 0$ and $\varphi = \alpha$ (Fig. 2.2.1). The diffracted field satisfies the equation

$$\nabla^2 u + k^2 u = 0 \tag{5.2.1}$$

and the boundary condition

$$u = 0 \quad \text{for} \quad \varphi = 0; \alpha \tag{5.2.2}$$

or

$$\frac{\partial u}{\partial n}=0 \quad \text{for} \quad \varphi=0;\alpha. \tag{5.2.3}$$

A solution to Eqn. (5.2.1) will be sought in the form

$$u=u_g+u_d, \tag{5.2.4}$$

where u_g is the field according to geometric optics; u_d is the supplemental diffracted field. Furthermore, let us introduce the step functions

$$\epsilon_1=\begin{cases}1 & \text{for} \quad \varphi\leqslant\pi+\varphi_0 \quad \text{(region 1)}\\ 0 & \text{for} \quad \varphi>\pi+\varphi_0,\end{cases} \tag{5.2.5}$$

$$\epsilon_2=\begin{cases}1 & \text{for} \quad \varphi\leqslant\pi-\varphi_0 \quad \text{(region 2)}\\ 0 & \text{for} \quad \varphi>\pi-\varphi_0.\end{cases} \tag{5.2.6}$$

When $\varphi_0<\alpha-\pi$ the solution for the function u_g can be assumed to be of the form

$$u_g=\epsilon_1 e^{-ikr\cos(\varphi-\varphi_0)}\mp\epsilon_2 e^{-ikr\cos(\varphi+\varphi_0)}. \tag{5.2.7}$$

Here, the minus sign corresponds to the boundary condition of Eqn. (5.2.2) and the plus sign corresponds to the boundary condition of Eqn. (5.2.3).

We are interested in finding the asymptotic solution to Eqn. (5.2.1) when $kr\gg 1$. Therefore, a function u_d will be sought in the form

$$u_d=W(r,\varphi)e^{ikr}. \tag{5.2.8}$$

The coordinate r is measured along rays originating at the "luminous line"— the edge of the wedge (see Chapter 20 in reference [43]). The exponential term $\exp[ikr]$ determines the basic phase progression of the wave along the ray. The slowly changing factor $W(r,\varphi)$ is the amplitude of the diffracted wave corresponding to the ray and smoothing discontinuities of u_g.

Substituting Eqn. (5.2.8) into Eqn. (5.2.1) produces

$$\frac{1}{r}\frac{\partial}{\partial r}\left(r\frac{\partial W}{dr}\right)+2ik\frac{\partial W}{dr}+\frac{1}{r^2}\frac{\partial^2 W}{d\varphi^2}+\frac{ik}{r}W=0. \tag{5.2.9}$$

Discarding the first term, i.e., neglecting longitudinal diffusion, we obtain the parabolic diffusion equation

$$\frac{\partial^2 W}{d\varphi^2}+2ikr^2\frac{\partial W}{dr}+ikrW=0. \tag{5.2.10}$$

Since the complete field $u = u_g+u_d$ is continuous everywhere and satisfies the boundary conditions given by Eqns. (5.2.2) and (5.2.3), $W(r,\varphi)$ must satisfy the following relationships:

$$\left.\begin{aligned} &W\left[r,(\pi-\varphi_0)+0^+\right]-W\left[r,(\pi-\varphi_0)+0^-\right]=-1,\\ &W\left[r,(\pi+\varphi_0)+0^+\right]-W\left[r,(\pi+\varphi_0)+0^-\right]=1,\end{aligned}\right\} \tag{5.2.11}$$

$$W(r,0)=W(r,\alpha)=0 \tag{5.2.12}$$

for the boundary conditions of Eqn. (5.2.2) and

$$\left.\begin{aligned} &W\left[r,(\pi-\varphi_0)+0^+\right]-W\left[r,(\pi-\varphi_0)+0^-\right]=1,\\ &W\left[r,(\pi+\varphi_0)+0^+\right]-W\left[r,(\pi+\varphi_0)+0^-\right]=1,\end{aligned}\right\} \tag{5.2.13}$$

$$\left.\frac{\partial}{\partial\varphi}W(r,\varphi)\right|_{\varphi=0;\alpha}=0 \tag{5.2.14}$$

for the boundary conditions of Eqn. (5.2.3). In Eqns. (5.2.11) and (5.2.13) we note that $0^+ = \lim(0+\epsilon)$ and $0^- = \lim(0-\epsilon)$ where ϵ is a positive value tending to zero ($\epsilon \to 0$).

In order to make the problem single-valued for the elliptic form of Eqn. (5.2.1), two conditions are imposed on the function u: the radiation condition (for $kr \to \infty$) and the boundary condition at the edge of the wedge (for $kr \to 0$). Due to the formulation of Eqn. (5.2.8) and the parabolic equation of Eqn. (5.2.10), the radiation condition is automatically satisfied, and the boundary condition at the edge is of the form

$$W(r,\varphi) \quad \text{remains finite as} \quad kr \to 0\,. \tag{5.2.15}$$

This makes the problem single-valued (see, § 5.3 below).

§5.3 Solution of the Parabolic Equation

This section will examine initially the solution to the parabolic equation for conditions of Eqns. (5.2.11) and (5.2.12). Here, $W(r, \varphi)$ is assumed to be of the form

$$W(r,\varphi) = W_0 + W_1 \tag{5.3.1}$$

where

$$W_0 = \epsilon_2 - \epsilon_1 \tag{5.3.2}$$

is a piecewise constant function of φ that satisfies the conditions of Eqn. (5.2.11) and approaches zero at the faces of the wedge. This function can be expanded as a Fourier series

$$W_0 = -\frac{4}{\alpha}\sum_{s=1} \frac{1}{\nu_s} \sin\nu_s\pi \sin\nu_s\varphi_0 \sin\nu_s\varphi\,, \quad \nu_s = s\frac{\pi}{\alpha}\;, \tag{5.3.3}$$

which is an odd periodic function with period 2α. It follows that $W_1(r, \varphi)$ will be a continuous function that approaches zero at the faces of the wedge. It is also possible to show that $\partial W_1(r, \varphi)/\partial\varphi$ is continuous, but $\partial^2 W_1(r, \varphi)/\partial\varphi^2$ possesses a discontinuity of the first kind at $\varphi = \pi \pm \varphi_0$.

Noticing that

$$\frac{\partial^2 W}{\partial \varphi^2} = \frac{\partial^2 W_1}{\partial \varphi^2}, \quad \frac{\partial W}{\partial r} = \frac{\partial W_1}{\partial r}$$

and substituting Eqn. (5.3.1) into Eqn. (5.2.10), we obtain

$$\frac{\partial^2 W_1}{d\varphi^2} + 2ikr^2 \frac{\partial W_1}{dr} + ikrW_1 = -ikrW_0 . \tag{5.3.4}$$

Furthermore, assuming that

$$W_1 = \sum_{s=1}^{\infty} R_{1s}(r) \sin\nu_s \pi \sin\nu_s \varphi_0 \sin\nu_s \varphi \tag{5.3.5}$$

and incorporating Eqn. (5.3.3), we find that

$$2ikr^2 \frac{\partial R_{1s}}{\partial r} + \left(ikr - \nu_s^2\right) R_{1s} = 4\frac{ikr}{\alpha \nu_s} . \tag{5.3.6}$$

Now, W will be sought in the form

$$W(r,\varphi) = \sum_{s=1}^{\infty} R_s(r) \sin\nu_s \pi \sin\nu_s \varphi_0 \sin\nu_s \varphi . \tag{5.3.7}$$

It is evident that

$$R_s(r) = R_{1s}(r) - \frac{4}{\alpha \nu_s}, \tag{5.3.8}$$

which satisfies

$$2ikr^2 \frac{\partial R_s}{\partial r} + \left(ikr - \nu_s^2\right) R_s = \frac{4}{\alpha} \nu_s . \tag{5.3.9}$$

Taking into account the condition given by Eqn. (5.2.15) we find that

$$R_s(r)=\frac{4}{i\alpha}\sqrt{\frac{2}{kr}}e^{i\frac{\nu_s^2}{2kr}}\int_{\frac{\nu_s}{\sqrt{2kr}}}^{\infty}e^{-it^2}dt\,, \tag{5.3.10}$$

and subsequently,

$$W=\frac{4}{i\alpha}\sqrt{\frac{2}{kr}}\sum_{s=1}^{\infty}\sin\nu_s\pi\sin\nu_s\varphi_0\sin\nu_s\varphi e^{i\frac{\nu_s^2}{2kr}}\int_{\frac{\nu_s}{\sqrt{2kr}}}^{\infty}e^{-it^2}dt\,. \tag{5.3.11}$$

It is not difficult to see that for $kr \ll 1$

$$W=W_0+O(kr),\quad u_g=-W_0+O(kr), \tag{5.3.12}$$

therefore,

$$u=u_g+u_d=O(kr), \tag{5.3.13}$$

whereas the exact solution to Eqn. (5.2.1) satisfying the boundary condition given by Eqn. (5.2.2) leads to the following estimation

$$u=O\left[(kr)^{\pi/\alpha}\right],\quad 0<\varphi<\alpha\,. \tag{5.3.14}$$

Examination of Eqns. (5.3.13) and (5.3.14) reveal that the field behavior predicted by the parabolic equation given by Eqn. (5.2.10) and by the exact wave equation given by Eqn. (5.2.1) is different. According to the parabolic equation, the field vanishes more quickly with the observation point approaching the edge ($r = 0$) of the wedge ($\alpha > \pi$) and vanishes more slowly as $r \to 0$ inside the horn ($\alpha < \pi$), when we compare it with the exact solution of the wave equation.

Again returning to Eqn. (5.3.11), it is possible to rewrite it as

$$W(r,\varphi)=\frac{4}{i\alpha}\sqrt{\frac{2}{kr}}\int_0^\infty e^{-it^2}dt\sum_{s=1}^{\infty}e^{-it\nu_s\sqrt{2/kr}}\sin\nu_s\pi\sin\nu_s\varphi_0\sin\nu_s\varphi\,. \tag{5.3.15}$$

Summing the series, we obtain

$$W(r,\varphi)=w(r,\varphi-\varphi_0)-w(r,\varphi+\varphi_0), \tag{5.3.16}$$

where

$$w(r,\psi)=-\frac{i}{2\alpha}\int_0^{0^+-i\infty}e^{-i\frac{kr}{2}\tau^2}\left(\frac{\sin\frac{\pi+\psi}{n}}{\cos\frac{\tau}{n}-\cos\frac{\pi+\psi}{n}}+\frac{\sin\frac{\pi-\psi}{n}}{\cos\frac{\tau}{n}-\cos\frac{\pi-\psi}{n}}\right)d\tau \tag{5.3.17}$$

and

$$n=\frac{\alpha}{\pi}.$$

Utilizing the formula

$$\frac{\sin\frac{\pi+\psi}{n}}{\cos\frac{\tau}{n}-\cos\frac{\pi+\psi}{n}}+\frac{\sin\frac{\pi-\psi}{n}}{\cos\frac{\tau}{n}-\cos\frac{\pi-\psi}{n}}=$$

$$=-\frac{\sin\frac{\pi}{n}}{\cos\frac{\pi}{n}-\cos\frac{\psi+\tau}{n}}-\frac{\sin\frac{\pi}{n}}{\cos\frac{\pi}{n}-\cos\frac{\psi-\tau}{n}},$$

we can write $w(r,\psi)$ as

$$w(r,\psi)=\frac{\sin\frac{\pi}{n}}{2n\pi i}\int_{0^+-i\infty}^{0^-+i\infty} e^{-i\frac{kr}{2}\xi^2}\frac{d\xi}{\cos\frac{\pi}{n}-\cos\frac{\psi+\xi}{n}}. \tag{5.3.18}$$

For all intents and purposes, using an analogous method, it is possible to show that the solution to the parabolic equation, satisfying Eqns. (5.2.13), (5.2.14), and (5.2.15) is equal to

$$W=\frac{4}{i\alpha}\sqrt{\frac{2}{kr}}\sum_{s=1}^{\infty}\sin\nu_s\pi\cos\nu_s\varphi_0\cos\nu_s\varphi e^{i\frac{\nu_s^2}{2kr}}\int_{\frac{\nu_s}{\sqrt{2kr}}}^{\infty} e^{-it^2}\,dt, \tag{5.3.19}$$

which can be transformed into the form

$$W(r,\varphi)=w(r,\varphi-\varphi_0)+w(r,\varphi+\varphi_0), \tag{5.3.20}$$

where the function $w(r,\psi)$ is again determined by Eqn. (5.3.18).

Equations (5.3.16) and (5.3.18) were first derived by Mal'uzhinets from the solution of functional equations related to the parabolic equation method [91]. He compared them with the solution of the exact (elliptic) wave equation

$$W^{exact}=\frac{\sin\frac{\pi}{n}}{2\pi in}\times$$

$$\times\int_{0^+-i\infty}^{0^-+i\infty}\left(\frac{e^{-2ikr\sin^2\frac{\xi}{2}}}{\cos\frac{\pi}{n}-\cos\frac{\xi+\varphi-\varphi_0}{n}}-\frac{e^{-2ikr\sin^2\frac{\xi}{2}}}{\cos\frac{\pi}{n}-\cos\frac{\xi+\varphi+\varphi_0}{n}}\right)d\xi, \tag{5.3.21}$$

and found that their asymptotic expressions for the edge waves were identical:

$$W(r,z) = W^{exact}(r,z) =$$

$$= \left(\frac{\frac{1}{n}\sin\frac{\pi}{n}}{\cos\frac{\pi}{n} - \cos\frac{\varphi-\varphi_0}{n}} - \frac{\frac{1}{n}\sin\frac{\pi}{n}}{\cos\frac{\pi}{n} - \cos\frac{\varphi+\varphi_0}{n}} \right) \frac{e^{i\frac{\pi}{4}}}{\sqrt{2\pi kr}}. \tag{5.3.22}$$

This asymptotic expression is valid far from the geometrical optics boundaries, where $(kr)^{1/2}|\cos[(\varphi \pm \varphi_0)/2]| \gg 1$. The following section will present a more detailed asymptotic investigation of the solution of the parabolic equation.

§5.4 Asymptotic Expansion for $w(r, \psi)$

The asymptotic expansion for $w(r, \psi)$ will be found for $kr \gg 1$ by refining the method of steepest descent (see, Pauli [36]), which assumes that the poles of the integrand can approach at the saddle point. To do this, $w(r, \psi)$ will be written in the form

$$w(r,\psi) = \frac{\sin\frac{\pi}{n}}{n\pi\sqrt{2}} e^{i\frac{\pi}{4}} \int_{-\infty}^{\infty} e^{-krt^2} \frac{f(t,\psi)}{t^2 - it_0^2} dt \tag{5.4.1}$$

where

$$f(t,\psi) = \frac{t^2 - it_0^2}{\cos\frac{\pi}{n} - \cos\frac{\psi+\zeta}{n}}, \quad n = \frac{\alpha}{\pi}, \tag{5.4.2}$$

$$t_0 = \sqrt{2}\cos\psi, \quad \zeta = \sqrt{2}te^{-i\frac{\pi}{4}}. \tag{5.4.3}$$

Substituting a Taylor series representation for $f(r,\psi)$

$$f(t,\psi)=\sum_{m=0}^{\infty}a_m t^m$$

into Eqn. (5.4.1) we obtain the asymptotic series for $kr \gg 1$

$$w(r,\psi)=\frac{1}{\cos\frac{\psi}{2}}\sum_{m=0}^{\infty}\Gamma\left(m+\frac{1}{2}\right)\frac{F_m(\tau)}{(kr)^m}A_m(\psi), \tag{5.4.4}$$

where

$$F_m(\tau)=\tau^{2m}e^{-i\tau^2}\int_{\tau}^{\infty\,\mathrm{sgn}\,\tau}\frac{e^{ix^2}}{x^{2m}}dx, \tag{5.4.5}$$

$$\tau=\sqrt{2kr}\cos\frac{\psi}{2}, \tag{5.4.6}$$

$$\left.\begin{aligned} F_0(\tau)&=e^{-i\tau^2}\int_{\tau}^{\infty\,\mathrm{sgn}\,\tau}e^{ix^2}dx \\ F_1(\tau)&=\tau+2i\tau^2e^{-i\tau^2}\int_{\tau}^{\infty\,\mathrm{sgn}\,\tau}e^{ix^2}dx \end{aligned}\right\}, \tag{5.4.7}$$

and $\Gamma(m)$ is the Gamma function. It is evident that

$$\frac{F_m(\tau)}{(kr)^m}\sim O\left[(kr)^{-m-\frac{1}{2}}\right] \quad \text{for} \quad |\tau|\gg 1, \quad m=0,1,2\dots. \tag{5.4.8}$$

In addition, from the identity

$$\int_0^\infty \frac{e^{-krt^2}}{t^2 - it_0^2} t^{2m} dt = \frac{1}{\sqrt{2}} \Gamma\left(m + \frac{1}{2}\right) \frac{F_m(\tau)}{(kr)^m \cos\frac{\psi}{2}}, \tag{5.4.9}$$

it follows that

$$\frac{F_m(\tau)}{(kr)^m \cos\frac{\psi}{2}} \sim O\left[(kr)^{-m+\frac{1}{2}}\right] \quad \text{for} \quad \psi = \pi, \quad m = 1, 2\ldots \tag{5.4.10}$$

is valid as $kr \to \infty$.

The coefficients $A_m(\psi)$ of the expansion are determined by

$$A_m(\psi) = \frac{\sin\frac{\pi}{n}}{n\pi} \frac{e^{i\frac{\pi}{4}}}{(2m)!} \frac{\partial^{2m} f(t,\psi)}{\partial t^{2m}}\Bigg|_{t=0}, \tag{5.4.11}$$

where

$$A_0(\psi) = \frac{2}{n\pi} \frac{\sin\frac{\pi}{n}\cos^2\frac{\psi}{2}}{\cos\frac{\pi}{n} - \cos\frac{\psi}{n}} e^{-i\frac{\pi}{4}}, \tag{5.4.12}$$

$$A_1(\psi) = \frac{\frac{1}{n}\sin\frac{\pi}{n}}{\cos\frac{\pi}{n} - \cos\frac{\psi}{n}} \frac{e^{i\frac{\pi}{4}}}{\pi} \times$$

$$\times \left\{1 + \frac{2}{n^2} \frac{\cos^2\frac{\psi}{2}\cos\frac{\psi}{n}}{\cos\frac{\pi}{n} - \cos\frac{\psi}{n}} - \frac{4}{n^2}\left(\frac{\sin\frac{\psi}{n}\cos\frac{\psi}{2}}{\cos\frac{\pi}{n} - \cos\frac{\psi}{n}}\right)^2\right\}, \tag{5.4.13}$$

$$A_1(\pi) = -\frac{e^{i\frac{\pi}{4}}}{n2\pi}\cot\frac{\pi}{n}. \tag{5.4.14}$$

Equation (5.4.4) will be compared with the asymptotic expansion of Sommerfeld's solution,

$$u_d e^{-ikr} = w(r,\varphi-\varphi_0) \mp w(r,\varphi+\varphi_0) \tag{5.4.15}$$

found by Pauli [36]. It turns out, that the first term in Pauli's asymptotic series for the function $w(r,\psi)$ is the same as in Eqn. (5.4.4). However, all the other terms are different. For example, Pauli's coefficient for $A_1^{Pauli}(\psi)$ contains an additional term in comparison to Eqn. (5.4.13):

$$A_1^{Pauli}(\psi) = -\frac{i}{4}A_0(\psi) + \frac{\frac{1}{n}\sin\frac{\pi}{n}}{\cos\frac{\pi}{n}-\cos\frac{\psi}{n}}\frac{e^{i\frac{\pi}{4}}}{\pi}\times$$

$$\times\left\{1+\frac{2}{n^2}\frac{\cos^2\frac{\psi}{2}\cos\frac{\psi}{n}}{\cos\frac{\pi}{n}-\cos\frac{\psi}{n}}-\frac{4}{n^2}\left(\frac{\sin\frac{\psi}{n}\cos\frac{\psi}{2}}{\cos\frac{\pi}{n}-\cos\frac{\psi}{n}}\right)^2\right\}. \tag{5.4.16}$$

It is evident that the values for $A_1(\pi)$ given by Eqns. (5.4.13) and (5.4.16) are identical.

It follows that the asymptotic expression,

$$w(r,\psi) = \frac{2}{n}\frac{\sin\frac{\pi}{n}\cos\frac{\psi}{2}}{\cos\frac{\pi}{n}-\cos\frac{\psi}{n}}\frac{e^{-i\frac{\pi}{4}}}{\sqrt{\pi}}e^{-2ikr\cos^2\frac{\psi}{2}}\int_{\sqrt{2kr}\cos\frac{\psi}{2}}^{\infty\,\mathrm{sgn}\cos\frac{\psi}{2}} e^{iq^2}\,dq, \tag{5.4.17}$$

describes both the solution of Sommerfeld and the solution of the parabolic equation with identical error. The error of Eqn. (5.4.17) is on the order of $(kr)^{-1/2}$ when $\psi \approx \pi$ and is on the order of $(kr)^{-3/2}$ for $(kr)^{1/2}|\cos[\psi/2]| \gg 1$.

We want to emphasize that the above error estimation relates only to the first term in the asymptotic expansion of Eqn. (5.4.4) for the solution of the parabolic equation. The accuracy of the entire solution given by Eqn. (5.4.4) is higher. To show this, let us consider the difference between Eqn. (5.4.4) and the exact asymptotic solution provided by Pauli [36].

$$\Delta w = w(r,\psi) - w^{Pauli}(r,\psi) = \tag{5.4.18}$$
$$= \frac{i}{4} A_0(\psi)\Gamma(3/2)\frac{F_1(\tau)}{kr\cos(\psi/2)} + O\left(\frac{F_2(\tau)}{(kr)^2\cos(\psi/2)}\right).$$

This quantity determines the error of the parabolic equation method. According to Eqns. (5.4.8) and (5.4.10), this error given by Eqn. (5.4.18)is on the order of $(kr)^{-3/2}$ both at $(\psi = \pi)$ and away from the shadow boundary, where $[kr]^{1/2}|\cos[(\varphi \pm \varphi_0)/2]| \gg 1$. It is easy to see that the error Δw is also on the order of $(kr)^{-3/2}$ in the vicinity of the shadow boundary where $kr|\cos[(\varphi \mp \varphi_0)/2] \ll 1$.

We have investigated the asymptotic behavior of $W(r, \varphi)$ utilizing its integral representation. If we were not interested in the behavior of $W(r, \varphi)$ in the effective diffusion zone and wanted to know only the angular distribution of the cylindrical wave diffracted from the edge of the wedge, it can be determined immediately from the series. When $kr \gg S \gg 1$, the sum of the S first terms of Eqn. (5.3.11) is given by

$$W = \frac{2}{i\alpha}\sqrt{\frac{2\pi}{kr}}e^{-i\frac{\pi}{4}}\sum_{s=1}^{S}\sin\nu_s\pi\sin\nu_s\varphi_0\sin\nu_s\varphi + O\left(\frac{1}{kr}\right). \tag{5.4.19}$$

Imagining here that $S = \infty$ and summing the series using Borel's method, we obtain Eqn. (5.3.16), in which

$$w(r,\psi) = \frac{\frac{1}{n}\sin\frac{\pi}{n}}{\cos\frac{\pi}{n} - \cos\frac{\psi}{n}}\frac{e^{i\frac{\pi}{4}}}{\sqrt{2\pi kr}}. \tag{5.4.20}$$

Using an analogous method, the series of Eqn. (5.3.19) can be transformed into Eqn. (5.3.20), with $w(r,\varphi)$ given by Eqn. (5.4.20).

§5.5 Reflection Method

The parabolic equation of Eqn. (5.2.10) possesses a partial solution

$$V(\tau)=e^{-i\tau^2}\frac{e^{-i\frac{\pi}{4}}}{\sqrt{\pi}}\int_{\tau}^{\infty\operatorname{sgn}\tau} e^{iq^2}\,dq, \tag{5.5.1}$$

where $\tau=(kr/2)^{1/2}(\varphi-\phi)$. This solution, by itself, is related to the diffraction of a plane wave from the wedge with black boundaries. The field resulting from the diffraction of a perfectly conducting wedge can be thought of as resulting from the multibounce reflection of waves given by Eqn. (5.5.1) from the boundary of the wedge. In this case, using the primary function $V(\tau)$ the incident wave is selected setting $\phi=\pi+\varphi_0$ and the reflected wave is selected setting $\phi=\pi-\varphi_0$. The reflection method leads us to the expression

$$W(r,\varphi)=w(r,\varphi-\varphi_0)\mp w(r,\varphi+\varphi_0),$$

where

$$\begin{aligned} w(r,\psi)=&V\left(\sqrt{kr/2}(\psi-\pi)\right)+\\ &+\sum_{m=1}^{\infty}\left\{V\left(\sqrt{kr/2}(\psi-\pi+2m\alpha)\right)+V\left(\sqrt{kr/2}(\psi-\pi-2m\alpha)\right)\right\}-\\ &-V\left(\sqrt{kr/2}(\psi+\pi)\right)-\\ &-\sum_{m=1}^{\infty}\left\{V\left(\sqrt{kr/2}(\psi+\pi+2m\alpha)\right)+V\left(\sqrt{kr/2}(\psi+\pi-2m\alpha)\right)\right\}. \end{aligned} \tag{5.5.2}$$

It is possible to show that Eqns. (5.3.18) and (5.5.2) are equivalent. To do this, one modifies Eqn. (5.3.17) by making the substitution

$$\frac{\sin\frac{x}{n}}{\cos\frac{\tau}{n}-\cos\frac{x}{n}}=\frac{1}{2}\left(\cot\frac{x+\tau}{2n}+\cot\frac{x-\tau}{2n}\right), \tag{5.5.3}$$

and using the identities

$$\frac{1}{2n}\cot\frac{x}{2n}=\frac{1}{x}+\sum_{m=1}^{\infty}\left(\frac{1}{x+2mn\pi}+\frac{1}{x-2mn\pi}\right), \tag{5.5.4}$$

$$\int_0^{\infty} e^{-i\frac{kr}{2}\tau^2}\frac{d\tau}{a^2-\tau^2}=\frac{\pi}{ia}V\left(\sqrt{\frac{kr}{2}}a\right). \tag{5.5.5}$$

Let us investigate the asymptotic properties of $w(r, \psi)$ proceeding from Eqn. (5.5.2). It is not difficult to see that if all the arguments of $V(\tau)$ are large, then

$$\begin{aligned} w(r,\psi)=\frac{e^{i\frac{\pi}{4}}}{\sqrt{2\pi kr}}\Bigg[\frac{1}{\psi-\pi}+\sum_{m=1}^{\infty}\left(\frac{1}{\psi-\pi+2m\alpha}+\frac{1}{\psi-\pi-2m\alpha}\right)- \\ -\frac{1}{\psi+\pi}-\sum_{m=1}^{\infty}\left(\frac{1}{\psi+\pi+2m\alpha}+\frac{1}{\psi+\pi-2m\alpha}\right)\Bigg]+O\left[(kr)^{-3/2}\right]. \end{aligned} \tag{5.5.6}$$

The series on the right side of this equation are easy to sum, and in so doing we arrive at Eqn. (5.4.20).

Provided $kr \gg 1$ we can present Eqn. (5.5.2) as

$$w(r,\psi)=V\left(\sqrt{\frac{kr}{2}}(\psi-\pi)\right)+\left(\frac{\frac{1}{n}\sin\frac{\pi}{n}}{\cos\frac{\pi}{n}-\cos\frac{\psi}{n}}-\frac{1}{\psi-\pi}\right)\frac{e^{i\frac{\pi}{4}}}{\sqrt{2\pi kr}}. \tag{5.5.7}$$

Here, the first term is transverse diffusion at the shadow boundary, independent of the reflective properties of the face of the wedge. The second term can be interpreted as the diffraction background. A similar representation of the field is well known in the theory of diffraction from convex bodies (see, especially [90] and [98]). Substituting $\psi = \pi \pm 0$ into Eqn. (5.5.7) we obtain

$$\mathrm{w}(r, \pi \pm 0) = \pm\frac{1}{2} - \frac{1}{2n}\cot\frac{\pi}{n}\frac{e^{i\frac{\pi}{4}}}{\sqrt{2\pi kr}}. \quad (5.5.8)$$

The discontinuity of w at $\psi = \pi$ guarantees the continuity of the total field, $u = u_g + u_d$.

It should be noted that due to its accuracy and generality, Eqn. (5.5.7) does not yield Eqn. (5.4.17). Moreover, Eqn. (5.4.17) may be considered as a "stenographic" description of Eqn. (5.5.7), which due to its form, is a more physically meaningful solution.

Thus, if one accepts the parabolic equation as well-founded, then fundamental asymptotic Eqns. (5.4.20) and (5.5.7) can be derived much more simply than by means of the exact approach. Still, a bigger payoff is obtained if one proceeds from the diffraction of plane waves to a more complicated problem — the diffraction of cylindrical waves.

§5.6 Transverse Diffusion and Diffraction of Cylindrical Waves at a Wedge

According to the laws of geometrical optics, the source of a cylindrical wave, oriented along a line r_0, φ_0 exterior to a wedge, excites a field given by

$$u_g = \epsilon_1 \frac{e^{ikR_1}}{\sqrt{kR_1}} \mp \epsilon_2 \frac{e^{ikR_2}}{\sqrt{kR_2}}, \quad (5.6.1)$$

where

$$R_1 = \sqrt{r^2 + r_0^2 - 2rr_0\cos(\varphi - \varphi_0)}, \quad (5.6.2)$$

$$R_2 = \sqrt{r^2 + r_0^2 - 2rr_0 \cos(\varphi + \varphi_0)}, \qquad (5.6.2, \text{cont.})$$

and ϵ_1 and ϵ_2 are determined by Eqns. (5.2.5) and (5.2.6). Here, the upper sign corresponds to the boundary conditions of Eqn. (5.2.2), and the lower sign refers to those of Eqn. (5.2.3).

The supplemental diffracted field for $kr \gg 1$ is assumed to be of the form

$$u_d = \frac{W(r,\varphi)}{\sqrt{k(r+r_0)}} e^{ik(r+r_0)}. \qquad (5.6.3)$$

As before, it is evident that W satisfies Eqns. (5.2.12), (5.2.14), and (5.2.15) and possesses discontinuities indicated by Eqns. (5.2.11) and (5.2.13). Furthermore, it follows from this that $R_1 = r+r_0$ when $\phi = \pi + \varphi_0$ and $R_2 = r+r_0$ when $\phi = \pi - \varphi_0$.

Substituting $W/[k(r+r_0)]^{1/2}$ into the parabolic differential equation (5.2.10), we obtain a differential equation for W,

$$\frac{\partial^2 W}{d\varphi^2} + 2ikr^2 \frac{\partial W}{dr} + ik\frac{rr_0}{r+r_0} W = 0. \qquad (5.6.4)$$

With the aid of a variable substitution

$$\rho = \frac{rr_0}{r+r_0}, \qquad (5.6.5)$$

we arrive at

$$\frac{\partial^2 W(\rho,\varphi)}{d\varphi^2} + 2ik\rho^2 \frac{\partial W(\rho,\varphi)}{d\rho} + ik\rho W(\rho,\varphi) = 0, \qquad (5.6.6)$$

completely agreeing with Eqn. (5.2.10).

Thus, $W(\rho, \varphi)$ satisfies the same conditions and same differential equation as does $W(r, \varphi)$ studied in the preceding section. Therefore, the solution to the problem of diffraction of cylindrical waves from an edge is obtained directly from the solution of the problem of the diffraction of plane waves by replacing r with $\rho = rr_0/(r+r_0)$ in the expressions for $W(r, \varphi)$ in §§ 5.3, 5.4, and 5.5. Physically, this solution is a reasonable approximation of the exact solution when the conditions

$$kr \gg 1, \quad kr_0 \gg 1 \tag{5.6.7}$$

are satisfied. In this case, as was observed previously, we do not need to utilize an exact solution of the parabolic equation, and $w(\rho, \varphi)$ can be approximated by either Eqn. (5.4.17) of Eqn. (5.5.7).

The results obtained above are related to the case when one face of the wedge is illuminated ($\varphi_0 < \alpha - \pi$). If both faces are illuminated then the equations require some modification.

The results of this chapter show that the parabolic equation allows one to obtain asymptotic solutions to problems on diffraction from bodies with edges. The next chapter will use this method to investigate more complicated problems — multibounce diffraction of edge waves of current on a dipole and a strip.

§5.7 Additional Comments

- As shown in this chapter, the asymptotic behavior of the field predicted by the parabolic equation is incorrect near the edge [see Eqn. (5.3.13)], while it is correct far from the edge [see Eqns. (5.4.4), (5.4.12) and the text following Eqn. (5.4.15)]. There is interesting physics behind this observation. Actually, this result shows that the edge and corner/vertex waves are created within some finite vicinity of an edge and corner/vertex. The size of this vicinity can be comparable with a wavelength. The contribution of an infinitesimal vicinity of edges and corners to the scattered field is very small and cannot significantly affect its magnitude. This observation agrees with known results based on numerical solutions for wedges with rounded edges. If the radius of the edge curvature is less than one tenth of a wavelength, the edge wave from the rounded wedge is nearly identical with the edge wave from the perfectly sharp wedge. The phenomenon of "illuminous" edges on illuminated objects was understood by Sommerfeld who emphasized that this effect merely results from an extrapolation made by our eyes (see Chapter 20 in [43]).

- Numerical solutions of parabolic equations in the wedge diffraction problems are presented in [97–100].
- Additional references related to diffraction at impedance wedges can be found in [101, 102].
- Extension of the parabolic equation method for regular and irregular waveguides is presented in [103]. Here, a reference mode approach is suggested that allows one to apply the parabolic equation at greater distances than the ordinary parabolic equation.
- An interesting connection between the parabolic equation and the Helmholtz equations is found in [104]. It is shown that in planar stratified waveguides with perfectly conducting walls, the exact solution of the elliptic type (Helmholtz) equation can be represented as an integral from the solution of the parabolic equation.

6

Current Waves on Thin Conductors and Strips

This chapter will utilize the parabolic equation method to analyze the diffraction from a thin cylindrical conductor (dipole) and a strip. The possibility of solving this problem was previously mentioned in [96]. The theory of dipoles is widely examined in the professional literature. The general physical properties of current waves on dipoles were studied in [105–108]. This chapter will show that the results obtained in [105–108] can be derived more simply by using the parabolic equation. In addition, these results include near field calculations, and in the limit as $a \rightarrow \infty$ (a is the radius of the dipole), also describe the current waves on a strip.

It should be noted that the field near the surface of a conductor consists of different types of waves. In this case, there are waves traveling from a source of electromotive force and waves due to multiple diffraction at the ends of the conductor. Therefore, determining the total scattered field using the parabolic equation method leads to the sequential solution of a series of mathematical problems. Thus, for instance, to find the current on a transmitting dipole, it is necessary to determine the current $\mathbf{J}_0$ excited by the point source located on an infinitely long conductor (§ 6.1), and then the edge waves of the current, etc. (§ 6.2). This problem possesses a special property that it is connected with the specifics of the parabolic equation, i.e., waves traveling opposite directions are described by different differential equations.

§6.1 Excitation of an Infinite Conductor by a Point Source

Let there be an infinitely long ($-\infty \leqslant z \leqslant \infty$) cylindrical conductor of radius a excited by an external source

$$E_z^e = \mathcal{E}\delta(z), \tag{6.1.1}$$

located at its surface ($r = a$). An expression for the field close to the conductor will be sought in the form

$$A_z = W(r,z)e^{ik|z|}, \tag{6.1.2}$$

where A_z is a component of the retarded vector potential. The function W must satisfy

$$\frac{\partial^2 W}{\partial r^2} + \frac{1}{r}\frac{\partial W}{\partial r} + 2ik\,\mathrm{sgn}\,z\frac{\partial W}{\partial z} + \frac{\partial^2 W}{\partial z^2} = 0. \tag{6.1.3}$$

Neglecting the last term, we arrive at the parabolic equation

$$\frac{\partial^2 W}{\partial r^2} + \frac{1}{r}\frac{\partial W}{\partial r} + 2ik\,\mathrm{sgn}\,z\frac{\partial W}{\partial z} = 0, \tag{6.1.4}$$

which is assumed to have a solution of the form

$$W(r,z) = \int_C f(u)H_0^{(1)}\left(\sqrt{2ku}\,r\right)e^{-iu|z|}du, \tag{6.1.5}$$

where $H_0^{(1)}$ is the zeroth-order Hankel function of the first kind and $\Im\mathrm{m}\{\sqrt{u}\} \geqslant 0$.

The boundary condition

$$E_z + E_z^e = 0 \quad \text{for} \quad r = a \quad , \tag{6.1.6}$$

written in the form

$$\frac{d^2}{dz^2}A_z + k^2 A_z = ik\mathcal{E}\delta(z) \text{ for } r=a \tag{6.1.7}$$

leads to the equation

$$\int_C \left(u^2 - 2ku\right) f(u) H_0^{(1)}\left(\sqrt{2ku}\, r\right) e^{-iu|z|} du = -ik\mathcal{E}\delta(z). \tag{6.1.8}$$

Noting

$$\delta(z) = \frac{1}{2\pi}\int_{-\infty}^{\infty} e^{-iu|z|} du, \tag{6.1.9}$$

we find from Eqn. (6.1.8) that

$$f(u) = -\frac{ik}{2\pi}\frac{\mathcal{E}}{\overset{\frown}{u}\left(u - \underset{\smile}{2k}\right) H_0^{(1)}\left(\sqrt{2ku}\, a\right)}. \tag{6.1.10}$$

The contour of integration proceeds along the real axis in the positive direction and skirts above the point $u = 0$ and below the point $u = 2k$. This behavior of the contour is dictated by two reasons.

First, the point $u = 0$ belongs to the branch cut $-i\infty \le u \le 0$ and must be skirted above. With this branch cut, we choose for the function $\sqrt{u}$ the branch where $\Im\mathrm{m}\{\sqrt{u}\} \geqslant 0$. This condition is necessary to ensure the convergence of the integrals in Eqns. (6.1.5) and (6.1.8).

Secondly, to determine the contour behavior near the point $u = 2k$, one should refer to Eqn. (6.1.7). It has the solution

$$A_z(a,z) = \text{const}\, e^{ik|z|}. \tag{6.1.11}$$

Taking into account the relationship

$$\frac{d^2}{dz^2}|z| = 2\delta(z) \tag{6.1.12}$$

we find

$$A_z(a,z) = \frac{1}{2}\mathcal{E}e^{ik|z|}, \tag{6.1.13}$$

and

$$W(a,z) = \frac{1}{2}\mathcal{E} = \text{const}. \tag{6.1.14}$$

Subsitution of Eqn. (6.1.10) into Eqn. (6.1.5) gives

$$W(a,z) = -\frac{ik}{2\pi}\mathcal{E}\int_{-\infty}^{\infty}\frac{1}{\widehat{u}(u-2k)}e^{-iu|z|}dz. \tag{6.1.15}$$

Referring to the Cauchy residue theorem, we find that this equation can be reduced to Eqn. (6.1.14) only on the condition that the contour of integration skirts below the point $u = 2k$.

Therefore, having determined the contour of integration, we obtain the final expression for the function $W(r,z)$:

$$W(r,z) = \frac{\mathcal{E}}{4\pi i}\int_C\left(\frac{1}{u-\underset{\smile}{2k}} - \frac{1}{\widehat{u}}\right)\frac{H_0^{(1)}\left(\sqrt{2ku}\,r\right)}{H_0^{(1)}\left(\sqrt{2ku}\,a\right)}e^{-iu|z|}du. \tag{6.1.16}$$

The current excited by the external source on the surface of the conductor will now be calculated. According to Eqn. (6.1.2) and $\mathbf{H} = \nabla\times\mathbf{A}$, the components of the magnetic field are

$$H_\varphi = -\frac{\partial A_z}{\partial r} = -\frac{\partial W}{\partial r} e^{ik|z|}, \quad H_r = H_z = 0. \tag{6.1.17}$$

The density of the surface current is determined by

$$j_z = \frac{c}{4\pi} H_\varphi \bigg|_{r=a^+} = -\frac{c}{4\pi} \frac{\partial W}{\partial r}\bigg|_{r=a^+} e^{e^{ik|z|}}. \tag{6.1.18}$$

Thus, the total current

$$J(z) = 2\pi a j_z \tag{6.1.19}$$

using Eqns. (6.1.16) and (6.1.18) is equal to

$$J(z) = S_0 \psi(|z|) e^{ik|z|}, \tag{6.1.20}$$

where

$$\psi(z) e^{ikz} = \frac{\Lambda(2ka)}{2\pi i} \int_{-\infty}^{\infty} \left(\frac{1}{u - \underset{\smile}{k}} - \frac{1}{u - \overset{\frown}{(-k)}} \right) \frac{e^{iuz}\, du}{\Lambda\left[\sqrt{2k(k-u)}\, a\right]}, \tag{6.1.21}$$

$$S_0 = \frac{c\mathcal{E}}{4\Lambda(2ka)}, \quad \Lambda(x) = \frac{H_0^{(1)}(x)}{x H_1^{(1)}(x)}. \tag{6.1.22}$$

The $\frown$ and $\smile$ symbols in the integrand of Eqn. (6.1.21) show that the contour of integration skirts below the pole at $u = k$ and above the pole at $u = -k$.

Notice the following properties of the function $\psi(z)$:

$$\psi(0) = 1, \quad \psi(\infty) = 0. \tag{6.1.23}$$

The first property can be proved by utilizing the Cauchy residue theorem if we close the contour of integration in Eqn. (6.1.21) in the lower half plane,

$\Im m(u) \leqslant 0$, where $u = \infty \exp(i\varphi)$ and $-\pi \leq \varphi \leq 0$. This extension of the integration contour is permissible since,

$$|\Lambda(x)| \to \frac{1}{|x|} \text{ with } |x| \to \infty. \tag{6.1.24}$$

In order to prove the second property, $\psi(\infty) = 0$, we deform the sections $-\infty \leq u < 0$ and $0 < u \leq \infty$ of the contour of integration to follow the rays given by $\arg(u) = \pi - \alpha$ and $\arg(u) = \alpha$, respectively, where for example, $0 < \alpha < \pi/4$. This deformation of the contour is also possible due to Eqn. (6.1.24). Along these rays, $\Im m(u) > 0$ and the integrand is zero, since $\exp(iu|z|) = 0$ when $|z| = \infty$. Besides, one can show that the integral of Eqn. (6.1.21) over the infintesimal arc that skirts below the branch point $u = k$ is zero as well.

Equations (6.1.20) and (6.1.21) are analogous to formulas obtained in [105] using the more complicated method of slowly varying amplitudes and can be transformed into the latter for $ka \ll 1$ when

$$\Lambda(2ka) \approx \ln \frac{i}{\gamma ka}, \quad \Lambda\left(\sqrt{2k(k-u)}\,a\right) \approx \ln \frac{2i}{\gamma a \sqrt{2k(k-u)}}, \tag{6.1.25}$$

where $\gamma = 1.718\ldots$. Here and below, we choose the principal value of the logarithm function, where $\ln(|z|e^{i\varphi}) = \ln|z| + i\varphi$, $\ln(1) = 0$, and $\ln(i) = i\pi/2$.

§6.2 Transmitting Dipole

We will now examine a finite length dipole ($z_1 \leqslant z \leqslant z_2$) excited by the external field given by Eqn. (6.1.1). It is evident that the point of excitation ($z = 0$) will cause the current of Eqn. (6.1.20) to propagate along the conductor. Let us call it J_0. Its arrival at the end of the conductor at $z = z_1$ creates a reflected wave, the field of which near the conductor for $z - z_1 \geqslant 0$ may be expressed as

$$A_z = W(r, z) e^{ik(z - z_1)}. \tag{6.2.1}$$

Using the parabolic equation

$$\frac{\partial^2 W}{\partial r^2}+\frac{1}{r}\frac{\partial W}{\partial r}+2ik\frac{\partial W}{\partial z}=0 \tag{6.2.2}$$

to solve for $W(r, z)$, we find that

$$W(r,z)=\int_C f(u)H_0^{(1)}\left(\sqrt{2ku}\,r\right)e^{-iu(z-z_1)}du. \tag{6.2.3}$$

From the boundary condition

$$E_z=0 \quad \text{for} \quad z-z_1>0, \quad r=a, \tag{6.2.4}$$

we obtain

$$W(a,z)=\text{const}, \quad \frac{\partial W(a,z)}{\partial z}=0 \quad \text{for} \quad z-z_1>0. \tag{6.2.5}$$

The current, reflected from $z=z_1$ is

$$J_1(z)=\frac{ca}{2}e^{ik(z-z_1)}\int_C \sqrt{2ku}\,f(u)H_1^{(1)}\left(\sqrt{2ku}\,a\right)e^{-iu(z-z_1)}du \tag{6.2.6}$$

and must satisfy the boundary condition at $z=z_1$ that

$$J_1(z_1)=-J_0(z_1)=-S_0\psi(-z_1)e^{-ikz_1}. \tag{6.2.7}$$

The conditions outlined by Eqns. (6.2.5) and (6.2.7) lead to the following system of functional equations

$$\left.\begin{aligned}&\int_C uf(u)H_0^{(1)}\left(\sqrt{2ku}\,a\right)e^{-iu(z-z_1)}du=0 \quad \text{for} \quad z-z_1>0,\\ &\frac{ca}{2}e^{ik(z-z_1)}\int_C \sqrt{2ku}\,f(u)H_1^{(1)}\left(\sqrt{2ku}\,a\right)e^{-iu(z-z_1)}du=\\ &\qquad\qquad =-S_0\psi(-z_1)e^{-ikz} \quad \text{for} \quad z-z_1<0.\end{aligned}\right\} \tag{6.2.8}$$

Note that independent of Eqn. (6.2.7),

$$J_1(z)=-J_0(z) \quad \text{for} \quad z-z_1<0, \tag{6.2.9}$$

which represents the absence of current for $z<z_1$. Noting Eqn. (6.2.8), one can obtain an approximate formulation of the problem that satisfies only Eqn. (6.2.7). The condition described by Eqn. (6.2.9) is satisfied only approximately, when $k|z_1|\gg 1$. This approximate formulation allows us to study the diffraction of the edge wave on a conductor of finite length, L. It is clear that the assumed approximations will hold true for sufficiently long conductors ($kL\gg 1$). This approximation is convenient because the resulting functional equations can be solved without factorization.

From Eqn. (6.2.8) we find that

$$f(u)=-\frac{2k}{c\pi i}\frac{\Lambda(2ka)S_0\psi(-z_1)e^{-ikz_1}}{u(u-2k)\mathrm{H}_0^{(1)}\left(\sqrt{2ku}\,a\right)}, \tag{6.2.10}$$

and thus,

$$J_1(z)=-J_0(z_1)\psi(z-z_1)e^{ik(z-z_1)}. \tag{6.2.11}$$

In a completely analogous manner it is possible to show that the current reflected at $z = z_2$ is equal to

$$J_2(z) = -J_0(z_2)\psi(z_2 - z)e^{ik(z_2 - z)}. \tag{6.2.12}$$

Thus, all of the waves of current on the transmitting dipole are expressed as a function of ψ. Therefore, the total current can be assumed to be of the form

$$J(z) = S_0\left\{\psi(|z|)e^{ik|z|} + A_1\psi(z - z_1)e^{ik(z - z_1)} + A_2\psi(z_2 - z)e^{ik(z_2 - z)}\right\}. \tag{6.2.13}$$

The constants A_1 and A_2 are determined from the boundary conditions at the ends of the conductor

$$J(z_1) = J(z_2) = 0, \tag{6.2.14}$$

and are equal to

$$\left.\begin{aligned} A_1 &= -\frac{1}{\mathcal{D}}\left[\psi(-z_1) - \psi(z_2)\psi(z_2 - z_1)e^{2ikz_2}\right]e^{-ikz_1}, \\ A_2 &= -\frac{1}{\mathcal{D}}\left[\psi(z_2) - \psi(-z_1)\psi(z_2 - z_1)e^{-2ikz_1}\right]e^{ikz_2}, \end{aligned}\right\} \tag{6.2.15}$$

where

$$\mathcal{D} = 1 - \psi^2(z_2 - z_1)e^{2ik(z_2 - z_1)} \tag{6.2.16}$$

represents the characteristic resonance of the dipole. When $z_2 - z_1 \approx n\lambda/2$ $(n = 1, 2, 3\ldots)$ and $\mathcal{D} \approx 0$, the dipole current is resonant.

§6.3 Excitation of a Semi-Infinite Conductor by a Plane Wave

Let a semi-infinite cylinder $(z_1 \leqslant z \leqslant \infty)$ be excited by the plane wave

$$E_z^e = E_{0z}e^{iwz}, \quad w = -k\cos\vartheta. \tag{6.3.1}$$

First, the current excited by this wave on an infinite conductor $(-\infty \leqslant z \leqslant \infty)$ will be calculated. The field near the surface will assumed to be of the form

$$A_z = W(r,z)e^{iwz}, \tag{6.3.2}$$

where

$$W(r,z) = \int_{-\infty}^{\infty} f(u)H_0^{(1)}\left(\sqrt{v^2+2wu}\,r\right)e^{-iuz}du, \quad v=\sqrt{k^2-w^2} \tag{6.3.3}$$

and is the solution to the parabolic equation

$$\left(\frac{\partial^2}{\partial r^2}+\frac{1}{r}\frac{\partial}{\partial r}+2iw\frac{\partial}{\partial z}+v^2\right)W(r,z)=0. \tag{6.3.4}$$

The boundary condition given by Eqn. (6.1.6) leads to

$$\int_{-\infty}^{\infty}\left(v^2+2wu-u^2\right)f(u)H_0^{(1)}\left(\sqrt{v^2+2wu}\,a\right)e^{-iuz}du = ikE_{oz}, \tag{6.3.5}$$

where

$$f(u) = \frac{ikE_{0z}}{v^2H_0^{(1)}(va)}\delta(u), \quad \Im\mathfrak{m}\{v\}\geqslant 0. \tag{6.3.6}$$

As a result, the total current on the conductor,

$$J(z) = \frac{ca}{2}e^{iwz}\int_{-\infty}^{\infty} f(u)\sqrt{v^2+2wu}\,H_1^{(1)}\left(\sqrt{v^2+2wu}\,a\right)e^{-iuz}du, \tag{6.3.7}$$

will be equal to

$$J(z) = Se^{iwz}, \tag{6.3.8}$$

where

$$S = E_{0z}\frac{icka}{2v}\frac{H_1^{(1)}(va)}{H_0^{(1)}(va)}, \quad v = k\sin\vartheta. \tag{6.3.9}$$

It is evident that such a wave exists on a semi-infinite conductor. Upon arriving at the end $z = z_1$ of the conductor, it excites an edge wave. The field of this edge wave and its corresponding current will once again be in the form of Eqns. (6.2.1), (6.2.3), and (6.2.6). The function $f(u)$ in these expressions is determined by the conditions

$$\left.\begin{aligned} &\frac{\partial}{\partial z}W(a,z)=0 &&\text{for} \quad z-z_1>0, \\ &J(z)=-Se^{iwz} &&\text{for} \quad z-z_1<0, \end{aligned}\right\} \tag{6.3.10}$$

which is the equivalent to the following system of equations

$$\left.\begin{aligned} &\int_C uf(u)H_0^{(1)}\left(\sqrt{2ku}\,a\right)e^{-iu(z-z_1)}du=0 \quad \text{for} \quad z-z_1>0, \\ &\frac{ca}{2}e^{ik(z-z_1)}\int_C \sqrt{2ku}\,f(u)H_1^{(1)}\left(\sqrt{2ku}\,a\right)e^{-iu(z-z_1)}du= \\ &\qquad\qquad = -Se^{i\omega z} \quad \text{for} \quad z-z_1<0. \end{aligned}\right\} \tag{6.3.11}$$

The contour, C, will be chosen such that it proceeds along the real axis in a positive direction, skirting above branch point at $u = 0$, and the below pole at $u = k-w$. Note that Eqns. (6.3.10) and (6.3.11) guarantee exactly, not approximately, zero current past the end of the semi-infinite conductor, i.e., at $r = a$, $z-z_1 < 0$. Thus, for this case, the diffraction of a plane wave from a semi-infinite conductor is solved in a rigorous fashion.

From Eqn. (6.3.11) we find that

$$f(u) = -\frac{Se^{iwz_1}}{c\pi i}\frac{k-w}{u(u-k+w)}\frac{\Lambda\left[\sqrt{2k(k-w)}a\right]}{H_0^{(1)}\left(\sqrt{2ku}\,a\right)}. \tag{6.3.12}$$

Therefore,

$$A_z = -\frac{Se^{iwz_1}}{c\pi i}\Lambda\left[\sqrt{2k(k-w)}a\right]\times$$

$$\times\int_{-\infty}^{\infty}\left(\frac{1}{u-\underset{\curvearrowleft}{k}}-\frac{1}{u-\overset{\frown}{w}}\right)\frac{H_0^{(1)}\left(\sqrt{2k(k-u)}r\right)}{H_0^{(1)}\left(\sqrt{2k(k-u)}a\right)}e^{iu(z-z_1)}du, \tag{6.3.13}$$

and

$$J(z) = -S\psi_-(z-z_1)e^{ik(z-z_1)+iwz_1}, \tag{6.3.14}$$

where

$$\psi_-(z)e^{ikz} = \frac{\Lambda\left[\sqrt{2k(k-w)}a\right]}{2\pi i}\times$$

$$\times\int_{-\infty}^{\infty}\left(\frac{1}{u-\underset{\curvearrowleft}{k}}-\frac{1}{u-\overset{\frown}{w}}\right)\frac{e^{iuz}du}{\Lambda\left[\sqrt{2k(k-u)}a\right]}. \tag{6.3.15}$$

Equation (6.3.14) describes the edge wave of the current. The total current on the conductor is given by

$$J(z) = S\left[e^{iwz} - \psi_-(z-z_1)e^{ik(z-z_1)+iwz_1}\right]. \tag{6.3.16}$$

Using this same method, it is possible to show that a plane wave given by Eqn. (6.3.1) excites a current given by

$$J(z)=S\left[e^{iwz}-\psi_{+}(z_2-z)e^{ik(z_2-z)+iwz_2}\right] \tag{6.3.17}$$

on a semi-infinite conductor ($-\infty \leqslant z \leqslant z_2$), where

$$\psi_{+}(z)e^{ikz}=\frac{\Lambda\left[\sqrt{2k(k+w)a}\right]}{2\pi i}\times$$
$$\times\int_{-\infty}^{\infty}\left(\frac{1}{u-k}-\frac{1}{u-(-w)}\right)\frac{e^{iuz}du}{\Lambda\left[\sqrt{2k(k-u)a}\right]}. \tag{6.3.18}$$

Now, some useful relationships for ψ and $\psi_{\pm}$ will be obtained. First of all, it is evident that they exhibit the following properties

$$\psi_{+}(0)=\psi_{-}(0)=\psi(0)=1, \tag{6.3.19}$$

$$\left.\begin{aligned}\psi_{+}(z)&=1 &&\text{for}\quad w=-k,\\ \psi_{+}(z)&=\psi(z) &&\text{for}\quad w=k,\end{aligned}\right\} \tag{6.3.20}$$

$$\left.\begin{aligned}\psi_{-}(z)&=1 &&\text{for}\quad w=k,\\ \psi\ (z)&=\psi(z) &&\text{for}\quad w=-k,\end{aligned}\right\} \tag{6.3.21}$$

$$\psi_{+}(\infty)=\psi_{-}(\infty)=\psi(\infty)=0. \tag{6.3.22}$$

Equations (6.3.19) and (6.3.22) can be proved in a manner identical to what was used for Eqn. (6.1.21). The expressions found on the second lines of Eqns. (6.3.20) and (6.3.21) are obvious. In order to prove the equations in the first lines of Eqns. (6.3.20) and (6.3.21), let us shift the contour of integration in Eqns. (6.3.15) and (6.3.18) a little bit downward in the vicinity of the pole $u = \pm w$ and intersect this pole. According to the Cauchy residue theorem, the function $\psi_{\pm}(z)\exp(ikz)$ will be equal to $\exp(\mp iwz)$ plus the integral over the shifted contour. As $w \to \mp k$, this integral approaches zero and one obtains the equality $\psi_{\pm}(z)\exp(ikz) = \exp(ikz)$, which means that $\psi_{\pm}(z) = 1$ if $w = \mp k$.

Furthermore, let us calculate the integral

$$I = \int_{z>0}^{\infty} \psi(\zeta) e^{i(k-w)\zeta} d\zeta .$$

Referring Eqn. (6.1.19), it is possible write this in the form

$$I = e^{-iwz} \frac{\Lambda(2ka)}{2\pi} \int_{-\infty}^{\infty} \left(\frac{1}{u-k} - \frac{1}{u-(-k)} \right) \frac{1}{u-w} \frac{e^{iu\zeta} du}{\Lambda\left[\sqrt{2k(k-u)}a\right]} .$$

With the aid of a simple transformation we obtain

$$\int_{z>0}^{\infty} \psi(\zeta) e^{i(k-w)\zeta} d\zeta = \tag{6.3.23}$$
$$= \frac{e^{i(k-w)z}}{i(k+w)} \left\{ \psi(z) - \frac{2k}{k-w} \frac{\Lambda(2ka)}{\Lambda\left[\sqrt{2k(k-w)}a\right]} \psi_{-}(z) \right\} .$$

Analogously,

$$\int_{-\infty}^{z_1} \psi_{+}(z_2 - \zeta) e^{-i(k-w_0)\zeta} d\zeta = \tag{6.3.24}$$

$$= \frac{e^{i(w_0-k)z_1}}{i(w+w_0)} \left\{ \psi_+(z_2-z_1) - \frac{k+w}{k-w_0} \frac{\Lambda\left[\sqrt{2k(k+w)a}\right]}{\Lambda\left[\sqrt{2k(k-w_0)a}\right]} \psi_-^0(z_2-z_1) \right\} \quad (6.3.24, \text{cont.})$$

and

$$\int_{z_2}^{\infty} \psi_-(\zeta - z_2) e^{i(k+w_0)\zeta} d\zeta =$$
$$= \frac{ie^{i(k+w_0)z_2}}{(w+w_0)} \left\{ \psi_-(z_2-z_1) - \frac{k-w}{k+w_0} \frac{\Lambda\left[\sqrt{2k(k-w)a}\right]}{\Lambda\left[\sqrt{2k(k+w_0)a}\right]} \psi_+^0(z_2-z_1) \right\}, \quad (6.3.25)$$

where $\psi_\pm^0$ is obtained from the expression for $\psi_\pm$ by substituting w_0 for w. Equations (6.3.23), (6.3.24), and (6.3.25) will be used in Chapter 7 for the investigation of the fields scattered by passive dipoles.

§6.4 Passive Dipole

This section will examine the diffraction of a plane wave given by Eqn. (6.3.1) from a conducting cylinder of finite length ($z_1 \leqslant z \leqslant z_2$). The first-order edge waves were calculated in the previous section. Subsequently, one can find the second-order edge waves created by the diffraction of the first-order waves from the ends of the conductor.

The first-order edge wave of the current, traveling from the end at $z = z_2$ is equal to

$$J(z) = -S\psi_+(z_2 - z) e^{ik(z_2-z)+iwz_2}. \quad (6.4.1)$$

The field and the current of the second-order edge wave close to the conductor (for $z-z_1 \geqslant 0$, $r-a \ll z-z_1$) will once again be sought in the form of Eqns. (6.2.1), (6.2.3), and (6.2.6). The unknown function $f(u)$ must be chosen such that the following conditions are satisfied

$$\left.\begin{aligned} &\frac{\partial}{\partial z}W(a,z)=0 && \text{for} \quad z-z_1>0, \\ &J(z)=S\psi_+(z_2-z_1)e^{ik(z_2-z)+iwz_2} && \text{for} \quad z-z_1<0. \end{aligned}\right\} \tag{6.4.2}$$

Having found $f(u)$, one can determine the value of the field

$$A_z=\frac{2}{c}\left[S\psi_+(z_2-z_1)e^{ik(z_2-z_1)+iwz_2}\right]\frac{\Lambda(2ka)}{2\pi i}\times$$
$$\times\int_{-\infty}^{\infty}\left(\frac{1}{u-\underset{\smile}{k}}-\frac{1}{u-\widehat{(-k)}}\right)\frac{H_0^{(1)}\left(\sqrt{2k(k-u)}r\right)}{H_0^{(1)}\left(\sqrt{2k(k-u)}a\right)}e^{iu(z-z_1)}du \tag{6.4.3}$$

and the current

$$J(z)=\left[S\psi_+(z_2-z_1)e^{ik(z_2-z_1)+iwz_2}\right]\psi(z-z_1)e^{ik(z-z_1)} \tag{6.4.4}$$

of the second-order edge wave.

In an identical manner, it is possible to show that the first-order edge wave of the current

$$J(z)=-S\psi_-(z-z_1)e^{ik(z-z_1)+iwz_1}, \tag{6.4.5}$$

experiences a reflection at $z=z_2$ that excites the field

$$A_z=\frac{2}{c}\left[S\psi_-(z_2-z_1)e^{ik(z_2-z_1)+iwz_1}\right]\frac{\Lambda(2ka)}{2\pi i}\times$$
$$\times\int_{-\infty}^{\infty}\left(\frac{1}{u-\underset{\smile}{k}}-\frac{1}{u-\widehat{(-k)}}\right)\frac{H_0^{(1)}\left(\sqrt{2k(k-u)}r\right)}{H_0^{(1)}\left(\sqrt{2k(k-u)}a\right)}e^{iu(z_2-z)}du \tag{6.4.6}$$

and the current

$$J(z)=\left[S\psi_{-}(z_2-z_1)e^{ik(z_2-z_1)+iwz_1}\right]\psi(z_2-z)e^{ik(z_2-z)}. \tag{6.4.7}$$

Returning to § 6.2, we see that all the following edge waves of the current are expressed using the function, ψ. Thus, the total current on a passive dipole can be expressed as

$$\begin{aligned}J(z)=S\Big\{&e^{iwz}-\psi_{-}(z-z_1)e^{ik(z-z_1)+iwz_1}-\psi_{+}(z_2-z)e^{ik(z_2-z)+iwz_2}+\\&+B_1\psi(z-z_1)e^{ik(z-z_1)}+B_2\psi(z_2-z)e^{ik(z_2-z)}\Big\}.\end{aligned} \tag{6.4.8}$$

The constants B_1 and B_2 are determined from the condition that at the ends of the conductor,

$$J(z_1)=J(z_2)=0, \tag{6.4.9}$$

and are equal to

$$\left.\begin{aligned}B_1&=\frac{e^{i(k+w)z_2}}{\mathcal{D}}\left[\psi_{+}(L)-\psi_{-}(L)\psi(L)e^{i(k-w)L}\right]e^{-ikz_1},\\B_2&=\frac{e^{-i(k-w)z_1}}{\mathcal{D}}\left[\psi_{-}(L)-\psi_{+}(L)\psi(L)e^{i(k+w)L}\right]e^{ikz_2},\end{aligned}\right\} \tag{6.4.10}$$

where $L=z_2-z_1$, and $\mathcal{D}$ is the resonant denominator given by Eqn. (6.2.16).

Taking into account Eqns. (6.3.20) and (6.3.21), we find that the induced current equals zero at grazing incidence ($\vartheta=0$, $\vartheta=\pi$). The physical reason for that is clear: in this case the incident wave does not have a E_z component and, therefore, cannot create a z-directed current. Another, but equivalent explanation is the following: the H_φ component of the incident wave creates equal but opposite z-directed currents on the two sides of the conductor (φ and $\pi+\varphi$), which sum to zero total current. Recall that the φ components of the current are negligible for thin conductors ($ka \ll 1$).

The expressions found above for the current on transmitting and passive dipoles differ from analogous expressions obtained using more complicated methods in [105, 106], only in terms of $\psi_{\pm}$ and ψ. However, for thin dipoles ($ka \ll 1$) our expressions and those in [105, 106] are equivalent. Their accuracy and other properties are examined in detail in [107], therefore, we will not dwell on these issues further. We only will note that these expressions possess a minimum accuracy near the location of the source and near the ends of the conductor, where the amplitude of the excited waves fails to change sufficiently slowly. In particular, Eqn. (6.1.20) which represents the current excited on an infinite conductor by a point source, possesses a relative error on the order of $[2\ln(i/\gamma ka)]^{-2}$ when $kz \ll 1$. When $kz \to \infty$, the error of this expression approaches zero. An identical error is exhibited by Eqn. (6.3.14) which represents the current excited on a semi-infinite conductor by a plane wave.

It is also interesting to observe that the expressions for $\psi_{\pm}$ and ψ, obtained with the parabolic equation method, are preferable to the corresponding expressions found in [105, 106]. First, they differ from the latter in that they do not posses *parasitic* poles, the residue of which produces waves possessing no physical meaning. Second, in the limit as $a \to \infty$, they accurately describe the waves of current on a strip.

Note that according to the more rigorous approach of [108], the amplitudes of all the reflected waves will be determined by different functions; however, the difference between them decreases with an increase in the number of reflections. The functions $\psi_{\pm}(z)$ and $\psi(z)$ only approximately describe these waves of current but permit their summation to be obtained in closed form.

In conclusion, we can provide the useful approximate expression

$$\psi_{\pm}(z) = \frac{2\ln\left[2i/\gamma a\sqrt{2k(k\pm w)}\right]}{\ln\left[2iz/\gamma ka^2\right] - E\left[(k\pm w)z\right]e^{-i(k\pm w)z}}, \tag{6.4.11}$$

$$w=-k\cos\vartheta,\quad E(y)=-\int_{y}^{\infty}\frac{e^{it}}{t}dt, \tag{6.4.11, cont.}$$

which follows from the variational principle [107] and is well described in tables of functions. Equation (6.4.11) has an asymptotic form given by

$$\psi_{\pm}(z)\sim\frac{2\ln\left[\sqrt{2}i/\gamma a\sqrt{k(k\pm w)}\right]}{\ln\left[2iz/\gamma ka^{2}\right]}\left[1-\frac{e^{-i(k\pm w)z}}{\ln\left[2iz/\gamma ka^{2}\right]}\int_{(k\pm w)z}^{\infty}\frac{e^{it}}{t}dt\right] \tag{6.4.12}$$

for $(k\pm w)z\gg 1$, and

$$\psi_{\pm}(z)\sim 1-\frac{1}{2\ln\left[\sqrt{2}i/\gamma a\sqrt{k(k\pm w)}\right]}\times$$
$$\times\left\{\ln\left[\gamma(k\pm w)z/i\right]+e^{-i(k\pm w)z}\int_{(k\pm w)z}^{\infty}\frac{e^{it}}{t}dt\right\} \tag{6.4.13}$$

for $kz\ll 1$, the relative error of which does not exceed a value on the order of $[2\ln(i/\gamma ka)]^{-2}$. It can be shown that these equations may be also obtained from Eqns. (6.3.15) and (6.3.18). In fact, for $ka\ll 1$ Eqns. (6.3.15) and (6.3.18) can be written as

$$\psi_{\pm}(z)=\frac{\ln\left[\sqrt{2}i/\gamma a\sqrt{k(k\pm w)}\right]}{i\pi}\int_{-\infty}^{\infty}\left(\frac{1}{t}-\frac{1}{t+k\pm w}\right)\frac{e^{itz}dt}{\ln\left[2/\gamma^{2}ka^{2}t\right]}, \tag{6.4.14}$$

where the contour of integration skirts above the poles at $t=-(k\pm w)$, above the branch point at $t=2/(\gamma^{2}ka^{2})$, and below the point $t=0$. Introducing the notation

$$I(x) = -\frac{1}{2\pi i}\int_{-\infty}^{\infty}\frac{e^{iux}du}{u\ln\left[\gamma^2 u/2\right]}, \tag{6.4.15}$$

where the contour of integration skirts above the branch point at $u = 2/\gamma^2$ and below the point $u = 0$, we obtain

$$\psi_{\pm}(z) = 2\ln\frac{\sqrt{2i}}{\gamma a\sqrt{k(k\pm w)}}\left[I\left(\frac{z}{ka^2}\right) + e^{-i(k\pm w)z}\int_z^{\infty} I_s'\left(\frac{s}{ka^2}\right)e^{i(k\pm w)s}ds\right] \tag{6.4.16}$$

where

$$I_s'\left(\frac{s}{ka^2}\right) = \frac{\partial}{\partial s}I\left(\frac{s}{ka^2}\right).$$

In reference [105], it is shown that for $x \gg 1$

$$I(x) = \frac{1}{\ln\left[2ix/\gamma\right]} + O\left[\frac{1}{\left[\ln(2ix/\gamma)\right]^3}\right]. \tag{6.4.17}$$

Substituting this expression into Eqn. (6.4.16) and taking into account that the basic contribution to the integral is provided by the vicinity of the lower limit (with $(k+w)z \gg 1$), we arrive at Eqn. (6.4.12).

Furthermore, it should be noted that in agreement with reference [106], Eqn. (6.4.14) satisfies the integral equation

$$\psi_{\pm}(z) + \frac{1}{g_{\pm}(0) + g(0)}\int_0^z \ln\frac{\gamma(k\pm w)(z-\zeta)}{i}\psi_{\pm}'(\zeta)d\zeta = \\ = 1 - \frac{g_{\pm}(z) - g_{\pm}(0)}{g_{\pm}(0) + g(0)}, \tag{6.4.18}$$

where

$$\left.\begin{aligned}
g_{\pm}(z) &= \ln\frac{2z}{a} + e^{-i(k\pm w)z}\int_{z}^{\infty}\frac{e^{i(k\pm w)\sigma}}{\sigma}d\sigma,\\
g(z) &= \ln\frac{2z}{a} + e^{-2ikz}\int_{z}^{\infty}\frac{e^{2ik\sigma}}{\sigma}d\sigma,\\
g_{\pm}(0) &= \ln\frac{2i}{\gamma(k\pm w)a},\\
g(0) &= \frac{i}{\gamma ka}.
\end{aligned}\right\}\tag{6.4.19}$$

Hence, given $kz \ll 1$ and therefore the integral term of Eqn. (6.4.18) is negligible, we have discovered that the asymptotic representation of Eqn. (6.4.14) is Eqn. (6.4.13). In this manner, Eqn. (6.4.11) describes Eqns. (6.3.15) and (6.3.18), with a relative error not exceeding a value on the order of $[2\ln(i/\gamma ka)]^{-2}$.

§6.5 The Near Field

The parabolic equation method can effectively calculate not only the waves of current, but also the field near the dipole. Taking into consideration that the electric vector potential **A** for a thin dipole consists of only of an $\mathbf{A}_z$ component and utilizing the relationships

$$\mathbf{H} = \nabla\times\mathbf{A}, \quad \mathbf{E} = -\frac{1}{ik}\left(\nabla\nabla\cdot\mathbf{A} + k^2\mathbf{A}\right), \tag{6.5.1}$$

we find that

$$H_{\varphi} = -\frac{\partial A_z}{\partial r}, \quad E_r = -\frac{1}{ik}\frac{\partial^2 A_z}{\partial z\partial r}, \tag{6.5.2}$$

$$E_z = -\frac{1}{ik}\left(\frac{\partial^2}{\partial z^2} + k^2\right)A_z, \quad E_\varphi = H_r = H_z = 0. \tag{6.5.3}$$

Substituting the expression for $\mathbf{A}_z$ from the preceding section, and observing that $k(r-a) \ll 1$, we obtain

$$E_r = \frac{2}{r}q, \quad H_\varphi = \frac{2}{cr}J, \tag{6.5.4}$$

as obtained for a quasistationary field. Here, the value of J represents the total current on the dipole, and q, the linear charge density, is equal to

$$q = \frac{1}{ick}\frac{d}{dz}J(z). \tag{6.5.5}$$

The component E_z and energy characteristics are expediently calculated separately for waves of different directions. Using the approximation of the parabolic equation, one can write

$$E_z = \mp 2\frac{\partial W}{\partial z}e^{\pm ikz} \tag{6.5.6}$$

where the +/− sign denotes a wave traveling in the $+z/-z$ direction, respectively. Further noting that $W(r, z)$ obeys the relationship that

$$W(r,z) \approx W(a,z) + \frac{\partial W(a,z)}{\partial ka}k(r-a), \tag{6.5.7}$$

and enforcing the boundary condition on the surface of the conductor,

$$\frac{\partial W(a,z)}{\partial z} = 0, \tag{6.5.8}$$

we find that

$$E_z = 4ik\frac{r-a}{a}\left(\pm q - \frac{1}{c}J\right). \tag{6.5.9}$$

It follows that the power flux density,

$$\mathbf{P} = \frac{c}{8\pi}\Re\mathrm{e}\left\{\mathbf{E}\times\mathbf{H}^*\right\}, \tag{6.5.10}$$

possesses the components

$$P_r = \mp\frac{1}{2\pi c}\frac{r-a}{ra}\frac{d}{dz}|J|^2, \quad P_z = \pm\frac{1}{2\pi c r^2}|J|^2. \tag{6.5.11}$$

In deriving these equations, we utilized the relationships

$$\Re\mathrm{e}\left\{J^*\frac{dJ}{dz}\right\} = \frac{1}{2}\frac{d}{dz}|J|^2, \quad \Re\mathrm{e}\left\{qJ^*\right\} \approx \Re\mathrm{e}\left\{\pm\frac{1}{c}JJ^*\right\}. \tag{6.5.12}$$

Notice that the near field possesses a partially a reactive character, i.e., the field corresponding to the second term in Eqn. (6.5.9) does not transfer energy in the radial direction.

§6.6 Waves of Current on a Strip

The expressions obtained for the current on a dipole possess an additional advantage: they can calculate the current in the limit as $a \to \infty$, when the dipole is transformed into a strip. Transforming Eqns. (6.2.13) and (6.4.8) as $a \to \infty$, it is possible to calculate the current excited on a strip ($z_1 \leqslant z \leqslant z_2$) on one of its sides ($y = 0^+$). The total current on the strip is equal to

$$j_z = 2j_z\left(y = 0^+\right). \tag{6.6.1}$$

Utilizing this equation and introducing the notation

$$\lim_{a\to\infty}\psi(z)=\chi(z)=\frac{2}{\sqrt{\pi}}e^{-i\frac{\pi}{4}}e^{-2ikz}\int_{\sqrt{2kz}}^{\infty}e^{it^2}\,dt\,, \tag{6.6.2}$$

we can use Eqn. (6.2.13) to determine the current on the strip

$$j_z=-\frac{ick\mathcal{E}}{2\pi}\left[\chi(|z|)e^{ik|z|}+\tilde{A}_1\chi(z-z_1)e^{ik(z-z_1)}+\tilde{A}_2\chi(z_2-z)e^{ik(z_2-z)}\right] \tag{6.6.3}$$

being excited by the external field given by Eqn. (6.1.1) at $y=0^+$. Here,

$$\left.\begin{aligned}\tilde{A}_1&=-\frac{1}{\tilde{\mathcal{D}}}\left[\chi(-z_1)-\chi(z_2)\chi(z_2-z_1)e^{2ikz_2}\right]e^{-ikz_1}\,,\\ \tilde{A}_2&=-\frac{1}{\tilde{\mathcal{D}}}\left[\chi(z_2)-\chi(-z_1)\chi(z_2-z_1)e^{-2ikz_1}\right]e^{ikz_2}\,,\end{aligned}\right\} \tag{6.6.4}$$

$$\tilde{\mathcal{D}}=1-\chi^2(z_2-z_1)e^{2ik(z_2-z_1)}\,. \tag{6.6.5}$$

Analogously, from Eqn. (6.4.8) we can arrive at a solution for the current

$$\begin{aligned}j_z=\frac{c}{2\pi}\Big[&e^{iwz}-\chi_-(z-z_1)e^{ik(z-z_1)+iwz_1}-\chi_+(z_2-z)e^{ik(z_2-z)+iwz_2}+\\ &+\tilde{B}_1\chi(z-z_1)e^{ik(z-z_1)}+\tilde{B}_2\chi(z_2-z)e^{ik(z_2-z)}\Big],\end{aligned} \tag{6.6.6}$$

excited by the external field

$$E_z^e=E_{0z}e^{iwz}=\sin\vartheta e^{iwz}\,,\quad w=-k\cos\vartheta\,, \tag{6.6.7}$$

located at $y=0^{+}$, where

$$\lim_{a\to\infty}\psi_{\pm}(z)=\chi_{\pm}(z)=\frac{2}{\sqrt{\pi}}e^{-i\frac{\pi}{4}}e^{-i(k\pm w)z}\int_{\sqrt{(k\pm w)z}}^{\infty}e^{it^{2}}dt \qquad (6.6.8)$$

$$\left.\begin{aligned}\tilde{B}_1&=\frac{e^{i(k+w)z_2}}{\tilde{\mathcal{D}}}\left[\chi_{+}(L)-\chi_{-}(L)\chi(L)e^{i(k-w)L}\right]e^{-ikz_1},\\ \tilde{B}_2&=\frac{e^{-i(k-w)z_1}}{\tilde{\mathcal{D}}}\left[\chi_{-}(L)-\chi_{+}(L)\chi(L)e^{i(k+w)L}\right]e^{ikz_2},\end{aligned}\right\} \qquad (6.6.9)$$

with $L=z_2-z_1$. The approximations given by Eqns. (6.6.3) and (6.6.6) can also be obtained directly using the parabolic equation, expressed in Cartesian coordinates y and z.

It should be noted that the external field given by Eqn. (6.1.1) illuminating an infinite perfectly conducting plane produces a current at $y=0^{+}$ equal to

$$j_z=\frac{ck\mathcal{E}}{4\pi}H_0^{(1)}(k|z|). \qquad (6.6.10)$$

Comparing this expression with the first term of Eqn. (6.6.3), we see that for $k|z|\gg 1$ its error is on the order of $(k|z|)^{-3/2}$. Below (see, § 9.5), it is also shown that when $(k\pm w)L\gg 1$, the absolute error of Eqn. (6.6.6) does not exceed a value on the order of $(kL)^{-3/2}$.

It is interesting that the parabolic equation method yields the exact expressions given by Sommerfeld for the first-order edge wave of the current excited on a strip by a plane wave. Let us recall that they were obtained under the condition that $j_z=0$ past the edge of the half-plane (see, § 6.3). Moreover, it is possible to show that if the diffraction of an edge wave from a strip is solved by the parabolic equation method, rigorously satisfying the condition that $j_z=0$ outside of the strip, an exact expression for all the waves of current is obtained. In order to find the current on the strip $(-l\leqslant z\leqslant l, y=0)$, let us represent it in the form

$$\frac{2\pi}{c} j_z = j_0(z) + \sum_{n=1}^{\infty} \left[j_n^{(-)}(z+l) e^{ik(z+l)} + j_n^{(+)}(l-z) e^{ik(l-z)} \right], \tag{6.6.11}$$

where $j_0(z)$ is the current on the infinite plane and $j_n^{(\pm)}$ is the edge wave. Let us use the expression

$$A_{nz}^{(\mp)} = W_n^{(\mp)}(y, l \pm z) e^{ik(l \pm z)} \tag{6.6.12}$$

to describe the nth-order edge wave. The function $W_n^{(\mp)}$ must satisfy

$$\left(\frac{\partial^2}{\partial y^2} \pm 2ik \frac{\partial}{\partial z} + \frac{\partial^2}{\partial z^2} \right) W_n^{(\mp)} = 0, \tag{6.6.13}$$

and in the transverse diffusion approximation, it satisfies the equation

$$\left(\frac{\partial^2}{\partial y^2} \pm 2ik \frac{\partial}{\partial z} \right) W_n^{(\mp)} = 0. \tag{6.6.14}$$

The solution to Eqn. (6.6.14) is given by

$$W_n^{(\mp)}(y, l \pm z) = \frac{i}{\sqrt{2k}} \int_{-\infty}^{\infty} \frac{F_n^{(\mp)}(t)}{\sqrt{k - t}} e^{i\sqrt{2k(k-t)}|y| - i(k-t)(l \pm z)} dt, \tag{6.6.15}$$

where the contour of integration skirts beneath the branch point at $t = k$.

Noting that

$$j_n^{(\mp)}(l \pm z) = -\left. \frac{\partial A_{nz}^{(\mp)}}{\partial y} \right|_{y=0^+}, \tag{6.6.16}$$

and utilizing Eqns. (6.6.12) and (6.6.15), we find that

$$j_n^{(\mp)}(\zeta) = \int_{-\infty}^{\infty} F_n^{(\mp)}(t) e^{it\zeta} dt. \tag{6.6.17}$$

From the boundary condition $E_{nz}^{(\mp)} = 0$ on the conductor surface ($y = 0^+$), it follows that

$$W_n^{(\mp)}(0^+, \zeta) = \text{const}, \quad \frac{\partial}{\partial z} W(0^+, \zeta) = 0 \quad \text{for} \quad \zeta > 0. \tag{6.6.18}$$

This requirement and the condition that

$$j(z) = 0 \quad \text{for} \quad |z| > l \tag{6.6.19}$$

lead to a system of equations for $F_n^{(\mp)}(t)$:

$$\left.\begin{aligned}
&\int_{-\infty}^{\infty} \sqrt{k-t}\, F_1^{(\pm)}(t) e^{it\zeta} dt = 0 \quad \text{for} \quad \zeta > 0, \\
&\int_{-\infty}^{\infty} F_1^{(\pm)}(t) e^{it\zeta} dt = -j_0[\pm(l-\zeta)] \quad \text{for} \quad \zeta < 0, \\
&\int_{-\infty}^{\infty} \sqrt{k-t}\, F_n^{(\pm)}(t) e^{it\zeta} dt = 0 \quad \text{for} \quad \zeta > 0, \quad n = 2,3\ldots \\
&\int_{-\infty}^{\infty} \left[F_n^{(\pm)}(t) + F_{n-1}^{(\mp)}(-t) e^{-2ilt}\right] e^{it\zeta} dt = 0 \quad \text{for} \quad \zeta < 0, \quad n = 2,3\ldots
\end{aligned}\right\}. \tag{6.6.20}$$

It is interesting that the exact solution of the strip diffraction problem based on the elliptic form of Eqn. (6.6.13) leads to the system presented later by

Eqn. (8.3.10), which is equivalent to Eqn. (6.6.20). One should also note that the solutions of the elliptic and parabolic equations are identical only on the strip surface. When the observation point moves away from the strip, these two solutions differ. The results obtained can be explained in this way. First, it follows from the boundary conditions given by Eqn. (6.6.18) that Eqns. (6.6.13) and (6.6.14) agree on the surface of the strip. Second, both the exact solution and the approximate solution (based on the parabolic equation) essentially rely on the same condition given by Eqn. (6.6.19), which does not depend on the nature of the differential equations.

This brings up one more interesting point. The formulation of the problem based on Eqn. (6.6.20) differs from that given earlier in Chapter 5. The first approach (utilizing Eqn. (6.6.20)) yields an exact expression for the field only on the surface of the body, while the second method (Chapter 5) determines the first term of the asymptotic solution of the exact problem, describing the scattered field in the far zone of all space.

§6.7 Fundamental Results

- Chapters 5 and 6 researched edge waves using the parabolic equation method. Chapter 5 provided the asymptotic solution to the problem of wedge diffraction. It was shown that the parabolic equation described in ray (polar) coordinates allows us to obtain the first term of the asymptotic solution to the exact problem when $kr \gg 1$. Proceeding from the solution of an absolutely black wedge, § 5.5 used the method of reflections to describe the diffraction from a perfectly conducting wedge. Here, a new asymptotic representation is given by Eqn. (5.5.7) for the diffracted field which, in comparison with the well known formula of Pauli, is a more physically meaningful solution.
- Chapter 6 developed the parabolic equation method, which utilizes approximate ray coordinates for edge waves running over thin conductors and strips. This method is based on functional equations and permits a rather simple and elegant method of investigating the multiple diffraction of edge waves on dipoles — thin cylindrical conductors. The strong resonant properties of these bodies requires the study of higher-order edge waves. When $ka \ll 1$ the expressions for the current on dipoles transform to the well known Vainshtein functions [105, 106] and in the limit as $a \to \infty$, describe the current on a strip. Using the parabolic

equation method we also calculated the near field of a dipole and studied its physical structure (§ 6.5).

- The results obtained at the end of § 6.6 should also be mentioned. Here, it was shown that the solutions to the boundary-value problem for the elliptic Helmholtz equation and its parabolic form are identical on the surface of the strip. Thus, we have seen that the parabolic equation method can serve as an effective means for the asymptotic study of edge waves.
- Finally, we would like to emphasize a close connection between this chapter and the PTD current concept presented in references [3, 13, 64, 75] and used, here, in Chapters 2–4. The first terms in Eqns. (6.2.13) and (6.4.8) determine the current excited on an infinite wire and can be interpreted as modified *uniform* components of the current. All other terms relate to the edge waves and represent the *nonuniform* part of the current. This interpretation completely agrees with the concept of references [3, 13, 64, 75], especially in the case when a dipole is transformed into a strip, as was shown in § 6.6. The next chapter will study the radiation generated by the currents on a dipole and a strip.

§6.8 Additional References

- Some results of this chapter were published in [19].
- A new version of the parabolic equation is proposed in [103]. It is called the method of reference modes, and it is recommended for the investigation of wave propagation in irregular waveguides.
- The parabolic equation method was applied in [109] to study currents on thin wire dipoles with finite conductivity. In [110] these currents were used to calculate the far field scattered from such wires.
- The problem of a wire dipole with finite conductivity was investigated in [111] using a special modification of the Vainshtein function.
- A thin wire dipole is an example of an open resonator. Interesting applications of the parabolic equation to other types of resonators are presented in [112] and [113].

7

Radiation of Edge Waves: Theory Based on the Reciprocity Theorem

§7.1 Calculation of the Far Field

Earlier we examined edge waves of current on dipoles and strips. This chapter will now determine the field radiated by these currents in the far zone. This problem, in principle, can be solved with the use of retarded vector potentials by integrating the current on the surface of the body. However, it turns out that direct integration of the approximate current results in a scattered field that possesses a significant flaw — it fails to obey the reciprocity theorem.[1] Therefore, we need a special method that uses an approximation for the current to determine the radiated field without this flaw.

To accomplish this, a theory based on the reciprocity theorem will be developed. This theory will determine explicit expressions for the scattered field suitable for all directions of illumination and observation and will include an approximation for multiple diffraction.

Here, one should note that research on dipole scattering is commonly encountered in the literature. As a rule, the traditional approach initially determines the induced current on the conductor, and upon integration, the scattered field in the far zone. However, given the complexity of the problem, comparatively simple formulas are obtained that only describe the monostatic case perpendicular to the axis of the conductor. In the more gen-

[1]This topic will be considered in more detail in §10.5.

eral bistatic case, when the directions of incidence and observation are arbitrary, the expressions for the scattered field become complicated and are difficult to analyze. In addition, they do not satisfy the reciprocity theorem.

The fundamental results of the research presented below were published in references [13] and [14].

§7.2 Radiation from a Transmitting Dipole

The radiated field of a transmitting dipole can be calculated by integrating its current. However, this approach is pointless, because it was shown in the previous chapter that the accuracy of Eqn. (6.2.13) varies across the conductor and is poor both near its ends ($z = z_1$ and $z = z_2$) and also at the external source ($z = 0$). More exact results can be obtained using the reciprocity theorem [114]. The assumption will be made that the transmitting dipole ($r = a$, $-l \le z \le l$) is excited by the point source given by Eqn. (6.1.1). Equation (6.4.8) will be used for the current on the dipole excited by the auxiliary incident wave of Eqn (6.3.1). Then the reciprocity theorem can be applied to find that the far field radiated by a dipole is equal to

$$\left.\begin{aligned} E_\vartheta &= H_\varphi = \frac{\mathcal{E}}{2\sin\vartheta\Lambda[ka\sin\vartheta]}\frac{e^{ikR}}{R}f(\vartheta), \\ E_\varphi &= H_\vartheta = 0. \end{aligned}\right\} \tag{7.2.1}$$

where

$$\begin{aligned} f(\vartheta) = 1 - \psi_+(z_2)e^{ikz_2(1-\cos\vartheta)} - \psi_-(-z_1)e^{-ikz_1(1+\cos\vartheta)} + \\ + B_1\psi(-z_1)e^{-ikz_1} + B_2\psi(z_2)e^{ikz_2} \end{aligned} \tag{7.2.2}$$

and coefficients B_1 and B_2 are determined by Eqn. (6.4.10). The function $f(\vartheta)$ and the current given by Eqn. (6.4.8), which is induced by the auxiliary plane wave given by Eqn. (6.3.1), are related by

$$J(0) = Sf(\vartheta). \tag{7.2.3}$$

Equation (7.2.2) can be rearranged into a form more convenient for physical analysis,

$$f(\vartheta)=1-\psi_{+}(z_2)e^{ikz_2(1-\cos\vartheta)}-\psi_{-}(-z_1)e^{-ikz_1(1+\cos\vartheta)}+ \\ +C_1\psi_{+}(z_2-z_1)e^{ikz_2(1-\cos\vartheta)}+C_2\psi_{-}(z_2-z_1)e^{-ikz_1(1+\cos\vartheta)}, \tag{7.2.4}$$

with coefficients C_1 and C_2 given by

$$\left.\begin{aligned} C_1&=\frac{1}{\mathcal{D}}\left[\psi(-z_1)-\psi(z_2)\psi(z_2-z_1)e^{2ikz_2}\right]e^{-2ikz_1}, \\ C_2&=\frac{1}{\mathcal{D}}\left[\psi(z_2)-\psi(-z_1)\psi(z_2-z_1)e^{-2ikz_1}\right]e^{2ikz_2}. \end{aligned}\right\} \tag{7.2.5}$$

The function $f(\vartheta)$ satisfies the condition

$$f(0)=f(\pi)=0, \tag{7.2.6}$$

which indicates that the field radiated by a finite conductor in its longitudinal direction must be equal to zero. The resonant denominator $\mathcal{D}$ of Eqn. (7.2.5) is given by Eqn. (6.2.16).

Equation (7.2.4) has clear physical meaning. The first term (unity) describes the field of an infinitely long conductor excited by a point source. Propagating in the $\vartheta = 0$ direction, this field arrives at the end of the conductor at $z = z_2$ and is diffracted, creating the first-order edge wave (second term). In an analogous manner, another first-order edge wave (third term) originates at $z = z_1$. The last two terms in Eqn. (7.2.2) represent waves, arising as a result of subsequent diffraction (second-order, third-order, etc.).

Equation (7.2.4), which describes the radiation pattern, was also presented in [106]. The only difference is in the definition of ψ and $\psi_{\pm}$. In [106] they were found by the method of slowly varying wave amplitudes, while we determined them using the parabolic equation method.

§7.3 First and Second-Order Diffraction at a Passive Dipole

Let an electromagnetic wave traveling at an angle, ϑ_0, be incident upon a thin cylindrical conductor of length $L = z_2 - z_1$ and radius a (Fig. 7.3.1). For generality, assume that the electric field, $\mathbf{E}_0$, is oriented at an angle, α, relative to the plane of the illustration. Then its tangential component on the surface of the conductor is equal to

$$E_z^e = E_{0z} e^{iw_0 z}, \tag{7.3.1}$$

where

$$E_{0z} = E \sin\vartheta_0, \quad E = E_o \cos\alpha, \quad w_0 = -k\cos\vartheta_0. \tag{7.3.2}$$

The current, induced on the dipole by this field, was examined in § 6.4. The resulting expression given by Eqn. (6.4.8) displays relatively low accuracy near the ends of the conductor. Therefore, calculation of the scattered field by integrating the current may not provide satisfactory results. We will seek the scattering characteristics of a passive dipole based on the scattering pattern, which naturally stems from the preceding results. The incident plane wave diffracted from the ends of the dipole creates the first-order edge waves radiated into the surrounding space. Propagating along the conductor, each of these waves diffracts at the opposite end of the conductor and excites second-order edge waves. Subsequently, in turn, third-order edge waves are created, *etc.* In so doing, the diffraction of waves at the ends of the conductor occurs in the same way as at the end of a semi-infinite conductor. Thus, the first-order waves can be determined from the diffraction of a plane wave from a semi-infinite wire (z_1, ∞) and $(-\infty, z_2)$. The sum of these waves produces the first-order field

$$E_\vartheta^{(1)} = H_\varphi^{(1)} = -E\frac{e^{ikR}}{kR} F^{(1)}(\vartheta, \vartheta_0), \tag{7.3.3}$$

$$F^{(1)}(\vartheta,\vartheta_0) = \frac{i}{2}\frac{\cot(\vartheta/2)\cot(\vartheta_0/2)e^{-ikz_1(\cos\vartheta+\cos\vartheta_0)}}{(\cos\vartheta+\cos\vartheta_0)\Phi(-k\cos\vartheta, -k\cos\vartheta_0)} - \tag{7.3.4}$$

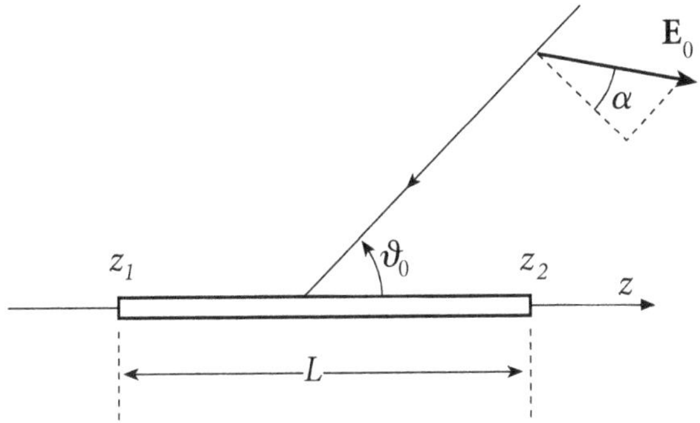

Figure 7.3.1: Incidence of a plane wave on a thin cylindrical conductor.

$$-\frac{i}{2}\frac{\tan(\vartheta/2)\tan(\vartheta_0/2)e^{-ikz_2(\cos\vartheta+\cos\vartheta_0)}}{(\cos\vartheta+\cos\vartheta_0)\Phi(k\cos\vartheta,k\cos\vartheta_0)}. \qquad (7.3.4, \text{cont.})$$

The function $\Phi(w, w_0)$ can be calculated using a rigorous solution to the semi-infinite wire conductor (see, [106] § 3 and [107]) and in this case satisfies the relationship

$$\Phi(w,w_0)\Phi(-w,-w_0)=\ln(2i/\gamma va)\ln(2i/\gamma v_0 a),$$
$$v=\sqrt{k^2-w^2}, \quad v_0=\sqrt{k^2-w_0^2}. \qquad (7.3.5)$$

However, a exact expression for Φ will not be necessary below.

The second-order edge wave propagating from $z=z_2$ is excited due to diffraction of the first-order current

$$S\psi_{-}^{0}(z-z_1)e^{iw_0z_1+ik(z-z_1)} \qquad (7.3.6)$$

where $\psi_{\mp}^{0}$ is obtained from $\psi_{\mp}(z)$ by substituting ϑ with ϑ_0. To find the second-order wave we must first determine the external field located at the infinite conductor ($-\infty\leqslant z\leqslant\infty$) that excites the current between $z_1\leqslant z\leqslant\infty$ given by Eqn. (7.3.6).

To do this, let us study the current induced on an infinite conductor by the external field

$$E_z^e = \hat{E}_{0z} e^{iw_0 z} \epsilon(z - z_1), \quad \epsilon(z) = \begin{cases} 1 & \text{for} \quad z < 0 \\ 0 & \text{for} \quad z > 0 \end{cases}. \tag{7.3.7}$$

It will be assumed that w_0 possesses a small negative imaginary part ($\Im\mathfrak{m}\{w_0\} \leqslant 0$). The value $\hat{E}_{0z} e^{iw_0\zeta} d\zeta$ represents the point source, which according to Eqn. (6.1.20) creates a current equal to

$$\frac{c\hat{E}_{0z}}{4\ln(i/\gamma ka)} \psi(|z-\zeta|) e^{iw_0\zeta + ik|z-\zeta|} d\zeta \tag{7.3.8}$$

on an infinite conductor ($ka \ll 1$). Therefore, according to the superposition principle, the total current in the region $z_1 \leqslant z \leqslant \infty$ induced by the external field is

$$J(z) = \frac{c\hat{E}_{0z}}{4\ln(i/\gamma ka)} \int_{-\infty}^{z_1} \psi(z-\zeta) e^{iw_0\zeta + ik(z-\zeta)} d\zeta = \\ = \frac{c\hat{E}_{0z}}{4\ln(i/\gamma ka)} e^{iw_0 z} \int_{z-z_1}^{\infty} \psi(\zeta) e^{i(k-w_0)\zeta} d\zeta \,. \tag{7.3.9}$$

Making use of Eqn. (6.3.23), we find that

$$J(z) = \frac{c\hat{E}_{0z} e^{iw_0 z_1}}{8ik\sin^2(\vartheta_0/2)\ln[(i/\gamma ka)]} \psi(z-z_1) e^{ik(z-z_1)} - \\ - \frac{c\hat{E}_{0z} e^{iw_0 z_1}}{2ik\sin^2\vartheta_0 \ln[i/\gamma ka\cos(\vartheta_0/2)]} \psi_-^0(z-z_1) e^{ik(z-z_1)} \,. \tag{7.3.10}$$

Thus, it has been shown that the external field given by Eqn. (7.3.7) excites both ψ_{-}^{0} and ψ waves. The excitation of the "pure" wave, ψ_{-}^{0}, evidently requires a complementary external field

$$E_z^e = \mathcal{E}_1 \delta(z - z_1), \tag{7.3.11}$$

such that

$$\begin{aligned} &\frac{c\hat{E}_{0z} e^{iw_0 z_1}}{8ik \sin^2(\vartheta_0/2) \ln(i/\gamma ka)} \psi(z - z_1) e^{ik(z-z_1)} + \\ &\qquad + \frac{c\mathcal{E}_1}{4\ln(i/\gamma ka)} \psi(z - z_1) e^{ik(z-z_1)} = 0 . \end{aligned} \tag{7.3.12}$$

Thus,

$$\mathcal{E}_1 = -\frac{\hat{E}_{0z} e^{iw_0 z_1}}{2k \sin^2(\vartheta_0/2)} . \tag{7.3.13}$$

In order for the sum of the external fields given by Eqns. (7.3.7) and (7.3.11) to create the current of Eqn. (7.3.6), the following equality must be satisfied

$$\begin{aligned} &\frac{c\hat{E}_{0z}}{2ik \sin^2 \vartheta_0 \ln[i/\gamma ka \cos(\vartheta_0/2)]} = S = \\ &\qquad = \frac{i\omega E_{0z}}{2k^2 \sin^2 \vartheta_0 \ln[2i/\gamma ka \sin \vartheta_0]} , \end{aligned} \tag{7.3.14}$$

and therefore,

$$\hat{E}_{0z} = -E_{0z} \frac{\ln[i/\gamma ka \cos(\vartheta_0/2)]}{\ln[2i/\gamma ka \sin \vartheta_0]} . \tag{7.3.15}$$

Thus, in order for an infinite conductor to support the excitation of a wave of current given by Eqn. (7.3.6) for $z > z_1$, it is necessary to apply an external field equal to

$$E_z^e = E_{0z}\frac{\ln\left[i/\gamma ka\cos(\vartheta_0/2)\right]}{\ln\left[2i/\gamma ka\sin\vartheta_0\right]}\left[\frac{e^{iw_0z_1}\delta(z-z_1)}{2ik\sin^2(\vartheta_0/2)}-\epsilon(z-z_1)e^{iw_0z}\right], \quad (7.3.16)$$

where

$$\epsilon(z)=\begin{cases}1 & \text{for} \quad z<0,\\ 0 & \text{for} \quad z>0.\end{cases}$$

Using an identical approach, it is possible to show that the external field,

$$E_z^e = E_{0z}\frac{\ln\left[i/\gamma ka\sin(\vartheta_0/2)\right]}{\ln\left[2i/\gamma ka\sin\vartheta_0\right]}\times$$
$$\times\left[\frac{e^{iw_0z_2}}{2ik\cos^2(\vartheta_0/2)}\delta(z_2-z)-\epsilon(z_2-z)e^{iw_0z}\right], \quad (7.3.17)$$

will excite a current equal to

$$-S\psi_+^0(z_2-z)e^{iw_0z_2+ik(z_2-z)} \quad (7.3.18)$$

on an infinite conductor (for $z < z_2$).

Let us now examine the diffraction of the wave given by Eqn. (7.3.6) from a semi-infinite conductor $(-\infty, z_2)$ using the reciprocity theorem [114]

$$\iiint\left(\mathbf{j}_1^e\cdot\mathbf{E}_2+\mathbf{j}_2^m\cdot\mathbf{H}_1\right)dV=0. \quad (7.3.19)$$

Here, $\mathbf{j}_1^e = -i\omega\mathbf{p}_1\delta(\mathbf{R}-\mathbf{R}')$ is the current of an infinitesimal dipole with moment $\mathbf{p}_1$ located at Point 1 with coordinates (R, ϑ), $\mathbf{H}_1$ is its magnetic field

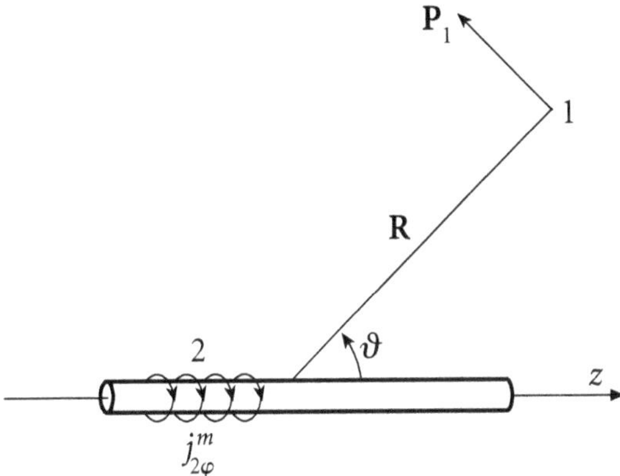

Figure 7.3.2: Excitation of a semi-infinite conductor $(-\infty, z_2)$ by the current, $\mathbf{j}^m_{2\varphi}$.

on the surface of the conductor at the location of the external magnetic current $\mathbf{j}^{\mathrm{m}}_2$, and $\mathbf{E}_2$ is the field created by the magnetic current at Point 1 (Fig. 7.3.2).

The external current, $\mathbf{j}^m_2$, is determined by the well known expression

$$\mathbf{j}^m_2 = -\frac{c}{4\pi}\mathbf{n}\times\mathbf{E} \tag{7.3.20}$$

as a function of $\mathbf{E}$ on the conductor surface. Enforcing the boundary condition

$$E_z + E^e_z = 0 \tag{7.3.21}$$

we have

$$j^m_{2\varphi} = -\frac{c}{4\pi}E^e_z\,. \tag{7.3.22}$$

Determining the dipole moment $\mathbf{p}_1$ as a function of its field in free space (at point $x = y = z = 0$),

$$E'_{0z} = -k^2 p_1 \frac{e^{ikR}}{R}\sin\vartheta\,, \tag{7.3.23}$$

and using the magnetic field $H_{1\varphi}$ to determine the total current,

$$J = \frac{ca}{2} H_{1\varphi}, \tag{7.3.24}$$

induced by a dipole on the conductor, we can use the reciprocity theorem to determine that

$$E_{2\vartheta} = H_{2\varphi} = \frac{k^2 \sin\vartheta}{i\omega E'_{0z}} \frac{e^{ikR}}{R} \int_{-\infty}^{z_2} E_z^e J(z) dz. \tag{7.3.25}$$

If $\mathbf{p}_1$ is located at a distance $R \gg z_2 - z_0$ $(z_2 - z_0 > z_2 - z_1)$, then the field it radiates along the length $z_2 - z_0$ of an infinite conductor $(-\infty, z_2)$ is approximately planar. Then, the current density induced on this part of the conductor can be determined by

$$J(z) = S'\left[e^{iwz} - e^{iwz_2}\psi_+(z_2 - z)e^{ik(z_2 - z)}\right], \tag{7.3.26}$$

where

$$S' = \frac{i\omega E'_{0z}}{2k^2 \sin^2\vartheta \ln(2i/\gamma ka \sin\vartheta)}, \quad w = -k\cos\vartheta. \tag{7.3.27}$$

The value of z_0 is chosen such that at a distance $z_2 - z_0$ from the end of the conductor the reflected wave of current is essentially zero $(\psi_+(z_2 - z_0) \approx 0)$. Substituting Eqn. (7.3.26) into the right half of Eqn. (7.3.25) and using Eqn. (7.3.16) for E_z^e, we obtain

$$E_{2\vartheta} = H_{2\varphi} = \frac{1}{2\sin\vartheta \ln(2i/\gamma ka \sin\vartheta)} \frac{e^{ikR}}{R} \times$$
$$\times \int_{-\infty}^{z_2} E_z^e\left[e^{iwz} - \psi_+(z_2 - z)e^{iwz_2}e^{ik(z_2 - z)}\right]dz = \tag{7.3.28}$$

$$= \frac{1}{2\sin\vartheta\ln(2i/\gamma ka\sin\vartheta)}\times$$

$$\times\left\{\mathcal{E}_1\left[e^{iwz_1} - \psi_+(z_2 - z_1)e^{iwz_2}e^{ik(z_2-z_1)}\right]+\right. \qquad (7.3.28, \text{cont.})$$

$$\left. +\hat{E}_{oz}\frac{e^{i(w+w_0)z_1}}{i(w+w_0)} - \hat{E}_{0z}e^{i(k+w)z_2}\int\limits_{-\infty}^{z_1}\psi_+(z_2 - z)e^{ik(k-w_0)z}dz\right\}\frac{e^{ikR}}{R}.$$

An important property of this relationship is that the integration is carried out not over the entire conductor $(-\infty, z_2)$, but only over $(-\infty, z_1)$ where $\psi_+(z_2-z)$ describes the current with sufficient accuracy. This integral is evaluated using Eqn. (6.3.24). The resultant field, radiated by the semi-infinite conductor $(-\infty, z_2)$, is equal to

$$E_{2\vartheta} = H_{2\varphi} = \frac{1}{2\sin\vartheta\ln(2i/\gamma ka\sin\vartheta)}\frac{e^{ikR}}{R}\times$$

$$\times\left\{\mathcal{E}_1\left[e^{iwz_1} - \psi_+(L)e^{iwz_2+ikL}\right] - \frac{\hat{E}_{oz}e^{i(w+w_0)z_1}}{ik(\cos\vartheta+\cos\vartheta_0)}+\right.$$

$$+\frac{\hat{E}_{0z}e^{ikL}e^{i(w_0z_1+wz_2)}}{ik(\cos\vartheta+\cos\vartheta_0)}\times \qquad (7.3.29)$$

$$\left.\times\left[\psi_+(L) - \frac{\sin^2(\vartheta/2)\ln[i/\gamma ka\sin(\vartheta/2)]}{\cos^2(\vartheta_0/2)\ln[i/\gamma ka\cos(\vartheta_0/2)]}\psi_-^0(L)\right]\right\}.$$

The terms in the braces possessing the $\exp[iwz_2]$ phase factor correspond to the second-order wave diffracted from the end of the conductor at $z = z_2$. Using Eqns. (7.3.13) and (7.3.15), this wave can be rewritten as

$$E_{\vartheta}^{(2)}(z_2) = H_{\varphi}^{(2)}(z_2) = \frac{\mathcal{E}^{(2)}(z_2)}{2\sin\vartheta \ln(2i/\gamma ka\sin\vartheta)} \frac{e^{ikR}}{R} e^{-ikz_2\cos\vartheta}, \tag{7.3.30}$$

where

$$\mathcal{E}^{(2)}(z_2) = \frac{4iEe^{iw_0 z_1 + ikL}}{k\sin\vartheta_0(\cos\vartheta + \cos\vartheta_0)\ln(2i/\gamma ka\sin\vartheta_0)} \times$$
$$\times\left\{\cos^2\frac{\vartheta_0}{2}\cos^2\frac{\vartheta}{2}\ln[i/\gamma ka\cos(\vartheta_0/2)]\psi_+(L) - \right. \tag{7.3.31}$$
$$\left. -\sin^2\frac{\vartheta_0}{2}\sin^2\frac{\vartheta}{2}\ln[i/\gamma ka\sin(\vartheta/2)]\psi_-^0(L)\right\}.$$

Here $L = z_2 - z_1$.

Similarly, we can find the second-order edge wave radiating from the end at $z = z_1$. This requires the examination of the diffraction of the first-order wave given by Eqn. (7.3.18) diffracted at $z = z_1$ from a semi-infinite conductor ($z_1 \leqslant z \leqslant \infty$). In this case, the reciprocity theorem yields

$$E_{2\vartheta} = H_{2\varphi} = \frac{k^2\sin\vartheta}{i\omega E'_{0z}} \frac{e^{ikR}}{R} \int_{z_1}^{\infty} E_z^e J(z)dz, \tag{7.3.32}$$

whereupon substituting Eqn. (7.3.17) and the current

$$J(z) = \frac{i\omega E'_{0z}}{2k^2\sin^2\vartheta\ln(2i/\gamma ka\sin\vartheta)}\left[e^{iwz} - \psi_-(z - z_1)e^{iwz_1 + ik(z - z_1)}\right], \tag{7.3.33}$$

gives the field radiated by a semi-infinite conductor (z_1, ∞). The wave radiated at the end of the conductor is a second-order wave equal to

$$E_{\vartheta}^{(2)}(z_1)=H_{\varphi}^{(2)}(z_1)=\frac{\mathcal{E}^{(2)}(z_1)}{2\sin\vartheta\ln(2i/\gamma ka\sin\vartheta)}\frac{e^{ikR}}{R}e^{-ikz_1\cos\vartheta}, \tag{7.3.34}$$

where

$$\mathcal{E}^{(2)}(z_1)=\frac{4iEe^{iw_0z_2+ikL}}{k\sin\vartheta_0(\cos\vartheta+\cos\vartheta_0)\ln(2i/\gamma ka\sin\vartheta_0)}\times$$

$$\times\left\{\sin^2\frac{\vartheta_0}{2}\sin^2\frac{\vartheta}{2}\ln[i/\gamma ka\sin(\vartheta_0/2)]\psi_-(L)-\right. \tag{7.3.35}$$

$$\left.-\cos^2\frac{\vartheta_0}{2}\cos^2\frac{\vartheta}{2}\ln[i/\gamma ka\cos(\vartheta/2)]\psi_+^0(L)\right\}.$$

However, this expression can be written using symmetry relationships, replacing z_2 with z_1, ϑ with $\pi-\vartheta$, and ϑ_0 with $\pi-\vartheta_0$ in Eqns. (7.3.30) and (7.3.31).

§7.4 Multiple Diffraction of Edge Waves

The second-order edge waves given by Eqns. (7.3.30) and (7.3.34) are waves radiating from the ends of the semi-infinite conductors $(-\infty, z_2)$ and (z_1, ∞). If an infinite conducting line is excited by an external field,

$$E_z^e=\mathcal{E}_2^{(2)}\delta(z-z_2), \tag{7.4.1}$$

where

$$\mathcal{E}_2^{(2)}=\mathcal{E}^{(2)}(z_2)\Big|_{\vartheta=\pi}=E\frac{2i\ln(i/\gamma ka)\psi_-^0(L)}{k\sin\vartheta_0\ln(2i/\gamma ka\sin\vartheta_0)}e^{ikL-ikz_1\cos\vartheta_0}, \tag{7.4.2}$$

a spherical wave is created that coincides with Eqn. (7.3.30) for $\vartheta\approx\pi$. Due to the excitation of an infinite wire by a voltage source

$$E_z^e=\mathcal{E}_1^{(2)}\delta(z-z_1), \tag{7.4.3}$$

where

$$\mathcal{E}_1^{(2)} = \mathcal{E}^{(2)}(z_1)\Big|_{\vartheta=0} = E\frac{2i\ln(i/\gamma ka)\psi_+^0(L)}{k\sin\vartheta_0\ln(2i/\gamma ka\sin\vartheta_0)}e^{ikL-ikz_2\cos\vartheta_0}, \tag{7.4.4}$$

a wave rises which coincides with Eqn. (7.3.34) for $\vartheta \approx 0$.

It is not difficult to show that on an infinite wire the external fields given by Eqns. (7.4.1) and (7.4.3) really excite current waves equivalent to the second-order current waves described in Eqn. (6.4.8) by the first terms in the brackets of Eqn. (6.4.10) when we replace $1/\mathcal{D}$ by 1. This substitution means that these waves correspond to the first terms of the geometrical progression of $1/\mathcal{D}$. Therefore, it is possible to consider the third-order waves as edge waves radiated by semi-infinite conductors (z_1, ∞) and$(-\infty, z_2)$ when they are excited by external fields given by Eqns. (7.4.1) and (7.4.3), respectively. Equations (7.3.25) and (7.3.32) show that for the given excitation, the conductor (z_1, ∞) radiates a total field given by

$$E_\vartheta = H_\varphi = \frac{\mathcal{E}_2^{(2)}}{2\sin\vartheta\ln(2i/\gamma ka\sin\vartheta)}\frac{e^{ikR}}{R}\left[e^{iwz_2} - \psi_-(L)e^{iwz_1+ikL}\right], \tag{7.4.5}$$

and the total field radiated by conductor $(-\infty, z_2)$ is

$$E_\vartheta = H_\varphi = \frac{\mathcal{E}_1^{(2)}}{2\sin\vartheta\ln(2i/\gamma ka\sin\vartheta)}\frac{e^{ikR}}{R}\left[e^{iwz_1} - \psi_+(L)e^{iwz_2+ikL}\right]. \tag{7.4.6}$$

As a result, the third-order waves, radiating from z_1 and z_2 are

$$E_\vartheta^{(3)}(z_1) = H_\varphi^{(3)}(z_1) = \frac{\mathcal{E}^{(3)}(z_1)}{2\sin\vartheta\ln(2i/\gamma ka\sin\vartheta)}\frac{e^{ikR}}{R}e^{-ikz_1\cos\vartheta}, \tag{7.4.7}$$

$$E_\vartheta^{(3)}(z_2) = H_\varphi^{(3)}(z_2) = \frac{\mathcal{E}^{(3)}(z_2)}{2\sin\vartheta\ln(2i/\gamma ka\sin\vartheta)}\frac{e^{ikR}}{R}e^{-ikz_2\cos\vartheta}, \tag{7.4.8}$$

where

$$\left.\begin{aligned}\mathcal{E}^{(3)}(z_1)&=-\mathcal{E}_2^{(2)}\psi_-(L)e^{ikL}\,,\\ \mathcal{E}^{(3)}(z_2)&=-\mathcal{E}_1^{(2)}\psi_+(L)e^{ikL}\,.\end{aligned}\right\}\tag{7.4.9}$$

In the directions toward the opposite ends of the conductor, these waves are equivalent to the radiation emitted by an infinite wire excited by the external fields

$$E_z^e=\mathcal{E}_1^{(3)}\delta(z-z_1),\quad \mathcal{E}_1^{(3)}=\mathcal{E}^{(3)}(z_1)\Big|_{\vartheta=0}\,,\tag{7.4.10}$$

$$E_z^e=\mathcal{E}_2^{(3)}\delta(z_2-z),\quad \mathcal{E}_2^{(3)}=\mathcal{E}^{(3)}(z_2)\Big|_{\vartheta=\pi}\,.\tag{7.4.11}$$

It follows that the fourth-order waves can, in turn, be described as edge waves radiated by semi-infinite conductors $(-\infty, z_2)$ and (z_1, ∞) due to their excitation by the external fields given by Eqns. (7.4.10) and (7.4.11). Using the reciprocity theorem it is easy to show that

$$\left.\begin{aligned}E_\vartheta^{(4)}(z_1)=H_\varphi^{(4)}(z_1)&=\frac{\mathcal{E}^{(4)}(z_1)}{2\sin\vartheta\ln(2i/\gamma ka\sin\vartheta)}\frac{e^{ikR}}{R}e^{-ikz_1\cos\vartheta}\,,\\ E_\vartheta^{(4)}(z_2)=H_\varphi^{(4)}(z_2)&=\frac{\mathcal{E}^{(4)}(z_2)}{2\sin\vartheta\ln(2i/\gamma ka\sin\vartheta)}\frac{e^{ikR}}{R}e^{-ikz_2\cos\vartheta}\,,\end{aligned}\right\}\tag{7.4.12}$$

where

$$\left.\begin{aligned}\mathcal{E}^{(4)}(z_1)&=-\mathcal{E}_2^{(3)}\psi_-(L)e^{ikL}\,,\\ \mathcal{E}^{(4)}(z_2)&=-\mathcal{E}_1^{(3)}\psi_+(L)e^{ikL}\,.\end{aligned}\right\}\tag{7.4.13}$$

Using this same method, we can determine the nth-order edge waves are

$$E_\vartheta^{(n)}(z_1)=H_\varphi^{(n)}(z_1)=\frac{\mathcal{E}^{(n)}(z_1)}{2\sin\vartheta\ln(2i/\gamma ka\sin\vartheta)}\frac{e^{ikR}}{R}e^{-ikz_1\cos\vartheta}\,,\tag{7.4.14}$$

$$E_{\vartheta}^{(n)}(z_2)=H_{\varphi}^{(n)}(z_2)=\frac{\mathcal{E}^{(n)}(z_2)}{2\sin\vartheta\ln(2i/\gamma ka\sin\vartheta)}\frac{e^{ikR}}{R}e^{-ikz_2\cos\vartheta}, \tag{7.4.15}$$

where

$$\left.\begin{aligned}\mathcal{E}^{(n)}(z_1)&=-\mathcal{E}_2^{(n-1)}\psi_-(L)e^{ikL},\\ \mathcal{E}^{(n)}(z_2)&=-\mathcal{E}_1^{(n-1)}\psi_+(L)e^{ikL}.\end{aligned}\right\} \tag{7.4.16}$$

and

$$\mathcal{E}_1^{(n)}=\mathcal{E}^{(n)}(z_1)\Big|_{\vartheta=0},\quad \mathcal{E}_2^{(n)}=\mathcal{E}^{(n)}(z_2)\Big|_{\vartheta=\pi}. \tag{7.4.17}$$

Thus, the field due to multiple diffraction (beginning with the second-order) can be written as

$$\begin{aligned}\sum_{n=2}^{\infty}\left[E_{\vartheta}^{(n)}(z_1)+E_{\vartheta}^{(n)}(z_2)\right]&=\frac{1}{2\sin\vartheta\ln(2i/\gamma ka\sin\vartheta)}\frac{e^{ikR}}{R}\times\\ &\times\left[\sum_{n=2}^{\infty}\mathcal{E}^{(n)}(z_1)e^{-ikz_1\cos\vartheta}+\sum_{n=2}^{\infty}\mathcal{E}^{(n)}(z_2)e^{-ikz_2\cos\vartheta}\right],\end{aligned} \tag{7.4.18}$$

where

$$\sum_{n=2}^{\infty}\mathcal{E}^{(n)}(z_1)=\mathcal{E}^{(2)}(z_1)+\left[\mathcal{E}_1^{(2)}\psi(L)e^{ikL}-\mathcal{E}_2^{(2)}\right]\frac{\psi_-(L)}{\mathcal{D}}e^{ikL}, \tag{7.4.19}$$

$$\sum_{n=2}^{\infty}\mathcal{E}^{(n)}(z_2)=\mathcal{E}^{(2)}(z_2)+\left[\mathcal{E}_2^{(2)}\psi(L)e^{ikL}-\mathcal{E}_1^{(2)}\right]\frac{\psi_+(L)}{\mathcal{D}}e^{ikL}, \tag{7.4.20}$$

and

$$\mathcal{D}=1-\psi^2(L)e^{2ikL}. \tag{7.4.21}$$

The functions $\mathcal{E}^{(2)}(z_{1,2})$ and $\mathcal{E}_{1,2}^{(2)}$ are determined by Eqns. (7.3.31), (7.3.35), (7.4.2), and (7.4.4). We will not print the cumbersome final expression for this field and will proceed to the calculation of the total field radiated by a dipole.

§7.5 Total Scattered Field

Prior to developing an expression for the scattering field, the following observation should be made. Equation (7.3.4) introduced the function, Φ, that satisfied Eqn. (7.3.5). It can be found using the factorization method. However, our examination of the series of waves diffracted from the ends of the conductor was approximate. Therefore, there is no need to utilize the rigorous expression for the first-order field given by Eqn. (7.3.4). Let us utilize the approximate expressions

$$\left.\begin{aligned}\Phi(-k\cos\vartheta,-k\cos\vartheta_0)&=\ln\frac{i}{\gamma ka\sin(\vartheta/2)\sin(\vartheta_0/2)},\\ \Phi(k\cos\vartheta,k\cos\vartheta_0)&=\ln\frac{i}{\gamma ka\cos(\vartheta/2)\cos(\vartheta_0/2)}\end{aligned}\right\} \tag{7.5.1}$$

that were obtained using the variational method and exhibit an accuracy sufficient for our needs (see, [106]). Strictly speaking, one can make use of the approximations provided by Eqn. (7.5.1) in combination with the rigorously derived Eqn. (7.3.5) to conclude that

$$\left.\begin{aligned}\frac{1}{\Phi(-k\cos\vartheta,-k\cos\vartheta_0)}&=\frac{\ln\left[i/\gamma ka\cos(\vartheta/2)\cos(\vartheta_0/2)\right]}{\ln(2i/\gamma ka\sin\vartheta)\ln(2i/\gamma ka\sin\vartheta_0)},\\ \frac{1}{\Phi(k\cos\vartheta,k\cos\vartheta_0)}&=\frac{\ln\left[i/\gamma ka\sin(\vartheta/2)\sin(\vartheta_0/2)\right]}{\ln(2i/\gamma ka\sin\vartheta)\ln(2i/\gamma ka\sin\vartheta_0)}.\end{aligned}\right\} \tag{7.5.2}$$

Then, the first-order field is equal to

$$E_{\vartheta}^{(1)} = H_{\varphi}^{(1)} = -\frac{iE}{2(\cos\vartheta + \cos\vartheta_0)\ln(2i/\gamma ka\sin\vartheta)\ln(2i/\gamma ka\sin\vartheta_0)} \times$$
$$\times\left\{\cot\frac{\vartheta}{2}\cot\frac{\vartheta_0}{2}\ln\left[\frac{i}{\gamma ka\cos(\vartheta/2)\cos(\vartheta_0/2)}\right]e^{-ikz_1(\cos\vartheta+\cos\vartheta_0)} - \right. \tag{7.5.3}$$
$$\left. -\tan\frac{\vartheta}{2}\tan\frac{\vartheta_0}{2}\ln\left[\frac{i}{\gamma ka\sin(\vartheta/2)\sin(\vartheta_0/2)}\right]e^{-ikz_2(\cos\vartheta+\cos\vartheta_0)}\right\}\frac{e^{ikR}}{R}.$$

The field scattered by the dipole is the sum of Eqns. (7.4.18) and (7.5.3)

$$E_{\vartheta} = H_{\varphi} = -E\frac{e^{ikR}}{kR}\mathrm{F}(\vartheta, \vartheta_0), \tag{7.5.4}$$

where

$$F(\vartheta, \vartheta_0) = \mathcal{F}(\vartheta, \vartheta_0)e^{ikl\cos\vartheta} + \mathcal{F}(\pi - \vartheta, \pi - \vartheta_0)e^{-ikl\cos\vartheta}, \tag{7.5.5}$$

$$\mathcal{F}(\vartheta, \vartheta_0) = \frac{2i\csc\vartheta\csc\vartheta_0}{(\cos\vartheta + \cos\vartheta_0)\ln(2i/\gamma ka\sin\vartheta)\ln(2i/\gamma ka\sin\vartheta_0)} \times$$
$$\times\left\{\cos^2\frac{\vartheta}{2}\cos^2\frac{\vartheta_0}{2}\ln\left[\frac{i}{\gamma ka\cos(\vartheta/2)\cos(\vartheta_0/2)}\right]e^{ikl\cos\vartheta_0} + \right.$$
$$+\left[\sin^2\frac{\vartheta}{2}\sin^2\frac{\vartheta_0}{2}\ln\left[\frac{i}{\gamma ka\sin(\vartheta_0/2)}\right]\psi_- - \right. \tag{7.5.6}$$
$$\left. -\cos^2\frac{\vartheta}{2}\cos^2\frac{\vartheta_0}{2}\ln\frac{i}{\gamma ka\cos(\vartheta/2)}\psi_+^0\right]e^{ik(L-l\cos\vartheta_0)} -$$
$$\left. -\frac{(\cos\vartheta + \cos\vartheta_0)\ln(i/\gamma ka)}{2\mathcal{D}}\left[\psi_+^0\psi e^{ikL-ikl\cos\vartheta_0} - \psi_-^0 e^{ikl\cos\vartheta_0}\right]\psi_- e^{2ikL}\right\},$$

where $\psi_\pm$, $\psi_\pm^0$, and ψ all possess the argument L. In addition, to maintain common conditions, let $z_2 = -z_1 = l$, such that $L = 2l$. This expression possesses a distinct physical quality. Effectively, the first term of Eqn. (7.5.6) corresponds to the first-order wave, the second term (in brackets) corresponds to the second-order wave, and the last term containing the resonant denominator $\mathcal{D}$ describes the series of waves that occur as a result of multiple diffraction. Resonance occurs when $L \approx n\lambda/2$ ($n = 1, 2, 3\ldots$) and $\mathcal{D} \approx 0$.

The other important property of $F(\vartheta, \vartheta_0)$ is that it obeys the reciprocity theorem, i.e., its value does not change when interchanging ϑ and ϑ_0. It can also be shown that the dipole does not radiate along its axis and does not scatter at grazing angles of incidence, i.e.,

$$F(0,\vartheta_0) = F(\pi,\vartheta_0) = F(\vartheta,0) = F(\vartheta,\pi) = 0. \qquad (7.5.7)$$

Equation (7.5.5), which describes the scattering pattern, was first published by the author in 1962 [14]. Later on, it was obtained using a more rigorous method based on the integral equations in reference [115]. It was shown that the error of Eqn. (7.5.5) did not exceed $[2\ln(i/\gamma ka)]^{-2}$ when $kl \gg 1$, and therefore, is small for thin dipoles when $ka \ll 1$.

Within this error, $\psi_\pm$ can be described by the approximation provided by Eqn. (6.4.11). With the aid of this expression, the scattering pattern in the specular direction can be written as

$$F(\pi-\vartheta_0,\vartheta_0) = -\frac{kL}{2\ln(2i/\gamma ka\sin\vartheta_0)} +$$

$$+kL\frac{(\psi_+^0)^2 E[2kL\sin^2(\vartheta_0/2)] + (\psi_-^0)^2 E[2kL\cos^2(\vartheta_0/2)]}{4[\ln(2i/\gamma ka\sin\vartheta_0)]^2} + \qquad (7.5.8)$$

$$+i\frac{\ln[2i/\gamma ka\sin\vartheta_0] - 1/2 + \phi(\vartheta_0) + \phi(\pi-\vartheta_0)}{[\sin\vartheta_0\ln(2i/\gamma ka\sin\vartheta_0)]^2},$$

where

$$\phi(\vartheta)=\left\{\frac{1}{2}\psi_+\cos^2(\vartheta/2)-\ln\left[\frac{i}{\gamma ka\sin(\vartheta/2)}\right]\right\}\psi_+ e^{ikL(1-\cos\vartheta)}-$$
$$-\frac{\ln[i/\gamma ka]}{\mathcal{D}}e^{2ikL}\left[\psi_+\psi e^{ikL(1-\cos\vartheta)}-\psi_-\right]\psi_+ . \tag{7.5.9}$$

At grazing angles of observation, when $\vartheta_0=0$, or $\vartheta_0=\pi$, it can be seen that $F(\pi,0)=F(0,\pi)=0$. Substituting $\vartheta_0=\pi/2$, we obtain

$$F(\pi/2,\pi/2)=-\frac{kL}{2\ln(2i/\gamma ka)}+kL\frac{\overline{\psi}^2E(kL)}{2[\ln(2i/\gamma ka)]^2}+$$
$$+\frac{i}{[\ln(2i/\gamma ka)]^2}\left[\ln\frac{2i}{\gamma ka}-\frac{1}{2}+\left(\frac{1}{2}\overline{\psi}-2\ln\frac{\sqrt{2i}}{\gamma ka}\right)\overline{\psi}e^{ikL}+\right. \tag{7.5.10}$$
$$\left.+2\overline{\psi}^2\frac{\ln(i/\gamma ka)}{1+\psi e^{ikL}}e^{2ikL}\right],$$

where

$$\overline{\psi}=\psi_\pm\big|_{\vartheta=\pi/2}, \tag{7.5.11}$$

which characterizes the reflected field for the normal angle of incidence.

The scattering pattern of Eqn. (7.5.5) was obtained by summing all waves created by multiple diffraction. This method is very straightforward but somewhat cumbersome. This result can be obtained more quickly if it is seen that the process of edge wave diffraction at the ends of a passive dipole, beginning with the third-order term, is identical to that of a transmitting dipole. Therefore, the expression for the scattering pattern of a passive dipole can be immediately expressed as

$$F(\vartheta,\vartheta_0)=\frac{2if(\vartheta,\vartheta_0)}{\sin\vartheta\sin\vartheta_0\ln(2i/\gamma ka\sin\vartheta)\ln(2i/\gamma ka\sin\vartheta_0)}, \tag{7.5.12}$$

where

$$f(\vartheta,\vartheta_0)=\frac{1}{(\cos\vartheta+\cos\vartheta_0)}\times$$

$$\times\{[f_1(\vartheta,\vartheta_0)+f_2(\vartheta,\vartheta_0)+\hat{C}_1\psi_-(L)]e^{ikl\cos\vartheta}- \tag{7.5.13}$$

$$-[f_1(\pi-\vartheta,\pi-\vartheta_0)+f_2(\pi-\vartheta,\pi-\vartheta_0)-\hat{C}_2\psi_+(L)]e^{-ikl\cos\vartheta}\}.$$

The functions f_1 and f_2 describe the first- and second-order edge waves. Therefore,

$$f_1(\vartheta,\vartheta_0)=\cos^2\frac{\vartheta}{2}\cos^2\frac{\vartheta_0}{2}\ln\left(\frac{i}{\gamma ka\cos(\vartheta/2)\cos(\vartheta_0/2)}\right)e^{ikl\cos\vartheta_0}, \tag{7.5.14}$$

$$f_2(\vartheta,\vartheta_0)=e^{ikL}\left\{\sin^2\frac{\vartheta}{2}\sin^2\frac{\vartheta_0}{2}\ln\left[\frac{i}{\gamma ka\sin(\vartheta_0/2)}\right]\psi_-(L)-\right.$$
$$\left.-\cos^2\frac{\vartheta}{2}\cos^2\frac{\vartheta_0}{2}\ln\left[\frac{i}{\gamma ka\cos(\vartheta/2)}\right]\psi_+^0(L)\right\}e^{-ikl\cos\vartheta_0}. \tag{7.5.15}$$

The terms containing $\psi_+(L)$ and $\psi_-(L)$ in Eqn. (7.5.13) are the sum of all edge second- and higher-order waves propagating from the ends $z=-l$ and $z=l$. The constants $\hat{C}_1$ and $\hat{C}_2$ are determined from the condition

$$f(0,\vartheta_0)=f(\pi,\vartheta_0)=0, \tag{7.5.16}$$

which leads to a system of equations

$$\left.\begin{aligned}\hat{C}_1\,\psi(L)e^{ikL}+\hat{C}_2&=\frac{1}{2}\ln\frac{i}{\gamma ka}\psi_+^0(L)e^{2ikL-ikl\cos\vartheta_0},\\ \hat{C}_1+\hat{C}_2\psi(L)e^{ikL}&=\frac{1}{2}\ln\frac{i}{\gamma ka}\psi_-^0(L)e^{2ikL+ikl\cos\vartheta_0}.\end{aligned}\right\}\tag{7.5.17}$$

Hence,

$$\left.\begin{aligned}\hat{C}_1&=-\frac{1}{2\mathcal{D}}e^{2ikL}\ln\frac{i}{\gamma ka}\left[\psi_+^0\psi e^{ikL-ikl\cos\vartheta_0}-\psi_-^0 e^{ikl\cos\vartheta_0}\right],\\ \hat{C}_2&=-\frac{1}{2\mathcal{D}}e^{2ikL}\ln\frac{i}{\gamma ka}\left[\psi_-^0\psi e^{ikL+ikl\cos\vartheta_0}-\psi_+^0 e^{-ikl\cos\vartheta_0}\right].\end{aligned}\right\}\tag{7.5.18}$$

Substituting the value for $f(\vartheta,\vartheta_0)$ found using this method into Eqn. (7.5.12), we obtain an expression which is in complete agreement with Eqn. (7.5.5).

§7.6 Short, Passive Dipole

The theory presented thus far on the diffraction of a plane wave from a thin cylindrical dipole is based on a series of physical considerations. A strong aspect of this theory is the property that the accuracy increases with the length of the dipole, since the current wave diffraction is more pronounced as the length increases. However, it is possible to show that for a short dipole, the length of which is small in comparison with a wavelength, the equations we have introduced exhibit satisfactory accuracy.

It is clear that a dipole, which is short in comparison with a wavelength, creates a scattered field equal to

$$E_\vartheta=H_\varphi=-k^2p_z\frac{e^{ikR}}{kR}\sin\vartheta,\tag{7.6.1}$$

where the dipole moment p_z can be calculated by means of solving the electrostatics problem and depends on the dimensions and form of the dipole. In accordance with reference [116], the dipole moment of a cylinder in a uniform electrostatic field E_z is equal to

$$p_z = \mathcal{D}(Z)\left(\frac{L}{2}\right)^3 E_z, \tag{7.6.2}$$

where $\mathcal{D}(Z)$ is a dimensionless function and $Z = L/2a$. For $Z \gg 1$, $\mathcal{D}(Z)$ can be calculated from the asymptotic series

$$\mathcal{D}(Z) = \frac{2}{3}\left(\frac{1}{\Omega_2} + \frac{0.977}{\Omega_2^3} + \ldots\right),$$
$$\Omega_2 = 2(\ln 4Z - 7/3). \tag{7.6.3}$$

If this series is limited to the first term, then

$$\mathcal{D}(Z) = \frac{1}{3(\ln 4Z - 7/3)}. \tag{7.6.4}$$

Results of a numerical calculation of Eqn. (7.6.4) show that this last expression provides good accuracy for $Z > 9$ [116].

Thus, the dipole moment of a dipole, short in comparison to a wavelength, is equal to

$$p_z = E\frac{L^3}{24}\frac{\sin\vartheta_0}{\ln(2L/a) - 7/3}\left\{1 + O\left[\left(\ln\frac{L}{a}\right)^{-2}\right]\right\}, \tag{7.6.5}$$

and its scattering pattern must be of the form

$$F(\vartheta, \vartheta_0) = \frac{k^3 L^3}{24}\frac{\sin\vartheta \sin\vartheta_0}{\ln[2L/a] - 7/3}\left\{1 + O\left[\left(\ln\frac{L}{a}\right)^{-2}\right]\right\}. \tag{7.6.6}$$

This section will find the first two terms of the series for the dipole scattering pattern F in degrees of $1/\ln(L/a)$ (as $\lambda \to \infty$) and compare them with Eqn. (7.6.6). To do this, we will limit ourselves to the case when $\vartheta = \vartheta_0 = \pi/2$ where F can be described more simply by Eqn. (7.5.10).

According to Eqn. (6.4.13) for small values of kz, $\psi(z)$ and $\bar{\psi}(z) = \psi_{\pm}\big|_{\varphi=\pi/2}$ can be written in the form

$$\left.\begin{aligned}\psi(z)&=1-\frac{g(z)-g(0)}{2g(0)}+O\left[\frac{1}{g^2(0)}\right],\\ \bar{\psi}(z)&=1-\frac{\bar{g}(z)-\bar{g}(0)}{2g(0)}+O\left[\frac{1}{g^2(0)}\right],\end{aligned}\right\} \tag{7.6.7}$$

where $g(z)$ is determined by Eqn. (6.4.19) and

$$\bar{g}(z)-\bar{g}(0)=\ln\frac{\gamma kz}{i}+e^{-ikz}\int_z^\infty\frac{e^{ik\sigma}}{\sigma}d\sigma. \tag{7.6.8}$$

Limiting the series expansion for $\psi(z)$ and $\bar{\psi}(z)$ to terms on the order of $(kz)^3$, we have

$$\begin{aligned}\psi(z)=1-\frac{1}{\ln(i/\gamma ka)}\Bigg[ikz\left(\ln\frac{2\gamma kz}{i}-1\right)+\\ +k^2z^2\left(\ln\frac{2\gamma kz}{i}-\frac{3}{2}\right)-i\frac{2}{3}k^3z^3\left(\ln\frac{2\gamma kz}{i}-\frac{11}{6}\right)\Bigg],\end{aligned} \tag{7.6.9}$$

$$\begin{aligned}\bar{\psi}(z)=1-\frac{1}{2\ln(i/\gamma ka)}\Bigg[ikz\left(\ln\frac{\gamma kz}{i}-1\right)+\\ +\frac{k^2z^2}{2}\left(\ln\frac{\gamma kz}{i}-\frac{3}{2}\right)-i\frac{k^3z^3}{6}\left(\ln\frac{\gamma kz}{i}-\frac{11}{6}\right)\Bigg].\end{aligned} \tag{7.6.10}$$

Moreover, the terms on the order of $[\ln(i/\gamma ka)]^{-2}$ were truncated in Eqns. (7.6.9) and (7.6.10). If these equations are substituted into

Eqn. (7.5.10) then it is necessary to omit the terms on the order of $[\ln(i/\gamma ka)]^{-3}$. Therefore, $F(\pi/2, \pi/2)$ can be presented as

$$F(\pi/2,\pi/2)=\frac{i}{\left[\ln(2i/\gamma ka)\right]^2}\left\{\frac{ikL}{2}\left[\ln\frac{2i}{\gamma ka}-E(kL)\right]+\ln\frac{2i}{\gamma ka}-\frac{1}{2}+\right.$$

$$\left.+\frac{1}{2}e^{ikL}-2\overline{\psi}(L)\ln\frac{\sqrt{2}i}{\gamma ka}e^{ikL}+\frac{2\ln(i/\gamma ka)\overline{\psi}^2(L)}{1+\psi(L)e^{ikL}}e^{2ikL}\right\}. \tag{7.6.11}$$

Incorporating Eqns. (7.6.9) and (7.6.10) and omitting all intermediate steps, which were partly outlined in reference [13], we obtain

$$F(\pi/2,\pi/2)=\frac{k^3L^3}{24}\left\{\frac{1}{\ln(L/a)}+\frac{7/3-\ln 2}{\left[\ln(L/a)\right]^2}+O\left[\left(\ln\frac{L}{a}\right)^{-3}\right]\right\}. \tag{7.6.12}$$

This equation may be rewritten as

$$F(\pi/2,\pi/2)=\frac{k^3L^3}{24\left[\ln(2L/a)-7/3\right]}\left\{1+O\left[\left(\ln\frac{L}{a}\right)^{-2}\right]\right\}. \tag{7.6.13}$$

This completely agrees with Eqn. (7.6.6), which follows from [116].

These results confirm the correctness of our calculation for the scattering pattern given by Eqn. (7.5.5) and show that it is applicable to dipoles of any length.

§7.7 Results of Numerical Calculations

The function $F(\vartheta, \vartheta_0)$ can be used to calculate the total scattering cross section, S, and the bistatic radar cross section σ of a dipole. The total scattering cross section characterizes the power a passive dipole scatters in all directions and is equal to

$$S=\frac{\mathcal{P}}{p}, \tag{7.7.1}$$

where

$$p = \frac{c}{8\pi} E_0^2 \tag{7.7.2}$$

is the time-average power flux density of the incident wave, and

$$\mathcal{P} = \frac{c}{8\pi} \Re e \int \mathbf{E} \times \mathbf{H}^* \cdot d\mathbf{s} = \frac{1}{2} E \sin\vartheta_0 \Re e \int_{-l}^{l} J(z) e^{ikz\cos\vartheta_0} dz\,. \tag{7.7.3}$$

is the time-average total power of the field scattered by the dipole into surrounding space. The scattered far field in the $\vartheta = \pi - \vartheta_0$ direction can be written as

$$E_\vartheta = H_\varphi = -\frac{ik}{c} \sin\vartheta_0 \frac{e^{ikR}}{R} \int_{-l}^{l} J(z) e^{ikz\cos\vartheta_0} dz\,, \tag{7.7.4}$$

and

$$E_\vartheta = H_\varphi = -E \frac{e^{ikR}}{kR} F(\pi - \vartheta_0, \vartheta_0)\,. \tag{7.7.5}$$

Thus, having determined the integral

$$\int_{-l}^{l} J(z) e^{ikz\cos\vartheta_0} dz = \frac{cE}{ik^2 \sin\vartheta_0} F(\pi - \vartheta_0, \vartheta_0)\,, \tag{7.7.6}$$

we obtain

$$S = \frac{\lambda^2}{\pi} \cos^2\alpha\, \Im m\{F(\pi - \vartheta_0, \vartheta_0)\}\,. \tag{7.7.7}$$

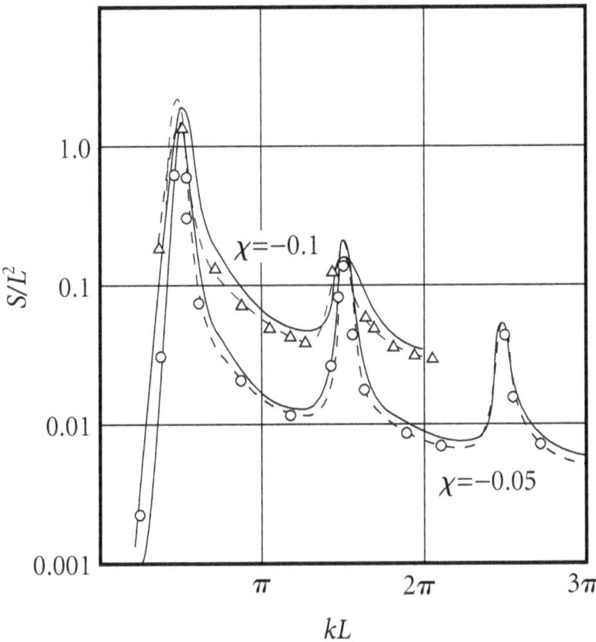

Figure 7.7.1: Dependence of the total scattering cross section on the length of the dipole for a normally incident plane wave. The solid line is constructed from the Leontovich and Levin solution. The dashed line is constructed from Eqn. (7.7.7). The circular and triangular icons represent the exact values calculated by Vainshtein.

Figure 7.7.1 presents S/L^2 (for $\alpha = 0, \vartheta_0 = \pi/2$) for dipoles for different values of $\chi = 1/[2\ln(ka)]$ ($\chi = -0.05$ and $\chi = -0.1$). The solid curves were calculated using expressions from the well known work of Leontovich and Levin [117]. The dashed curves were calculated using Eqn. (7.7.7). The circular and triangular icons denote exact values of the numerical solution of the integral equations calculated by Vainshtein [118]. As Fig. 7.7.1 illustrates, the exact values are in good agreement with the curves calculated using Eqns. (7.7.7) and (7.5.10). The Leontovich and Levin expressions applied to the same dipole are somewhat worse, but nevertheless, still entirely satisfactory.

By definition, the bistatic radar cross section σ is equal to

$$\sigma(\vartheta, \vartheta_0) = \frac{\mathcal{P}_r 4\pi R^2}{p}, \tag{7.7.8}$$

where p is given by Eqn. (7.7.2), and

$$\mathcal{P}_r = \frac{c}{8\pi}|E_\vartheta|^2 = \frac{c}{8\pi}\frac{|F(\vartheta, \vartheta_0)|^2}{k^2 R^2}E^2 \tag{7.7.9}$$

represents the power flux density scattered by the dipole in the ϑ direction. Thus,

$$\sigma(\vartheta, \vartheta_0) = \frac{\lambda^2}{\pi}\cos^2\alpha\,|F(\vartheta, \vartheta_0)|^2. \tag{7.7.10}$$

If the receiving antenna operates at the same polarization as the transmitting antenna, then the bistatic RCS is equal to

$$\sigma_\alpha(\vartheta, \vartheta_0) = \frac{\lambda^2}{\pi}\cos^4\alpha\,|F(\vartheta, \vartheta_0)|^2. \tag{7.7.11}$$

In the monostatic case when the receiving and transmitting antennas coincide, but with arbitrary polarization, the scattering properties of the dipole are characterized by the average value given by

$$\overline{\sigma}(\vartheta) = \frac{1}{2\pi}\int_0^{2\pi}\sigma_\alpha(\vartheta, \vartheta)d\alpha = \frac{3\lambda^2}{8\pi}|F(\vartheta, \vartheta_0)|^2. \tag{7.7.12}$$

Figure 7.7.2 plots $\overline{\sigma}(\pi/2)$ as a function of frequency for $L = 5$ cm and $L/(2a) = 452$. The solid lines represent the exact values of Vainshtein [118] and agree with the curve Lindroth [119] calculated by integrating the current. Lindroth obtained an expansion for the scattered field in negative orders of $\Omega_p = 2\ln(2L/a)$, which terminated at Ω_p^{-3}. The dashed curve was

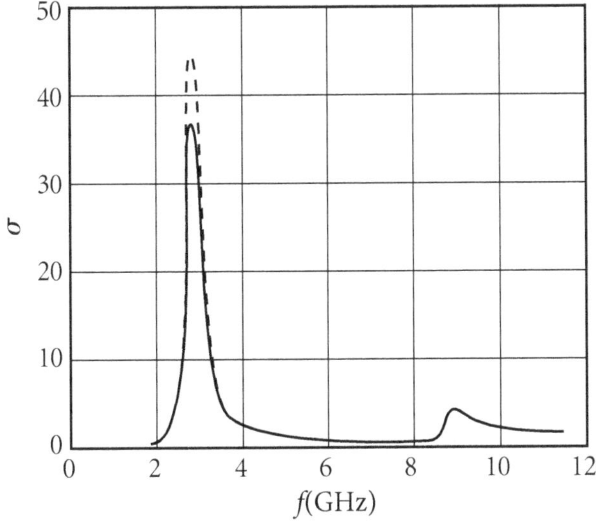

Figure 7.7.2: Dependence of the radar cross section of a dipole on the frequency in gigahertz for normal plane wave incidence. Reprinted from [14] with the permission of *Radiotekhnika i elektronika.*

constructed using Eqns. (7.7.12) and (7.5.10). It provides a somewhat higher value for $\bar{\sigma}$ near the first resonance (for $\lambda \approx 4l$) but at other frequencies practically overlays with the exact curve. Reference [13] contains a series of other graphs constructed using Eqns. (7.7.7), (7.7.12), and (7.5.8), and more completely characterizes the scattering properties of a dipole.

§7.8 Radiation of Edge Waves from a Strip

This section applies the theory of edge wave radiation to a strip. Since the solution of this problem uses fundamental theoretical elements that remain unchanged, we will attempt to present it briefly and limit ourselves to the formulation of the basic relationships.

Thus, let there be a plane wave

$$H_x = e^{-ik(z\cos\vartheta_0 + y\sin\vartheta_0)}, \quad E_x = 0, \quad E_z = -\sin\vartheta_0 e^{-ik(z\cos\vartheta_0 + y\sin\vartheta_0)}, \qquad (7.8.1)$$

diffracting at a perfectly conducting strip located in the $y = 0$ plane and bounded by $-l \leqslant z \leqslant l$ and $-\infty \leqslant x \leqslant \infty$. The first-order edge wave is determined from a solution to the diffraction from a half-plane (see, for instance, [13, 120]) and is described in the far zone (for $R \gg kl^2$) by

$$H_x = \frac{2}{(\cos\vartheta + \cos\vartheta_0)}\left[\sin\frac{\vartheta}{2}\sin\frac{\vartheta_0}{2}e^{-ikl(\cos\vartheta+\cos\vartheta_0)} - \right.$$
$$\left. -\cos\frac{\vartheta}{2}\cos\frac{\vartheta_0}{2}e^{ikl(\cos\vartheta+\cos\vartheta_0)}\right]\frac{e^{i\left(kr+\frac{\pi}{4}\right)}}{\sqrt{2\pi kr}}, \tag{7.8.2}$$

where $0 \leqslant \vartheta \leqslant \pi$ and $0 \leqslant \vartheta_0 \leqslant \pi$.

Our calculation of the higher-order edge waves will use the expression

$$-\frac{2\pi}{c}j_z = e^{-ikz\cos\vartheta_0} -$$
$$-\chi_-^0(z+l)e^{ik(z+l)+ikl\cos\vartheta_0} - \chi_+^0(l-z)e^{ik(l-z)-ikl\cos\vartheta_0} + \tag{7.8.3}$$
$$+\tilde{\mathcal{B}}_1\chi(l+z)e^{ik(z+l)} + \tilde{\mathcal{B}}_2\chi(l-z)e^{ik(l-z)},$$

which in agreement with Eqn. (6.6.6), determines the current excited on a strip by the incident wave given by Eqn. (7.8.1). The functions $\chi_\pm^0$ and χ are expressed in terms of Fresnel integrals using Eqns. (6.6.8) and (6.6.2), where w is replaced by $w_0 = -k\cos\vartheta_0$. Also, in accordance with Eqn. (6.6.9)

$$\left.\begin{aligned}\tilde{\mathcal{B}}_1 &= \frac{e^{ikL}}{\mathcal{D}}\left[\chi_+^0(L) - \chi_-^0(L)\chi(L)e^{i(k-w_0)L}\right]e^{iw_0l},\\ \tilde{\mathcal{B}}_2 &= \frac{e^{ikL}}{\mathcal{D}}\left[\chi_-^0(L) - \chi_+^0(L)\chi(L)e^{i(k+w_0)L}\right]e^{-iw_0l},\end{aligned}\right\} \tag{7.8.4}$$

where

$$\mathcal{D} = 1 - \chi^2(L)e^{2ikL}, \quad L = 2l, \quad w_0 = -k\cos\vartheta_0. \tag{7.8.5}$$

It is possible to show that for $z > -l$ the external field

$$E_z^e = -\frac{1}{2ik}\left(\cos\frac{\vartheta_0}{2}e^{-iw_0 l} - \tilde{\mathcal{B}}_1\right)\delta(z+l) + \epsilon(z+l)\sin^2\frac{\vartheta_0}{2}\cos\frac{\vartheta_0}{2}e^{iw_0 z},$$

$$\epsilon(z+l) = \begin{cases} 1 & \text{for} \quad z+l<0, \\ 0 & \text{for} \quad z+l>0, \end{cases} \qquad (7.8.6)$$

located at $y = \pm 0$ on an infinite perfectly conducting plane ($y = 0$) excites a total current given by

$$j_z = j_z(y=0^+) + j_z(y=0^-) = \frac{c}{2\pi}\left[\chi_-^0(z+l)e^{-iw_0 l} - \tilde{\mathcal{B}}_1\chi(z+l)\right]e^{ik(z+l)}, \qquad (7.8.7)$$

which is equivalent to the total edge wave current flowing along the strip from the end at $z = -l$. It follows that the field radiating into surrounding space from the edge of the strip at $z = l$ can be determined by the solving the excitation of the half-plane $-\infty \leqslant z \leqslant l$ by the external field of Eqn. (7.8.6). Utilizing the reciprocity theorem, we find that

$$H_x = \frac{2}{(\cos\vartheta + \cos\vartheta_0)}\frac{e^{i\left(kr+\frac{\pi}{4}\right)}}{\sqrt{2\pi kr}}e^{ik(L-l\cos\vartheta)} \times$$

$$\times\left\{\left[-\sin\frac{\vartheta}{2}\sin^2\frac{\vartheta_0}{2}\chi_-^0(L) + \cos^2\frac{\vartheta}{2}\cos\frac{\vartheta_0}{2}\chi_+(L)\right]e^{ikl\cos\vartheta_0} + \right.$$

$$+\left[\chi_-^0(L)\chi(L)e^{ik(1+\cos\vartheta)L} - \chi_+^0(L)\right]\times, \qquad (7.8.8)$$

$$\left.\times\frac{\cos\vartheta + \cos\vartheta_0}{2\mathcal{D}}\chi_+(L)e^{ik(L-l\cos\vartheta_0)}\right\}$$

$$0 \leqslant \vartheta \leqslant \pi, \quad 0 \leqslant \vartheta_0 \leqslant \pi.$$

Here, the first term in the braces is the second-order wave, and the second term is the sum of all remaining edge waves. From symmetry relationships it is clear that by replacing ϑ with $\pi-\vartheta$, and ϑ_0 with $\pi-\vartheta_0$, we obtain an expression for the waves radiated by the edge of the strip at $z = -l$.

As a result, the total field radiated by the strip can be written as

$$H_x = F(\vartheta,\vartheta_0)\frac{e^{i\left(kr+\frac{\pi}{4}\right)}}{\sqrt{2\pi kr}}, \quad 0\leqslant\vartheta\leqslant\pi, \quad 0\leqslant\vartheta_0\leqslant\pi, \tag{7.8.9}$$

where

$$F(\vartheta,\vartheta_0) = \mathcal{F}(\vartheta,\vartheta_0)e^{ikl\cos\vartheta} + \mathcal{F}(\pi-\vartheta,\pi-\vartheta_0)e^{-ikl\cos\vartheta}, \tag{7.8.10}$$

$$\begin{aligned}\mathcal{F}(\vartheta,\vartheta_0) = &-\frac{2\cos(\vartheta/2)\cos(\vartheta_0/2)}{\cos\vartheta+\cos\vartheta_0}e^{ikl\cos\vartheta_0} + \\ &+\left[-\sin^2\frac{\vartheta}{2}\sin\frac{\vartheta_0}{2}\chi_-(L)+\cos\frac{\vartheta}{2}\cos^2\frac{\vartheta_0}{2}\chi_+^0(L)\right]\times \\ &\times\frac{2e^{ikL}}{\cos\vartheta+\cos\vartheta_0}e^{-ikl\cos\vartheta_0} + \\ &+\frac{e^{i2kL}}{\mathcal{D}}\left[-\chi_-^0(L)+\chi_+^0(L)\chi(L)e^{ikL(1-\cos\vartheta_0)}\right]\chi_-(L)e^{ikl\cos\vartheta_0}.\end{aligned} \tag{7.8.11}$$

The function $F(\vartheta,\vartheta_0)$ satisfies the reciprocity theorem and Eqn. (7.5.7). Note that when $\vartheta = \pi-\vartheta_0$, i.e., in the specular direction, both terms in Eqn. (7.8.10) approach infinity, but their sum remains finite:

$$F(\pi-\vartheta_0,\vartheta_0) = -\frac{1}{\sin\vartheta_0} - 2ikl\sin\vartheta_0 + \beta(\vartheta_0) + \beta(\pi-\vartheta_0), \tag{7.8.12}$$

and

$$\beta(\vartheta_0)=\left[\left(\frac{\cos^3(\vartheta_0/2)}{\sin\vartheta_0}+\sin\frac{\vartheta_0}{2}+2ikl\sin\vartheta_0\cos\frac{\vartheta_0}{2}\right)\chi_+^0(L)+\right.$$

$$\left.+2e^{-i\frac{\pi}{4}}\sqrt{\frac{kl}{\pi}}\cos^2\frac{\vartheta_0}{2}\right]e^{ikL(1-\cos\vartheta_0)}+ \qquad (7.8.13)$$

$$+\frac{e^{2ikL}}{\mathcal{D}}\left[\chi_+^0(L)\chi(L)e^{ikL(1-\cos\vartheta)}-\chi_-^0(L)\right]\chi_+^0(L).$$

For a plane wave at normal incidence ($\vartheta_0=\pi/2$) we find that

$$F(\pi/2,\pi/2)=-1-2ikL+2\overline{\beta}e^{ikL}, \qquad (7.8.14)$$

where

$$\overline{\beta}=\frac{1}{2\sqrt{2}}\left[(3+2ikL)\overline{\chi}(L)+2\sqrt{\frac{2kl}{\pi}}e^{-i\frac{\pi}{4}}\right]+\frac{e^{ikL}}{\mathcal{D}}\left[\chi(L)e^{ikL}-1\right]\overline{\chi}^2(L), \qquad (7.8.15)$$

$$\overline{\chi}(L)=\overline{\chi}_\pm^0(L)\Big|_{\vartheta_0=\frac{\pi}{2}}=\frac{2}{\sqrt{\pi}}e^{-i\frac{\pi}{4}-ikL}\int_{\sqrt{kL}}^{\infty}e^{it^2}dt. \qquad (7.8.16)$$

Using Eqn. (7.8.14) to calculate the total scattering cross section

$$S=-\frac{2}{k}\Im\mathfrak{m}\{F(\pi/2,\pi/2)\}, \qquad (7.8.17)$$

we obtain

$$S=2L\left\{1-2\sqrt{\frac{2}{\pi}}\frac{\cos(kL-\pi/4)}{(kL)^{3/2}}+\frac{2}{\pi}\frac{\cos 2kL}{(kL)^2}-\right. \qquad (7.8.18)$$

$$-\frac{1}{\pi}\sqrt{\frac{2}{\pi}}\frac{\cos(3kL+\pi/4)-(6\pi/4)\cos(kL+\pi/4)}{(kL)^{5/2}}+ \qquad (7.8.18, \text{cont.})$$

$$+O\left[(kL)^{-3}\right]\Bigg\}.$$

A comparison of this equation with Eqn. (31.5) from reference [121] shows that its relative error is on the order of $(kl)^{-5/2}$. Chapter 10 will examine the error of Eqn. (7.8.10) in more detail and show that it is more accurate than the corresponding expressions found in references [5] and [122].

§7.9 Conclusion

- This chapter constructed an approximate theory of the radiation of higher-order edge waves. This theory is based on the reciprocity theorem and provides a straightforward representation of the scattered field. It was used to calculate the scattering pattern for thin cylindrical conductors (dipoles) and strips. The expressions obtained are valid for arbitrary directions of observation and illumination, satisfy the reciprocity theorem, and permit calculations to be made with satisfactory accuracy, provided $kl \gg 1$.

- It is interesting that the scattering pattern determined here is correct for not only long ($kl \gg 1$) but also short ($kl \ll 1$) conducting dipoles. To examine this characteristic, numerical results were presented that agreed with data from other approximations (Leontovich and Levin [117], Lindroth [119]) and with the exact solution provided by Vainshtein [118].

- This chapter concludes the development of approximate methods described in the previous sections. These methods investigated the scattering from certain objects and provided a visual picture of multiple diffraction. Recall that Chapter 1 presents the theory of diffraction at black bodies. This theory reveals the physical nature of such phenomena as Fresnel diffraction and forward scattering. It also establishes the diffraction limit for reduction of the total scattering power, when reflecting objects are covered with radar absorbing materials.

- Chapters 2–7 presented the Physical Theory of Diffraction (PTD), its extension, and development. They present results from the author of reference [13], which became a bibliographic rarity a long time ago. PTD is extended here for bodies of revolution with concave elements. A modified PTD in combination with the parabolic equation method is developed for investigation of multiple diffraction. First, this method is analyzed in Chapter 5 on the example of the classic wedge diffraction problem. Then, in Chaper 6 it is used for investigation of the *nonuniform* [in PTD terminology] component of current excited in dipoles and strips. Chapter 7 develops a special PTD technique for the calculation of the far field radiated by these currents.
- However, all this does not eliminate the need to solve rigorously some particular problems to first, establish limits of the applicability of approximate methods, and second, to study more minute diffraction effects. To accomplish this goal, the following chapters develop the mathematical theory of diffraction, which provides a more detailed structural description of higher-order edge waves.

8

Functional and Integral Equations for Strip Diffraction (Neumann Boundary Problem)

§8.1 Asymptotic Solutions for Strip Diffraction

Chapters 6 and 7 approximated higher-order edge waves. Now, using strips as an example, this chapter will study the properties of these waves in more detail. There are several reasons that prompt us to investigate these problems. Presently, there is significant interest in studying the exact structure of weak signals reflected from various objects. Clearly, their physical nature is related to higher-order edge waves. Besides, diffraction not only from isolated scatterers but also from groups of bodies, is of important practical interest. Referring to duality theory, diffraction from a strip is equivalent to a slot formed by two half-planes and is a special case of diffraction from two bodies. It is evident that the diffraction interaction of two half-planes is also caused by edge waves. Understanding strip diffraction is useful for studying these issues.

The asymptotic solution to diffraction from a strip (slot) has been examined by many references. Integral equations were used by Millar, Westpfahl, Lüneburg, Kieburtz, and Stöckel [123–127] for the diffracted field in order to determine the first few terms of an asymptotic expansion as a function of $1/(kl)$, where $k = 2\pi/\lambda$ is the wave number and $2l$ is the width of the strip (slot). Applying asymptotic expressions for the Mathieu function, Hansen [128] extracted the asymptotic expressions related to Sommerfeld first-order edge waves from the exact solution of the boundary value problem in elliptical coordinates.

The asymptotic solution to the integral equation of the second kind for a planar screen with a slot was performed by Grinberg [129, 130]. The asymptotic solution of a "key" equation for the current on a strip was also obtained by Grinberg with an accuracy on the order of $(kl)^{-5/2}$ [131]. Further research on this equation was conducted by Kuritsyn [132] and Popov [133]. Khaskind, Vainshtein [122] and Fialkovskii [134] used the integral equation method to obtain the foundation and refinement of approximations previously introduced in [5]. References [5, 122, 134] obtained formulas applicable for arbitrary angles of incidence and observation but correctly describe only the first term of the asymptotic expansion for $kl \gg 1$.

The most complete asymptotic research on diffraction from a strip was given by Borovikov [135, 136] who provided the scattering in the far zone to within an accuracy of any given order of $(kl)^{-n}$ as $kl \to \infty$ $(n = 1, 2, 3\ldots)$. However, in this research there isn't a single expression applicable for arbitrary angles of incidence and observation, and the surface current is not investigated.

The theory developed below will obtain the asymptotic expressions for the current excited on a strip and for the scattered field in the far zone to within an accuracy of any given order of $(kl)^{-n/2}$ for $2kl \gg n$. These expressions are applicable for any angle of observation and illumination. Chapters 8–10 examine the Neumann boundary problem. The Dirichlet boundary problem is considered in Chapter 11. The fundamental results of this research were previously published in [18, 22–24].

§8.2 Symmetry of Edge Waves

Let a planar electromagnetic wave

$$H_x = e^{-ik\left(z\alpha_0 + y\sqrt{1-\alpha_0^2}\right)-i\omega t}, \quad E_x = 0, \quad \alpha_0 = \cos\vartheta_0 \tag{8.2.1}$$

excite a perfectly conducting strip, the orientation of which in free space is defined by $y = 0$, $-l \leqslant z \leqslant l$, and $-\infty \leqslant x \leqslant \infty$ (Fig. 8.2.1). The incident wave propagates at an angle ϑ_0 with respect to the z-axis $(0 \leqslant \vartheta_0 \leqslant \pi)$.

Let us assume that the surface current density excited on the strip is

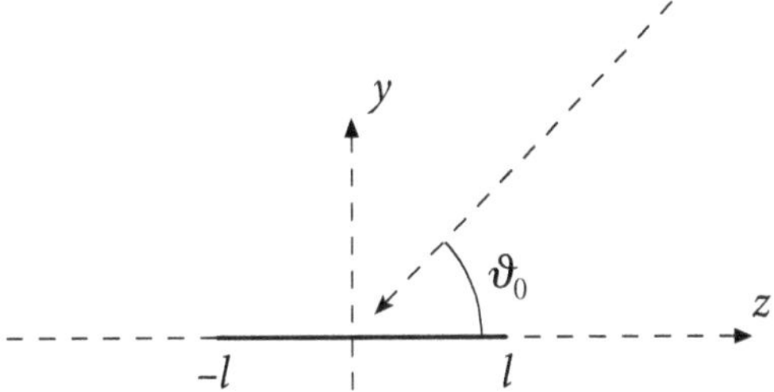

Figure 8.2.1: Segment of the z-axis ($-l \leq z \leq l$) depicting the cross section of a strip in the *yoz* plane.

$$j_z = \frac{c}{2\pi} J(z, \alpha_0), \tag{8.2.2}$$

$$J(z, \alpha_0) = j^{(0)}(z, \alpha_0) + J^{(-)}(z + l, \alpha_0) + J^{(+)}(l - z, \alpha_0), \tag{8.2.3}$$

where

$$j^{(0)}(z, \alpha_0) = -e^{-ikz\alpha_0} \tag{8.2.4}$$

is the current excited by the wave given by Eqn. (8.2.1) on an infinite perfectly conducting plane $y = 0$, $-\infty \leqslant z \leqslant \infty$. The values of $J^{(-)}$ and $J^{(+)}$ represent edge waves propagating from left to right and right to left, respectively.

Utilizing the notations introduced in Eqns. (8.2.2)–(8.2.4), a primed system of coordinates x', y, z' (Fig. 8.2.2) will be used to describe the current excited on a strip by a plane wave given by

$$H_{x'} = -e^{-ik\left(z'\alpha'_0 + y\sqrt{1-\alpha'^2_0}\right) - i\omega t}, \quad E_{x'} = 0, \quad \alpha'_0 = \cos\vartheta'_0 . \tag{8.2.5}$$

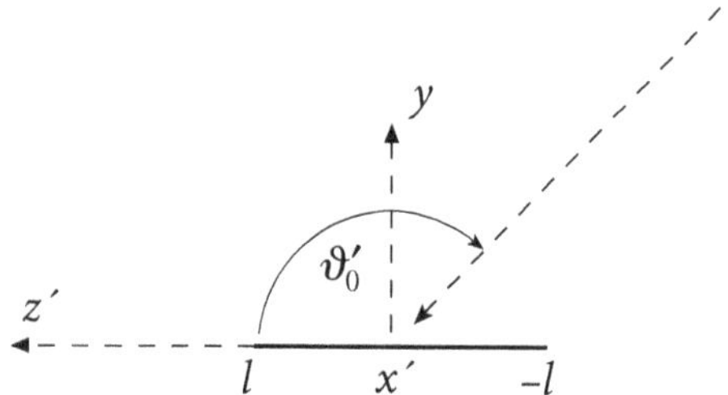

Figure 8.2.2: Cross section of a strip described in x', y, z' coordinates

It is evident that

$$\frac{2\pi}{c} j_{z'} = -j^{(0)}(z',\alpha_0') - J^{(-)}(z'+l,\alpha_0') - J^{(+)}(l-z',\alpha_0') \tag{8.2.6}$$

or

$$\frac{2\pi}{c} j_{z'} = -j^{(0)}(z,\alpha_0) - J^{(+)}(z+l,-\alpha_0) - J^{(-)}(l-z,-\alpha_0). \tag{8.2.7}$$

Noticing further that

$$j_{z'} = -j_z ,$$

and comparing Eqn. (8.2.2) with Eqn. (8.2.7) we find that

$$J^{(+)}(z,\alpha_0) = J^{(-)}(z,-\alpha_0). \tag{8.2.8}$$

Using this symmetry property of edge waves, Eqn. (8.2.3) can be written as

$$J(z,\alpha_0) = j^{(0)}(z,\alpha_0) + \tilde{J}(z+l,\alpha_0) + \tilde{J}(l-z,-\alpha_0). \tag{8.2.9}$$

Similarly, the scattering pattern of the edge wave also obeys the symmetry property

$$\Phi^{(+)}(\alpha,\alpha_0)=\Phi^{(-)}(-\alpha,-\alpha_0), \quad \alpha=\cos\vartheta . \tag{8.2.10}$$

Subsequently, the scattered far field (for $y \geqslant 0$) is

$$H_x=\Phi(\alpha,\alpha_0)\frac{e^{i(kr+\pi/4)}}{\sqrt{2\pi kr}}, \tag{8.2.11}$$

where

$$\Phi(\alpha,\alpha_0)=\tilde{\Phi}(\alpha,\alpha_0)e^{i\kappa\alpha}+\tilde{\Phi}(-\alpha,-\alpha_0)e^{-i\kappa\alpha}, \tag{8.2.12}$$

and the values of κ, α, and r are determined by

$$\kappa=kl, \tag{8.2.13}$$

$$z=r\alpha, \quad y=r\sqrt{1-\alpha^2}, \quad \alpha=\cos\vartheta . \tag{8.2.14}$$

Here, Eqn. (8.2.13) introduced a new parameter κ which should not be confused with the wave number k. These symmetry properties significantly facilitate the formulation and solution of the problem.

§8.3 Formulation and Solution of the Functional Equations

Let us present $\tilde{J}$ as a series of edge waves,

$$\tilde{J}(z,\alpha_0)-\sum_{n=1}^{\infty} j_n(z,\alpha_0) \tag{8.3.1}$$

excited on the edges of a strip. Furthermore, let us postulate that

$$j_n(z,\alpha_0)=\int_{-\infty}^{\infty} F_n(t,k\alpha_0)e^{izt}\,dt \tag{8.3.2}$$

and designate $\mathbf{E}_n(z, y, \alpha_0)$ as the electric vector radiated by $j_n(z, \alpha_0)$. The component of $\mathbf{E}_n$ tangential to the strip can be found from Eqns. (1.2.2) and (1.2.3) and is equal to

$$E_{nz} = -\frac{1}{ik}\left(\frac{\partial^2}{\partial z^2} + k^2\right)A_{nz}, \tag{8.3.3}$$

where $\mathbf{A}_n$ is the retarded vector potential and

$$A_{nz} = \frac{i}{2}\int_{-\infty}^{\infty} \mathrm{j}_n(\zeta, \alpha_0) H_0^{(1)}\left(k\sqrt{y^2 + (z-\zeta)^2}\right)d\zeta. \tag{8.3.4}$$

Substituting the equation

$$H_0^{(1)}\left(k\sqrt{y^2+(z-\zeta)^2}\right) = \frac{1}{\pi}\int_{-\infty}^{\infty}\frac{e^{i[vy+w(z-\zeta)]}}{v}dw$$

$$v = \sqrt{k^2 - w^2}, \quad \Im\mathrm{m}\{v\} \geqslant 0, \quad y \geqslant 0, \quad v = k \ \text{ for } \ w = 0 \tag{8.3.5}$$

we obtain

$$A_{nz} = i\int_{-\infty}^{\infty} \mathrm{F}_n(w, k\alpha_0)\frac{e^{i(vy+wz)}}{v}dw. \tag{8.3.6}$$

Thus,

$$E_{nz}(z, 0, \alpha_0) = -\frac{1}{k}\int_{-\infty}^{\infty}\sqrt{k^2 - t^2}\,F_n(t, k\alpha_0)e^{izt}dt. \tag{8.3.7}$$

As a result, the conditions

$$J(z,\alpha_0)=0, \quad \text{for} \quad |z|>l \tag{8.3.8}$$

and

$$E_{nz}(z,0,\alpha_0)=0, \quad \text{for} \quad z>0 \tag{8.3.9}$$

lead us to the following recursive system of functional equations:

$$\left.\begin{aligned}
&\int_{-\infty}^{\infty} F_1(t,k\alpha_0)e^{izt}\,dt=-j^{(0)}(z-l,\alpha_0) && \text{for } z<0,\\
&\int_{-\infty}^{\infty}\Big[\underset{n=2,3,4,\ldots}{F_n(t,k\alpha_0)+F_{n-1}(-t,-k\alpha_0)e^{-2ilt}}\Big]e^{izt}\,dt=0 && \text{for } z<0,\\
&\int_{-\infty}^{\infty}\sqrt{k^2-t^2}\,\underset{n=1,2,3\ldots}{F_n(t,k\alpha_0)}e^{izt}\,dt=0 && \text{for } z>0,
\end{aligned}\right\} \tag{8.3.10}$$

where the contour of integration skirts over the branch point at $t = -k$ and under the branch point at $t = k$. It is also possible to let the contour of integration to be on the real axis in the complex t plane and k possess a small positive imaginary part.

The solution of Eqn. (8.3.10) is found using the factorization method technique, (the Wiener-Hopf–Fock method [96]). It follows from the last expression in Eqn. (8.3.10) that $\sqrt{k-t}\,F_n(t,k\alpha_0)$ is holomorphic in the upper half-plane of the complex variable t ($\Im\mathfrak{m}\{t\} > 0$) and vanishes as $|t| \to \infty$. The first expression in Eqn. (8.3.10) states that $F_1(t,k\alpha_0)$ is holomorphic in the lower half-plane ($\Im\mathfrak{m}\{t\} < 0$), vanishes with $|t| \to \infty$, and has a simple pole at $t = -k\alpha_0$. Therefore, $F_1(t,k\alpha_0)$ can be represented as

$$F_1(t,k\alpha_0)=\frac{A}{\sqrt{k-t}}\frac{1}{t+k\alpha_0}. \tag{8.3.11}$$

The constant A is found from the first expression of Eqn. (8.3.10). As a result, we have

$$F_1(t,k\alpha_0)=-\frac{1}{2\pi i}\sqrt{\frac{k+k\alpha_0}{k-t}}\frac{e^{i\kappa\alpha_0}}{t+\widehat{k\alpha_0}}, \tag{8.3.12}$$

The second expression in Eqn. (8.3.10) states that the function

$$F_n(t,k\alpha_0)+F_{n-1}(-t,-k\alpha_0)e^{-i2lt} \tag{8.3.13}$$

is holomorphic in the lower half-plane ($\Im\mathfrak{m}\{t\}<0$) and vanishes as $|t|\to\infty$. Besides, it is clear that the function

$$\sqrt{k-t}F_n(t,k\alpha_0)+\sqrt{k-t}F_{n-1}(-t,-k\alpha_0)e^{-i2lt}=h^-(t) \tag{8.3.14}$$

as well as that of Eqn. (8.3.13) is holomorphic in the lower half-plane and vanishes as $|t|\to\infty$. One can consider that the function

$$\sqrt{k-t}F_{n-1}(-t,-k\alpha_0)e^{-i2lt}=H(t) \tag{8.3.15}$$

is holomorphic on the strip $-\epsilon\le\Im\mathfrak{m}\{t\}\le\epsilon, -\infty\le\Re\mathfrak{e}\{t\}\le\infty$ and vanishes as $|t|\to\pm\infty$. Here, $\epsilon\ll\Im\mathfrak{m}\{k\}\ll 1$, and later we will set $\epsilon=0$. Therefore, this function can be represented as the sum [137, 138]

$$H(t)=H^+(t)+H^-(t) \tag{8.3.16}$$

with

$$H^+(t)=\frac{1}{2\pi i}\int_{-i\epsilon-\infty}^{-i\epsilon+\infty}\frac{H(u)}{u-t}du,\quad H^-(t)=-\frac{1}{2\pi i}\int_{i\epsilon-\infty}^{i\epsilon+\infty}\frac{H(u)}{u-t}du, \tag{8.3.17}$$

where $H^+(t)$ is holomorphic in the upper half-plane ($\mathfrak{Im}\{t\} > -\epsilon$) and $H^-(t)$ is holomorphic in the lower half-plane ($\mathfrak{Im}\{t\} < \epsilon$).

Now, Eqn. (8.3.14) can be written as

$$\sqrt{k-t}\,F_n(t,k\alpha_0)+H^+(t)+H^-(t)=h^-(t) \tag{8.3.18}$$

Since the function $\sqrt{k-t}\,F_n(t,k\alpha_0)$ is holomorphic in the upper half-plane, we find from Eqn. (8.3.18) that $h^-(t) = H^-(t)$ and

$$F_n(t,k\alpha_0)=-\frac{H^+(t)}{\sqrt{k-t}}. \tag{8.3.19}$$

Substituting Eqn. (8.3.15) into Eqn. (8.3.17) for $H^+(t)$ and setting $\epsilon = 0$, we finally obtain,

$$\underset{n=2,3,\ldots}{F_n(t,k\alpha_0)}=-\frac{1}{2\pi i\sqrt{k-t}}\int_{-\infty}^{\infty}\frac{\sqrt{k-u}}{u-\overset{\frown}{t}}F_{n-1}(-u,-k\alpha_0)e^{-i2lu}du. \tag{8.3.20}$$

Thus,

$$F_1(k\alpha,k\alpha_0)=-\frac{1}{2\pi ik}\sqrt{\frac{1+\alpha_0}{1-\alpha}}\,\frac{e^{i\kappa\alpha_0}}{\alpha+\widehat{\alpha_0}}, \tag{8.3.21}$$

$$F_2(k\alpha,k\alpha_0)=-\frac{e^{-i\kappa\alpha_0}}{k(2\pi i)^2}\sqrt{\frac{1-\alpha_0}{1-\alpha}}\int_{-\infty}^{\infty}\frac{\mathrm{h}(t)dt}{(t+\widehat{\alpha})(t-\widehat{\alpha_0})}, \tag{8.3.22}$$

$$\underset{n=3,4,\ldots}{F_n(k\alpha,k\alpha_0)}=-\frac{e^{-i\kappa\beta_n}}{k(2\pi i)^n}\sqrt{\frac{1-\beta_n}{1-\alpha}}\int_{-\infty}^{\infty}\frac{h(t_1)dt_1}{t_1-\widehat{\beta_n}}\int_{-\infty}^{\infty}\frac{h(t_2)dt_2}{t_2+\widehat{t_1}}\times \tag{8.3.23}$$

$$\times\int_{-\infty}^{\infty}\frac{h(t_3)dt_3}{t_3+\widehat{t_2}}\cdots\int_{-\infty}^{\infty}\frac{h(t_{n-2})dt_{n-2}}{t_{n-2}+\widehat{t_{n-3}}}\int_{-\infty}^{\infty}\frac{h(t_{n-1})dt_{n-1}}{(t_{n-1}+\widehat{t_{n-2}})(t_{n-1}+\widehat{\alpha})}, \qquad (8.3.23, \text{cont.})$$

where

$$h(t)=\sqrt{\frac{1+t}{1-t}}e^{iqt}, \qquad (8.3.24)$$

$$\beta_n=(-1)^n\alpha_0, \quad q=2\kappa=2kl, \qquad (8.3.25)$$

and the $\frown$ symbol indicates the contour of integration skirts above the pole. In evaluating the square root of $(1 \pm t)^{1/2}$ the branch is chosen such that it is equal to 1 for $t = 0$. In the integrals of Eqns. (8.3.22) and (8.3.23) the contour of integration skirts above the branch point at $t = -1$ and below the branch point at $t = 1$.

§8.4 Scattering Pattern and the Edge Wave Equation

According to Eqn. (8.3.6) the currents $j_n(z+l, \alpha_0)$ and $j_n(l-z, -\alpha_0)$ respectively radiate the fields

$$\left.\begin{aligned} A_{nz}&=i\int_{-\infty}^{\infty}F(w,k\alpha_0)e^{iw(z+l)}\frac{e^{ivy}}{v}dw, \\ A_{nz}&=i\int_{-\infty}^{\infty}F(-w,-k\alpha_0)e^{iw(z-l)}\frac{e^{ivy}}{v}dw, \end{aligned}\right\} y\geqslant 0. \qquad (8.4.1)$$

Given that

$$H_x = -\frac{\partial A_z}{\partial y}, \tag{8.4.2}$$

the scattered field for $y \geqslant 0$ is

$$H_x = e^{ik\left(-z\alpha_0 + y\sqrt{1-\alpha_0^2}\right)} - \\ -\int_{-\infty}^{\infty} \sum_{n=1}^{\infty} \left[F_n(w, k\alpha_0) e^{iwl} + F_n(-w, -k\alpha_0) e^{-iwl} \right] e^{i(wz+vy)} dw. \tag{8.4.3}$$

Setting

$w = k\cos\tau$, $v = k\sin\tau$, $z = r\cos\vartheta$, $y = r\sin\vartheta$,

then

$$H_x = \epsilon e^{ik\left(-z\alpha_0 + y\sqrt{1-\alpha_0^2}\right)} + \\ +k \int_{\vartheta+0-i\infty}^{\vartheta-0+i\infty} \sum_{n=1}^{\infty} \left[F_n(k\cos\tau, k\alpha_0) e^{ikl\cos\tau} + \right. \\ \left. + F_n(-k\cos\tau, -k\alpha_0) e^{-ikl\cos\tau} \right] e^{ikr\cos(\tau-\vartheta)} \sin\tau \, d\tau, \tag{8.4.4}$$

where

$$\epsilon = \begin{cases} 1 & \text{inside the region} \\ 0 & \text{outside the region} \end{cases} -l - y\cot\vartheta_0 \leqslant z \leqslant l - y\cot\vartheta_0, \; y \geqslant 0. \tag{8.4.5}$$

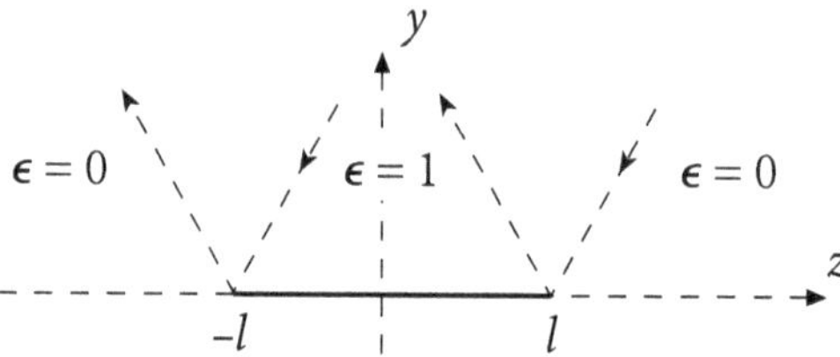

Figure 8.4.1: Region of space ($\epsilon = 1$) where strip diffraction is provided by the geometrical optics solution.

The region where $\epsilon = 1$, is illustrated in Fig. 8.4.1 and is the region corresponding to the geometrical optics reflected beam.

Applying the method of steepest descent [78] to Eqn. (8.4.4), the far field is determined to be

$$H_x = 2\pi ik\sqrt{1-\alpha^2}\sum_{n=1}^{\infty}\Big[F_n(k\alpha, k\alpha_0)e^{i\kappa\alpha} + F_n(-k\alpha, -k\alpha_0)e^{-i\kappa\alpha}\Big]\frac{e^{i(kr+\pi/4)}}{\sqrt{2\pi kr}}, \tag{8.4.6}$$

where $\alpha = \cos\vartheta$. Thus, F_n is the scattering pattern of the nth-order edge wave. An introduction of the functions

$$\Phi_n(\alpha, \alpha_0) = 2\pi ik\sqrt{1-\alpha^2}\,F_n(k\alpha, k\alpha_0) \tag{8.4.7}$$

and

$$\tilde{\Phi}(\alpha, \alpha_0) = \sum_{n=1}^{\infty}\Phi_n(\alpha, \alpha_0), \tag{8.4.8}$$

will allow the scattered field to be written in the form of Eqns. (8.2.11) and (8.2.12). According to Eqn. (8.3.20), Φ_n satisfies the recursion relationship

$$\left.\begin{aligned}\Phi_n(\alpha,\alpha_0)&=\frac{\sqrt{1+\alpha}}{2\pi i}\int_{-\infty}^{\infty}\frac{\Phi_{n-1}(u,-\alpha_0)}{\sqrt{1-u}\left(u+\widehat{\alpha}\right)}e^{iqu}du,\\ \Phi_n(-\alpha,-\alpha_0)&=\frac{\sqrt{1-\alpha}}{2\pi i}\int_{-\infty}^{\infty}\frac{\Phi_{n-1}(u,\alpha_0)}{\sqrt{1-u}\left(u-\widehat{\alpha}\right)}e^{iqu}du.\end{aligned}\right\}\tag{8.4.9}$$

It is evident that the series given by Eqn. (8.4.8) is a Neumann series for the following integral equation of the second kind

$$\tilde{\Phi}(\alpha,\alpha_0)=\Phi_1(\alpha,\alpha_0)+\frac{\sqrt{1+\alpha}}{2\pi i}\int_{-\infty}^{\infty}\frac{\tilde{\Phi}(u,-\alpha_0)}{\sqrt{1-u}\left(u+\widehat{\alpha}\right)}e^{iqu}du,\tag{8.4.10}$$

where

$$\Phi_1(\alpha,\alpha_0)=-\frac{\sqrt{1+\alpha}\sqrt{1+\alpha_0}}{\alpha+\widehat{\alpha}_0}e^{i\kappa\alpha_0}.\tag{8.4.11}$$

This equation [for $\tilde{\Phi}(-\alpha,\alpha_0)$] was first obtained by [124] using other methods and was utilized later in [122, 125, 134] for an asymptotic solution of the problem. However, the significance of this equation has remained unclear. Our derivation of this equation provides a clear physical meaning: the nth term in the Neumann series for Eqn. (8.4.10) is the scattering pattern of the nth-order edge wave. Therefore, Eqn. (8.4.10) may be expediently called the *edge wave equation*.

§8.5 Infinite Series for the Current and Its Properties

Equations (8.3.1), (8.3.2), (8.3.21), (8.3.22), and (8.3.23) can be used to represent the current in the form of a series

$$\tilde{J}(z,\alpha_0) = -e^{ikz}\sum_{n=1}^{\infty}(-1)^n \chi_n(z,-\beta_n)e^{iq(n-1)-i\kappa\beta_n}, \tag{8.5.1}$$

where

$$\beta_n = (-1)^n\alpha_0, \quad \chi_1(z,\alpha)e^{ikz} = -\frac{\sqrt{1+\alpha}}{2\pi i}\int_{-\infty}^{\infty}\frac{e^{ikzt}}{\sqrt{1-t}}\frac{dt}{t+\widehat{\alpha}}, \tag{8.5.2}$$

$$\chi_2(z,\alpha)e^{iq+ikz} = \frac{\sqrt{1+\alpha}}{(2\pi i)^2}\int_{-\infty}^{\infty}\frac{h(t_1)dt_1}{t_1+\widehat{\alpha}}\int_{-\infty}^{\infty}\frac{e^{ikzt_2}}{\sqrt{1-t_2}}\frac{dt_2}{(t_2+\widehat{t_1})}, \tag{8.5.3}$$

$$\begin{aligned}\underset{n=3,4,5,\ldots}{\chi_n(z,\alpha)e^{iq(n-1)+ikz}} &= (-1)^n\frac{\sqrt{1+\alpha}}{(2\pi i)^n}\int_{-\infty}^{\infty}\frac{h(t_1)dt_1}{t_1+\widehat{\alpha}}\int_{-\infty}^{\infty}\frac{h(t_2)dt_2}{t_2+\widehat{t_1}}\times \\ &\times\int_{-\infty}^{\infty}\frac{h(t_3)dt_3}{t_3+\widehat{t_2}}\cdots\int_{-\infty}^{\infty}\frac{h(t_{n-1})dt_{n-1}}{t_{n-1}+\widehat{t_{n-2}}}\int_{-\infty}^{\infty}\frac{e^{ikzt_n}}{\sqrt{1-t_n}}\frac{dt_n}{(t_n+\widehat{t_{n-1}})}.\end{aligned} \tag{8.5.4}$$

Let us transform χ_n into the form

$$\chi_1(z,\alpha) = \frac{\sqrt{1+\alpha}}{i\pi}\int_{1}^{1+i\infty}\frac{e^{ikz(t-1)}}{\sqrt{1-t}}\frac{dt}{t+\alpha} \quad \text{for} \quad z>0, \tag{8.5.5}$$

$$\chi_2(z,\alpha)e^{iq+ikz} = -\frac{\sqrt{1+\alpha}}{2(i\pi)^2}\int_1^{1+i\infty}\frac{h(t_1)dt_1}{t_1+\alpha}\int_{-\infty}^{\infty}\frac{e^{ikzt_2}}{\sqrt{1-t_2}}\frac{dt_2}{(t_2+\widehat{t_1})}, \qquad (8.5.6)$$

$$\underset{n=3,4,5,\ldots}{\chi_n(z,\alpha)}e^{iq(n-1)+ikz} = -\frac{\sqrt{1+\alpha}}{2(i\pi)^n}\int_1^{1+i\infty}\frac{h(t_1)dt_1}{t_1+\alpha}\int_1^{1+i\infty}\frac{h(t_2)dt_2}{t_2+t_1}\int_1^{1+i\infty}\frac{h(t_3)dt_3}{t_3+t_2}\cdots$$
$$\cdots\int_1^{1+i\infty}\frac{h(t_{n-1})dt_{n-1}}{t_{n-1}+t_{n-2}}\int_{-\infty}^{\infty}\frac{e^{ikzt_n}}{\sqrt{1-t_n}}\frac{dt_n}{(t_n+\widehat{t_{n-1}})}, \qquad (8.5.7)$$

where the contour of integration between the limits from 1 to $1+i\infty$ is on the left side of the branch cut $1 \to 1+i\infty$, where $(1-t)^{1/2} = |(1-t)^{1/2}|\exp(-i\pi/4)$ and $(t-1)^{1/2} = |(t-1)^{1/2}|\exp(-i3\pi/4)$.

From Eqns. (8.5.2), (8.5.6), and (8.5.7) it is easy to see that

$$\left.\begin{aligned}&\chi_1(z,\alpha)=e^{-ikz(1+\alpha)},\\&\chi_n(z,\alpha)=\chi_{n-1}(2l-z,\alpha)e^{-2ikz}\end{aligned}\right\}\text{for}\quad z\leqslant 0. \qquad (8.5.8)$$

From here, in particular, it follows that

$$\chi_1(0,\alpha)=1,\quad \chi_n(0,\alpha)=\chi_{n-1}(2l,\alpha). \qquad (8.5.9)$$

For $z > 0$, χ_n can be presented as

$$\chi_1(z,\alpha)=\frac{2}{\sqrt{\pi}}e^{-i\frac{\pi}{4}}e^{-ikz(1+\alpha)}\int_{\sqrt{kz(1+\alpha)}}^{\infty}e^{it^2}dt, \qquad (8.5.10)$$

$$\chi_2(z,\alpha)e^{iq+ikz}=\frac{\sqrt{1+\alpha}}{(i\pi)^2}\int_1^{1+i\infty}\frac{h(t_1)dt_1}{t_1+\alpha}\int_1^{1+i\infty}\frac{e^{ikzt_2}}{\sqrt{1-t_2}}\frac{dt_2}{(t_2+t_1)}, \tag{8.5.11}$$

$$\underset{n=3,4,\ldots}{\chi_n(z,\alpha)}e^{iq(n-1)+ikz}=\frac{\sqrt{1+\alpha}}{(i\pi)^n}\int_1^{1+i\infty}\frac{h(t_1)dt_1}{t_1+\alpha}\int_1^{1+i\infty}\frac{h(t_2)dt_2}{t_2+t_1}\int_1^{1+i\infty}\frac{h(t_3)dt_3}{t_3+t_2}\ldots$$
$$\ldots\int_1^{1+i\infty}\frac{h(t_{n-1})dt_{n-1}}{t_{n-1}+t_{n-2}}\int_1^{1+i\infty}\frac{e^{ikzt_n}}{\sqrt{1-t_n}}\frac{dt_n}{(t_n+t_{n-1})}. \tag{8.5.12}$$

The transformation of Eqn. (8.5.5) into Eqn. (8.5.10) is shown in § 8.8.

Let us now show that χ_n also exhibits the following property

$$\chi_1(z,-1)=1,\quad \chi_n(z,-1)=\chi_{n-1}(z,1),\quad n=2,3,\ldots. \tag{8.5.13}$$

The equality $\chi_1(z,-1)=1$ is evident. Therefore, one can immediately proceed to χ_n, where $n=2,3\ldots$. We will write χ_n for $z\geq 0$ in the form

$$\chi_n(z,\alpha)e^{iq(n-1)+ikz}=-\frac{\sqrt{1+\alpha}}{2(i\pi)^n}\times$$
$$\times\int_1^{1+i\infty}\frac{e^{ikzt_n}dt_n}{\sqrt{1-t_n}}\int_1^{1+i\infty}\frac{h(t_{n-1})dt_{n-1}}{t_{n-1}+t_n}\int_1^{1+i\infty}\frac{h(t_{n-2})dt_{n-1}}{t_{n-2}+t_{n-1}}\ldots \tag{8.5.14}$$
$$\ldots\int_1^{1+i\infty}\frac{h(t_2)dt_2}{t_2+t_3}\int_{-\infty}^{\infty}\frac{h(t_1)}{(t_1+\widehat{\alpha})}\frac{dt_1}{(t_1+\widehat{t_2})},$$

and examine the limit

$$p = \lim_{\alpha \to -1} \frac{\sqrt{1+\alpha}}{2\pi i}(-1) \int_{-\infty}^{\infty} \frac{h(t_1)}{(t_1 + \widehat{\alpha})} \frac{dt_1}{(t_1 + \widehat{t_2})} = \tag{8.5.15}$$

$$= -\frac{1}{2\pi i} \lim_{\alpha \to -1} \sqrt{1+\alpha} \int_{-\infty}^{\infty} \frac{\sqrt{(1+t_1)/(1-t_1)}\, e^{iqt_1}}{(t_1 + \widehat{\alpha})(t_1 + \widehat{t_2})} dt_1 .$$

We will transform the integral in Eqn. (8.5.15) into an integral over the left side of the branch cut $1 \to 1 + i\infty$ and separate it into the parts, such that

$$p = \lim_{\alpha \to -1} \frac{\sqrt{1+\alpha}}{i\pi} \left(\int_{1}^{1+i\delta} + \int_{1+i\delta}^{1+i\infty} \right),$$

where δ is a sufficiently small positive number. It is evident that

$$\lim_{\alpha \to -1} \frac{\sqrt{1+\alpha}}{i\pi} \int_{1+i\delta}^{1+i\infty} \frac{h(t_1)}{(t_1 + \alpha)} \frac{dt_1}{(t_1 + t_2)} = 0;$$

therefore, p can be written as

$$p = -\frac{\sqrt{2}e^{iq}}{2\pi i(1+t_2)} \lim_{\alpha \to -1} \sqrt{1+\alpha} \int_{-\infty}^{\infty} \frac{dt_1}{\sqrt{1-t_1}\,(t_1 + \widehat{\alpha})} . \tag{8.5.16}$$

Applying the Cauchy integral theorem to Eqn. (8.5.16), we obtain

$$p = \frac{\sqrt{2}e^{iq}}{1+t_2} . \tag{8.5.17}$$

As a result, we obtain

$$\chi_n(z,-1)=\chi_{n-1}(z,1), \quad n=2,3,\ldots$$

which was to be shown.

Using the established properties for χ_n, it is possible to show that the total current given by Eqn. (8.2.9) satisfies the conditions that

$$J(z,\alpha_0)=0 \quad \text{for} \quad |z|>l, \tag{8.5.18}$$

$$J(\pm l,\alpha_0)=0, \tag{8.5.19}$$

$$J(z,\pm 1)=0. \tag{8.5.20}$$

Equations (8.5.18) and (8.5.19) show that the current is equal to zero outside the strip and at its edges. Equation (8.5.20) follows from the fact that a scattered field does not result from an incident field of Eqn. (8.2.1) at grazing angles of incidence.

§8.6 Convergence of Infinite Series for the Current

This section will study the convergence of Eqn. (8.5.1). Accomplishing this requires the use Eqn. (8.5.12) for χ_n. We will change the contour of integration of this equation from the left side of the branch cut $1 \to 1+i\infty$ to a contour on the right side and will displace the contour of integration along the real axis. As a result, the following expression is obtained

$$\begin{aligned}\chi_n(z,\alpha)e^{iq(n-1)+ikz} &= \frac{\sqrt{1+\alpha}}{\pi^n}\int_1^\infty \sqrt{\frac{1+t_1}{t_1-1}}\frac{e^{iqt_1}dt_1}{(t_1+\alpha)}\int_1^\infty \sqrt{\frac{1+t_2}{t_2-1}}\frac{e^{iqt_2}dt_1}{(t_2+t_1)}\cdots \\ &\cdots\int_1^\infty \sqrt{\frac{1+t_{n-1}}{t_{n-1}-1}}\frac{e^{iqt_{n-1}}dt_{n-1}}{(t_{n-1}+t_{n-2})}\int_1^\infty \frac{e^{ikzt_n}}{\sqrt{t_n-1}}\frac{dt_n}{(t_n+t_{n-1})}.\end{aligned} \tag{8.6.1}$$

Setting $t_m = 1 + s_m$:

$$\chi_n(z,\alpha) = \frac{\sqrt{1+\alpha}}{\pi^n} \int_0^\infty \sqrt{\frac{2+s_1}{s_1}} \frac{e^{iqs_1}\,ds_1}{(s_1+\alpha+1)} \int_0^\infty \sqrt{\frac{2+s_2}{s_2}} \frac{e^{iqs_2}\,ds_2}{(s_2+s_1+2)} \cdots$$
$$\cdots \int_0^\infty \sqrt{\frac{2+s_{n-1}}{s_{n-1}}} \frac{e^{iqs_{n-1}}\,ds_{n-1}}{(s_{n-1}+s_{n-2}+2)} \int_0^\infty \frac{e^{ikzs_n}}{\sqrt{s_n}} \frac{ds_n}{(s_n+s_{n-1}+2)}. \quad (8.6.2)$$

Utilizing a well known technique [139] we set

$$\left.\begin{aligned} s_m &= x_m + p_m \frac{\pi}{q}, \quad m = 1,2,3,\ldots,n-1 \\ s_n &= x_n + p_n \frac{\pi}{kz}, \quad p_{m,n} = 0,1,2,3,\ldots. \end{aligned}\right\} \quad (8.6.3)$$

Then Eqn. (8.6.2) can be written as

$$\chi_n(z,\alpha) = \frac{\sqrt{1+\alpha}}{\pi^n} \int_0^{\pi/q} e^{iqx_1}dx_1 \int_0^{\pi/q} e^{iqx_2}dx_2 \ldots \int_0^{\pi/q} e^{iqx_{n-1}}dx_{n-1} \times$$
$$\times \sum_{p_1=0}^{\infty} \sum_{p_2=0}^{\infty} \cdots \sum_{p_{n-1}=0}^{\infty} (-1)^{p_1+p_2+\ldots+p_{n-1}} \times$$
$$\times y_1(p_1,\alpha) y_2(p_2,p_1) \ldots y_{n-1}(p_{n-1},p_{n-2}) \times \quad (8.6.4)$$
$$\times \int_0^{\frac{\pi}{kz}} e^{ikzx_n} \sum_{p_n=0}^{\infty} (-1)^{p_n} y_n(p_n,p_{n-1}) dx_n,$$

where

$$\left.\begin{aligned}
&y_1(p_1,\alpha)=\left(\frac{2+x_1+p_1\pi/q}{x_1+p_1\pi/q}\right)^{1/2}\frac{1}{x_1+p_1\pi/q+\alpha+1},\\
&\underset{i=2,3,\ldots,n-1}{y_i(p_i,p_{i-1})}=\left(\frac{2+x_i+p_i\pi/q}{x_i+p_i\pi/q}\right)^{1/2}\frac{1}{x_i+x_{i-1}+2+(p_i+p_{i-1})\pi/q},\\
&y_n(p_n,p_{n-1})=\left(x_n+p_n\frac{\pi}{kz}\right)^{-1/2}\frac{1}{x_n+x_{n-1}+2+p_{n-1}\pi/q+p_n\pi/kz}.
\end{aligned}\right\}\quad(8.6.5)$$

The terms in each series in Eqn. (8.6.4) alternate in sign and monotonically decrease with increasing order. Therefore, it is possible to write

$$\left.\begin{aligned}
&\sum_{p_n=0}^{\infty}(-1)^{p_n}y_n(p_n,p_{n-1})<y_n(0,p_{n-1}),\\
&\sum_{p_{n-1}=0}^{\infty}(-1)^{p_{n-1}}y_{n-1}(p_{n-1},p_{n-2})\sum_{p_n=0}^{\infty}(-1)^{p_n}y_n(p_n,p_{n-1})<\\
&\qquad<y_{n-1}(0,p_{n-2})\,y_n(0,0)
\end{aligned}\right\}\quad(8.6.6)$$

and

$$\begin{aligned}
&\sum_{p_1=0}^{\infty}\sum_{p_2=0}^{\infty}\cdots\sum_{p_n=0}^{\infty}(-1)^{p_1+p_2+\ldots+p_n}y_1(p_1,\alpha)y_2(p_2,p_1)\ldots y_n(p_n,p_{n-1})<\\
&\qquad<y_1(0,\alpha)\,y_2(0,0)\ldots y_n(0,0).
\end{aligned}\quad(8.6.7)$$

Noting that,

$$|\chi_n(z,\alpha)| \leqslant \frac{\sqrt{1+\alpha}}{\pi^n} \int_0^{\pi/q} dx_1 \int_0^{\pi/q} dx_2 \dots \int_0^{\pi/q} dx_{n-1} \int_0^{\frac{\pi}{kz}} dx_n \times$$

$$\times \left| \sum_{p_1=0}^{\infty} \sum_{p_2=0}^{\infty} \dots \sum_{p_n=0}^{\infty} (-1)^{p_1+p_2+\dots+p_n} \times \right. \tag{8.6.8}$$

$$\left. \times y_1(p_1,\alpha) y_2(p_2,p_1) \dots y_{n-1}(p_{n-1},p_{n-2}) y_n(p_n,p_{n-1}) \right|$$

and noting the inequality of Eqn. (8.6.7), we obtain

$$|\chi_n(z,\alpha)| < \frac{\sqrt{1+\alpha}}{\pi^n} \int_0^{\pi/q} dx_1 \int_0^{\pi/q} dx_2 \dots$$

$$\dots \int_0^{\pi/q} dx_{n-1} \int_0^{\frac{\pi}{kz}} dx_n \, y_1(0,\alpha) y_2(0,0) \dots y_n(0,0). \tag{8.6.9}$$

Here,

$$\int_0^{\frac{\pi}{kz}} y_n(0,0)\,dx_n = \int_0^{\frac{\pi}{kz}} \frac{dx_n}{\sqrt{x_n}\,(x_n+x_{n-1}+2)} = \frac{2}{\sqrt{2+x_{n-1}}} \tan^{-1} \sqrt{\frac{\pi}{kz(2+x_{n-1})}} <$$

$$< \frac{2}{\sqrt{2+x_{n-1}}} \tan^{-1} \sqrt{\frac{\pi}{2kz}} \;,$$

$$\int_0^{\frac{\pi}{q}} \frac{dx_i}{\sqrt{x_i}\left(x_i+x_{i-1}+2\right)} < \frac{2}{\sqrt{x_{i-1}+2}} \tan^{-1}\sqrt{\frac{\pi}{2q}},$$

$$\frac{\sqrt{1+\alpha}}{\pi}\int_0^{\frac{\pi}{q}} \frac{dx_1}{\sqrt{x_1}\left(x_1+\alpha+1\right)} < \frac{2}{\pi}\tan^{-1}\sqrt{\frac{\pi}{q(\alpha+1)}} < 1.$$

Therefore,

$$\left|\chi_n(z,\alpha)\right| < \frac{2}{\pi}\tan^{-1}\sqrt{\frac{\pi}{2kz}}\left(\frac{2}{\pi}\tan^{-1}\sqrt{\frac{\pi}{2q}}\right)^{n-2}. \tag{8.6.10}$$

Thus, the series determined by Eqn. (8.5.1) has a majorant geometric progression with a ratio of

$$\frac{2}{\pi}\tan^{-1}\sqrt{\frac{\pi}{2q}} < 1, \tag{8.6.11}$$

and therefore, converges absolutely and uniformly for any value of $q = 2kl > 0$.

§8.7 Integral Equation for the Current and Schwarzschild's Solution

8.7.1 Integral Equation Resulting from the Solution of Functional Equations (8.3.10)

A Fourier transform will be applied to Eqn. (8.3.20) using the expression

$$F_{n-1}\left(-u,-k\alpha_0\right) = \frac{1}{2\pi}\int_{-\infty}^{\infty} j_{n-1}\left(s,-\alpha_0\right)e^{isu}ds. \tag{8.7.1}$$

As a result, we obtain a recursive formula for the current

$$j_n(z,\alpha_0) = -\frac{1}{\pi}\int_{-\infty}^{\infty} K(s-2l,z)\, j_{n-1}(s,-\alpha_0)\,ds. \tag{8.7.2}$$

where

$$K(x,z) = \frac{1}{4\pi i}\int_{-\infty}^{\infty} \frac{e^{izt}}{\sqrt{k-t}}dt \int_{-\infty}^{\infty} \frac{\sqrt{k-u}}{u-t} e^{ixu}\,du. \tag{8.7.3}$$

Noting that

$$K(x,z) = 0 \quad \text{for} \quad x < 0, \tag{8.7.4}$$

and one can transform Eqn. (8.7.2) into the form

$$j_n(z,\alpha_0) = -\frac{1}{\pi}\int_{0}^{\infty} K(x,z)\, j_{n-1}(2l+x,-\alpha_0)\,dx. \tag{8.7.5}$$

It is evident that the series

$$j(z,\alpha_0) = \sum_{n=1}^{\infty} j_n(z,\alpha_0) \tag{8.7.6}$$

is a Neumann series for the integral equation of the second kind, i.e.,

$$j(z,\alpha_0) = j_1(z,\alpha_0) - \frac{1}{\pi}\int_{0}^{\infty} K(x,z)\, j(2l+x,-\alpha_0)\,dx. \tag{8.7.7}$$

8.7.2 Integral Equation Resulting from Schwarzschild's Solution

Schwarzschild [139] examined the diffraction of a plane wave

$$E_x = -e^{-iky-i\omega t}, \quad H_x = 0 \tag{8.7.8}$$

from a slot $-l \leqslant z \leqslant l$ in the plane of a perfectly conducting screen at $y = 0$. Setting the component of the scattered field, $E_x = u$ in the slot at $y = 0^-$, we obtain the following expression

$$u = \frac{1}{\pi}\int_0^\infty \left[G(r_0', \pi, r) + G(r_0, \pi, r)\right]\bar{u}(r)dr, \tag{8.7.9}$$

where

$$G(r_0, \pi, r) = 2\sqrt{\frac{r_0}{r}}\frac{e^{ik(r+r_0)}}{r+r_0}, \tag{8.7.10}$$

$$r_0' = z + l, \quad r_0 = l - z, \tag{8.7.11}$$

$$\left.\begin{aligned} &\bar{u}(r) = \sum_{n=1}^{\infty}\bar{u}_n(r), \quad \bar{u}_1(r) = 1, \\ &\bar{u}_{n+1}(r) = -\frac{1}{\pi}\int_0^\infty \sqrt{\frac{r+2l}{r'}}\frac{e^{ik(r+r'+2l)}}{r+r'+2l}\bar{u}_n(r')dr'. \end{aligned}\right\} \tag{8.7.12}$$

Equation (8.7.9) may be rewritten into a sum of edge waves

$$u(r) = \frac{1}{\pi}e^{ik(z+l)}\int_0^\infty \sqrt{\frac{l+z}{r}}\frac{e^{ikr}}{l+z+r}\bar{u}(r)dr + \tag{8.7.13}$$

$$+\frac{1}{\pi}e^{ik(l-z)}\int_0^{\infty}\sqrt{\frac{l-z}{r}}\frac{e^{ikr}}{l-z+r}\overline{u}(r)dr. \tag{8.7.13, cont.}$$

Applying Babinet's principle to this result, we find that at normal incidence ($\alpha_0 = 0$) the plane wave given by Eqn. (8.2.1) will excite the scattered field

$$H_x = u \quad \text{for} \quad y = 0^- \tag{8.7.14}$$

on the strip. Since,

$$H_x\big|_{y=0^+} = -H_x\big|_{y=0^-}, \tag{8.7.15}$$

the corresponding current equals

$$j_z = \frac{c}{2\pi}H_x\left(y=0^-\right) = \frac{c}{2\pi}u. \tag{8.7.16}$$

It follows that the value

$$\frac{2\pi}{c}j_z = u \tag{8.7.17}$$

can be presented in the form

$$u = j(z+l,0) + j(l-z,0), \tag{8.7.18}$$

where

$$j(z,0) = \frac{1}{\pi}e^{ikz}\int_0^{\infty}\sqrt{\frac{z}{x}}\frac{e^{ikx}}{x+z}\overline{u}(x)dx = \sum_{n=1}^{\infty}j_n(z,0). \tag{8.7.19}$$

Here,

$$j_n(z,0)=\frac{1}{\pi}e^{ikz}\int_0^\infty \sqrt{\frac{z}{x}}\frac{e^{ikx}}{x+z}\overline{u}_n(x)dx, \tag{8.7.20}$$

or

$$j_n(z,0)=-\frac{1}{\pi}\int_0^\infty \hat{K}(x,z)j_{n-1}(2l+x,0)dx, \tag{8.7.21}$$

where

$$\hat{K}(x,z)=\sqrt{\frac{z}{x}}\frac{e^{ik(z+x)}}{x+z}. \tag{8.7.22}$$

This indicates that Eqn. (8.7.19) is a Neumann series for integral equations of the second kind,

$$j(z,0)=j_1(z,0)-\frac{1}{\pi}\int_0^\infty \hat{K}(x,z)j(2l+x,0)dx, \tag{8.7.23}$$

where

$$j_1(z,0)=\chi_1(z,0)e^{ikz}=\frac{1}{\pi}\int_0^\infty \sqrt{\frac{z}{x}}\frac{e^{ik(z+x)}}{x+z}dx. \tag{8.7.24}$$

We will now show that Eqns. (8.7.7) and (8.7.23) are equivalent.

8.7.3 Equivalency of Kernels $K(x,z)$ and $\hat{K}(x,z)$

When $x \geqslant 0$ and $z \geqslant 0$ the kernel in Eqn. (8.7.3) can be written as

$$K(x,z)=\frac{1}{i\pi}\int_{k}^{k+i\infty}\frac{e^{izt}dt}{\sqrt{k-t}}\int_{k}^{k+i\infty}\frac{e^{ixu}\sqrt{k-u}du}{u-t}, \tag{8.7.25}$$

where the integration is evaluated to the left of the branch cut $k \to k+i\infty$. The first integral in Eqn. (8.7.3) is evaluated in terms of its principal value, and the contribution from the pole located on the branch cut is equal to zero. Choosing

$$t=k+i\frac{\tau_1}{z}, \quad u=k+i\frac{\tau_2}{x},$$

we obtain

$$K(x,z)=\frac{1}{\pi}\sqrt{\frac{z}{x}}e^{ik(z+x)}\int_0^\infty\int_0^\infty\frac{e^{-(\tau_1+\tau_2)}\sqrt{\tau_2}}{\sqrt{\tau_1}(z\tau_2-x\tau_1)}d\tau_1 d\tau_2\,. \tag{8.7.26}$$

Using the variables $t_1=\tau_1$ and $t_2=\tau_1+\tau_2$, this expression possesses the form

$$K(x,z)=\frac{1}{\pi}\sqrt{\frac{z}{x}}e^{ik(z+x)}\int_0^\infty e^{-t_2}dt_2\int_0^{t_2}\sqrt{\frac{t_2-t_1}{t_1}}\frac{dt_1}{zt_2-(z+x)t_1}. \tag{8.7.27}$$

Substituting $t_1=\eta t_2$, we find that

$$K(x,z)=-\frac{1}{\pi}\sqrt{\frac{z}{x}}\frac{e^{ik(z+x)}}{z+x}\int_0^1\sqrt{\frac{1-\eta}{\eta}}\frac{d\eta}{\eta-\beta}, \tag{8.7.28}$$

where

$$\beta = \frac{z}{z+x}. \tag{8.7.29}$$

In Eqn. (8.7.28), the function $[(1-\eta)/\eta]^{1/2}$ is real and positive. For further calculation, it is useful to continue analytically this function on the whole complex plane (η). Note that the functions $\eta^{1/2}$ and $(1-\eta)^{1/2}$ are double valued.To make them single-valued functions we introduce two branch cuts on the real axis ($\mathfrak{Im}\{\eta\}=0$): the branch cut $-\infty \le \mathfrak{Re}\{\eta\} \le 0$ for $\eta^{1/2}$ and the branch cut $-\infty \le \mathfrak{Re}\{\eta\} \le 1$ for $(1-\eta)^{1/2}$. For the function $\eta^{1/2}$, we choose the branch with positive real values in the interval $0 \le \eta \le \infty$, with positive imaginary values on the upper side of the branch cut $-\infty \le \eta \le 0$ and negative imaginary values on the lower side. For the function $(1-\eta)^{1/2}$, we choose the branch with positive imaginary values in the interval $1 \le \eta \le \infty$, with positive real values on the lower side of the branch cut $-\infty \le \eta \le 1$ and negative real values on the upper side. Each of the functions $\eta^{1/2}$ and $(1-\eta)^{1/2}$ is discontinuous on the line $-\infty \le \eta \le 0$, but their combination $[(1-\eta)/\eta]^{1/2}$ is continuous. Thus, we have constructed the function $[(1-\eta)/\eta]^{1/2}$ which is single valued and analytical in the complex plane (η) with the branch cut along the section $0 \le \eta \le 1$. This function is positive real on the lower side of this branch cut and it is negative real on its upper side. Using this property of the function $[(1-\eta)/\eta]^{1/2}$ one can represent the integral of Eqn. (8.7.28) in the form of an integral along the contour C (Fig. 8.7.1), enveloping the branch cut $0 \to 1$ in a counterclockwise direction

$$\int_0^1 \sqrt{\frac{1-\eta}{\eta}} \frac{d\eta}{\eta-\beta} = \frac{1}{2} \int_C \sqrt{\frac{1-\eta}{\eta}} \frac{d\eta}{\eta-\beta}. \tag{8.7.30}$$

Note also that the residue of this integral at the pole $\eta = \beta$ equals zero because this pole is located on the branch cut. Finally deforming the contour C to infinity, we obtain

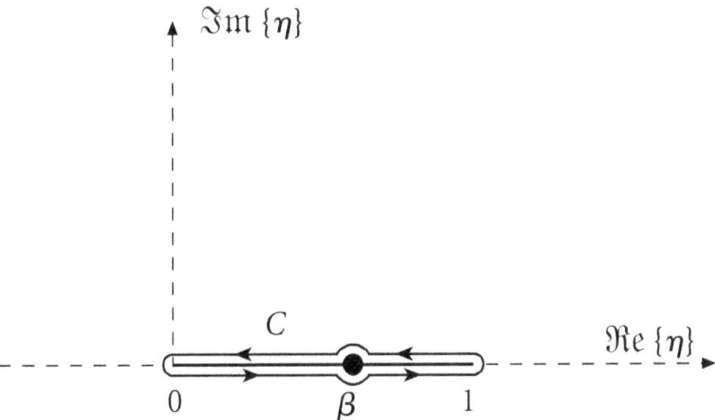

Figure 8.7.1: Contour of integration for the evaluation of Eqn. (8.7.30).

$$\frac{1}{2}\int_C \sqrt{\frac{1-\eta}{\eta}}\frac{d\eta}{\eta-\beta} = i^2\pi = -\pi \tag{8.7.31}$$

and as a result

$$K(x,z) = \sqrt{\frac{z}{x}}\frac{e^{ik(z+x)}}{z+x} = \hat{K}(x,z), \tag{8.7.32}$$

which was to be shown.

Thus, Schwarzschild's method [139] is, in essence, the method of edge waves. His solution is equivalent to our Eqn. (8.7.6) and differs only in the form written. Our solution is not only more distinctly physically meaningful, since scattered field is represented as a series of edge waves, but also is more convenient for asymptotic research.

In light of these results, it is necessary to mention a few words regarding the fortune of Schwarzschild's work. This classic work was, apparently, the first high frequency asymptotic analysis of diffraction theory and established a scientific topic almost a half century ahead of its time. In the 1950s, when this topic became a pressing issue, this work was almost forgotten (even in his native land of Germany). Thus, for instance, it was not referenced in the articles of Braunbek [140, 141], which are directly related to the same problem. In the following 15–20 years it again became acknowledged; however, despite the large amount of work investigating strip (slot) diffraction up to this point no connection had been established with Schwarzschild's research, which was considered only a phenomenon of sorts. Above, we have shown that Schwarzschild's solution, in essence, is equivalent to all solutions based on edge wave integral equations. Though Schwarzschild obtained the solution only in the case of the normal incidence, it is clear that his method is valid in the general case, for arbitrary directions of an incident wave.

§8.8 Transformation of Equation (8.5.2) into Equation (8.5.10)

According to Eqn. (8.5.2),

$$\chi_1(z,\alpha)e^{ikz} = -\frac{\sqrt{1+\alpha}}{2\pi i}\int_{-\infty}^{\infty}\frac{e^{ikzt}}{\sqrt{1-t}}\frac{dt}{t+\widehat{\alpha}}. \tag{8.8.1}$$

Here, $\alpha = \cos\vartheta, 0 \leqslant \vartheta \leqslant \pi, -1 \leqslant \alpha \leqslant 1, \sqrt{1+\alpha} = \sqrt{2}\cos(\vartheta/2) \geqslant 0$. The contour of integration is shown in Fig. 8.8.1. The line $1 \to 1+i\infty$ is the branch cut. The square root is defined as follows:

$$\sqrt{1-t} = i\sqrt{t-1} = i\left|\sqrt{t-1}\right|e^{i\varphi/2}.$$

At point A, $\varphi = 0$, $\sqrt{1-t} = i\left|\sqrt{t-1}\right|$.

At point B, $\varphi = -\pi$, $\sqrt{1-t} = i\left|\sqrt{t-1}\right|\exp(-i\pi/2) = \left|\sqrt{t-1}\right| \geqslant 0$.

At point C, $\varphi = -3\pi/2$, $\sqrt{1-t} = i\left|\sqrt{t-1}\right|\exp(-i3\pi/4) = \left|\sqrt{t-1}\right|\exp(-i\pi/4)$.

At point D, $\varphi = \pi/2$, $\sqrt{1-t} = i\left|\sqrt{t-1}\right|\exp(i\pi/4) = \left|\sqrt{t-1}\right|\exp(i3\pi/4)$.

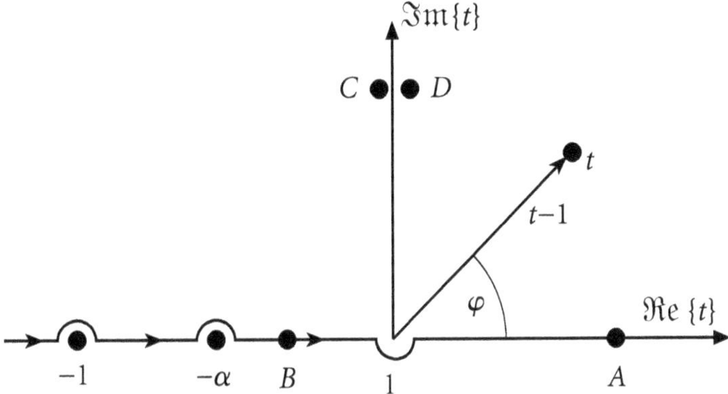

Figure 8.8.1: Contour of integration for Eqn. (8.8.1).

We note that

$$\sqrt{1-t}\Big|_C = -\sqrt{1-t}\Big|_D$$

and transform the contour of integration around the branch cut as shown in Fig. 8.8.2. Now, for $z \geq 0$

$$\frac{\sqrt{1+\alpha}}{2\pi i}\int_{-\infty}^{\infty}\frac{e^{ikz(t-1)}}{\sqrt{1-t}}\frac{dt}{t+\widehat{\alpha}} = -\frac{\sqrt{1+\alpha}}{\pi i}\int_{1}^{1+i\infty}\frac{e^{ikz(t-1)}}{i\sqrt{t-1}}\frac{dt}{t+\alpha},$$

where the last integral is taken along the left side ($\varphi = -3\pi/2$) of the branch cut. As a result,

$$\chi_1(z,\alpha) = \frac{\sqrt{1+\alpha}}{\pi i}\int_{1}^{1+i\infty}\frac{e^{i\zeta(t-1)}}{i\sqrt{t-1}}\frac{dt}{t+\alpha}, \quad \zeta = kz \tag{8.8.2}$$

with $\sqrt{t-1} = \left|\sqrt{t-1}\right|\exp(-i3\pi/4)$. Now, let us introduce a new variable, x,

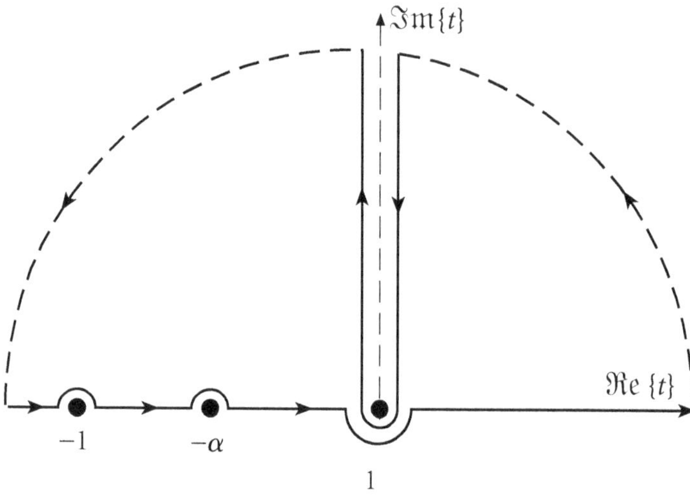

Figure 8.8.2: Contour of integration for transforming Eqn. (8.5.2) into Eqn. (8.5.10).

$$x^2 = t - 1, \quad x = \left|\sqrt{t-1}\right| e^{-i3\pi/4}$$

and obtain

$$\chi_1(z,\alpha) = -2\frac{\sqrt{1+\alpha}}{\pi} \int_0^{\infty e^{-i3\pi/4}} e^{i\zeta x^2} \frac{dx}{x^2+\alpha+1}. \tag{8.8.3}$$

After setting $s = -x$, we have

$$\chi_1(z,\alpha) = 2\frac{\sqrt{1+\alpha}}{\pi} \int_0^{\infty} e^{i\zeta s^2} \frac{ds}{s^2+\alpha+1}. \tag{8.8.4}$$

Next, substituting $\beta^2 = \alpha + 1$ leads to

$$\chi_1(z,\alpha) = 2\frac{\beta}{\pi}e^{-i\zeta\beta^2}\int_0^\infty e^{i\zeta(s^2+\beta^2)}\frac{ds}{s^2+\beta^2}. \tag{8.8.5}$$

Now, consider the integral

$$I(\zeta) = \int_0^\infty e^{i\zeta(s^2+\beta^2)}\frac{ds}{s^2+\beta^2}. \tag{8.8.6}$$

Its derivative equals

$$\frac{dI(\zeta)}{d\zeta} = i\int_0^\infty e^{i\zeta(s^2+\beta^2)}ds.$$

Since $I(\infty) = 0$, we obtain

$$I(\zeta) = i\int_\infty^\zeta d\zeta\int_0^\infty e^{i\zeta(s^2+\beta^2)}ds = i\int_\infty^\zeta e^{i\zeta\beta^2}d\zeta\int_0^\infty e^{\zeta is^2}ds =$$

$$= i\frac{\sqrt{\pi}}{2}e^{i\pi/4}\int_\infty^\zeta e^{i\zeta\beta^2}\frac{d\zeta}{\sqrt{\zeta}}. \tag{8.8.7}$$

After substituting $x^2 = \zeta\beta^2$, Eqn. (8.8.7) becomes.

$$I(\zeta)=\frac{\sqrt{\pi}}{\beta}e^{-i\pi/4}\int_{\sqrt{\zeta}\beta}^{\infty}e^{ix^2}\,dx \tag{8.8.8}$$

and

$$\chi_1(z,\alpha)=\frac{2}{\sqrt{\pi}}e^{-i\pi/4}e^{-ikz(1+\alpha)}\int_{\sqrt{kz(1+\alpha)}}^{\infty}e^{ix^2}\,dx, \tag{8.8.9}$$

which is identical to Eqn. (8.5.10).

9

Asymptotic Representation for the Current Density on a Strip

The preceding chapter showed that the solution to the problem of a strip are multiple integrals of a special type. This chapter studies the asymptotic properties of these integrals as $kl \to \infty$ and develops approximate expressions for the current.

§9.1 Lemmas on Asymptotic Series for Multiple Integrals

LEMMA 1. Let $\varphi(s_1, s_2, \ldots s_n)$ be an analytic function of variables $s_1, s_2, \ldots s_n$, regular in the neighborhood of the coordinate origin, where

$$\varphi(s_1, s_2, \ldots, s_n) = \sum_{m-0}^{\infty} \frac{1}{m!} \left[s_1 \frac{\partial}{\partial s_1} + s_2 \frac{\partial}{\partial s_2} + \ldots + s_n \frac{\partial}{\partial s_n} \right]^m \varphi(0, 0, \ldots, 0) \quad (9.1.1)$$

is its Taylor series representation. For $s_1, s_2, \ldots s_n \geqslant 0$ also let

$$\left|\varphi(s_1, s_2, \ldots, s_n)\right| < K e^{b\left(s_1^2 + s_2^2 + \ldots + s_n^2\right)}, \quad (9.1.2)$$

where K and b are positive numbers independent of $s_1, s_2, \ldots s_n$.

Then,

$$\int_0^\infty \ldots \int_0^\infty e^{-q\left(s_1^2+s_2^2+\ldots s_n^2\right)} \varphi(s_1, s_2, \ldots, s_n) ds_1 ds_2 \ldots ds_n \underset{q\to\infty}{\sim}$$

$$\underset{q\to\infty}{\sim} \sum_{m=0}^{\infty} \frac{1}{m!} \int_0^\infty \ldots \int_0^\infty e^{-q\left(s_1^2+s_2^2+\ldots s_n^2\right)} \times \tag{9.1.3}$$

$$\times \left[s_1 \frac{\partial}{\partial s_1} + s_2 \frac{\partial}{\partial s_2} + \ldots + s_n \frac{\partial}{\partial s_n} \right]^m \varphi(0,0,\ldots,0) ds_1 ds_2 \ldots ds_n .$$

Proof

Let us present the left side of Eqn. (9.1.3) in the form

$$\int_0^\infty \ldots \int_0^\infty e^{-q\left(s_1^2+s_2^2+\ldots s_n^2\right)} \times$$

$$\times \sum_{m=0}^{M-1} \frac{1}{m!} \left[s_1 \frac{\partial}{\partial s_1} + s_2 \frac{\partial}{\partial s_2} + \ldots + s_n \frac{\partial}{\partial s_n} \right]^m \varphi(0,0,\ldots,0) ds_1 ds_2 \ldots ds_n + R_M, \tag{9.1.4}$$

where

$$R_M = \int_0^\infty \ldots \int_0^\infty e^{-q\left(s_1^2+s_2^2+\ldots s_n^2\right)} ds_1 ds_2 \ldots ds_n \times$$

$$\times \left\{ \varphi(s_1, s_2, \ldots, s_n) - \sum_{m=0}^{M-1} \frac{1}{m!} \left[s_1 \frac{\partial}{\partial s_1} + s_2 \frac{\partial}{\partial s_2} + \ldots + s_n \frac{\partial}{\partial s_n} \right]^m \varphi(0,0,\ldots,0) \right\}. \tag{9.1.5}$$

It is possible to choose a constant $C > 0$ such that

$$\left|\varphi(s_1, s_2, \ldots, s_n) - \sum_{m=0}^{M-1} \frac{1}{m!}\left[s_1 \frac{\partial}{\partial s_1} + s_2 \frac{\partial}{\partial s_2} + \ldots + s_n \frac{\partial}{\partial s_n}\right]^m \varphi(0,0,\ldots,0)\right| \leqslant \tag{9.1.6}$$

$$\leqslant C(s_1 + s_2 + \ldots + s_n)^M e^{b\left(s_1^2 + s_2^2 + \ldots s_n^2\right)}.$$

Then,

$$|R_M| \leqslant C \int_0^\infty \ldots \int_0^\infty e^{-(q-b)\left(s_1^2 + s_2^2 + \ldots s_n^2\right)} (s_1 + s_2 + \ldots + s_n)^M ds_1 ds_2 \ldots ds_n =$$

$$= C(q-b)^{-\frac{M+n}{2}} \int_0^\infty \ldots \int_0^\infty e^{-\left(t_1^2 + t_2^2 + \ldots t_n^2\right)} (t_1 + t_2 + \ldots + t_n)^M dt_1 dt_2 \ldots dt_n = \tag{9.1.7}$$

$$= \frac{\text{const}}{(q-b)^{\frac{M+n}{2}}}.$$

Now, let us estimate the Mth term in the series of Eqn. (9.1.3):

$$a_M = \int_0^\infty \ldots \int_0^\infty e^{-q\left(s_1^2 + s_2^2 + \ldots s_n^2\right)} \left[s_1 \frac{\partial}{\partial s_1} + s_2 \frac{\partial}{\partial s_2} + \ldots + s_n \frac{\partial}{\partial s_n}\right]^M \times \tag{9.1.8}$$

$$\times \varphi(0,0,\ldots,0) ds_1 ds_2 \ldots ds_n =$$

$$= q^{-\frac{M+n}{2}} \int_0^\infty \dots \int_0^\infty e^{-\left(t_1^2+t_2^2+\dots t_n^2\right)} \left[t_1 \frac{\partial}{\partial t_1} + t_2 \frac{\partial}{\partial t_2} + \dots + t_n \frac{\partial}{\partial t_n} \right]^M \times$$

$$\times \varphi(0,0,\dots,0) dt_1 dt_2 \dots dt_n = \qquad (9.1.8, \text{cont.})$$

$$= \frac{\text{const}}{q^{(M+n)/2}}.$$

Therefore, it follows that

$$\lim_{q\to\infty} \frac{|R_M|}{|a_M|} = \text{const} \qquad (9.1.9)$$

and the lemma is proved. In a completely analogous manner we can prove another lemma which will be needed in the future.

Lemma 2. Let $\varphi(s_1, s_2, \dots s_n)$ be an analytic function of $s_1, s_2, \dots s_n$ possessing a Taylor series given by Eqn. (9.1.1) in the neighborhood of the coordinate origin. Also, let $q, \beta_1, \beta_2, \dots, \beta_n, K$, and b be real positive numbers, and the modulus φ satisfy Eqn. (9.1.2). Then,

$$\int_0^\infty \dots \int_0^\infty \frac{e^{-q\left(s_1^2+s_2^2+\dots s_n^2\right)}}{\left(s_1^2 - i\beta_1^2\right)\left(s_2^2 - i\beta_2^2\right)\dots\left(s_n^2 - i\beta_n^2\right)} \varphi(s_1, s_2, \dots, s_n) ds_1 ds_2 \dots ds_n \sim$$

$$\sim \sum_{m=0}^\infty \frac{1}{m!} \int_0^\infty \dots \int_0^\infty \frac{e^{-q\left(s_1^2+s_2^2+\dots s_n^2\right)} ds_1 ds_2 \dots ds_n}{\left(s_1^2 - i\beta_1^2\right)\left(s_2^2 - i\beta_2^2\right)\dots\left(s_n^2 - i\beta_n^2\right)} \times \qquad (9.1.10)$$

$$\times \left[s_1 \frac{\partial}{\partial s_1} + s_2 \frac{\partial}{\partial s_2} + \dots + s_n \frac{\partial}{\partial s_n} \right]^m \varphi(0,0,\dots,0)$$

as $q \to \infty$. Let us formulate this lemma in another form that is more convenient for our purposes.

Let $\varphi(t_1, t_2, \ldots t_n)$ be an analytic function of $t_1, t_2, \ldots t_n$ possessing in the neighborhood $t_1 = t_2 = \ldots = t_n = 1$ a Taylor series given by

$$\varphi(t_1, t_2, \ldots, t_n) = \\ = \sum_{m=0}^{\infty} \frac{1}{m!} \left[(t_1 - 1)\frac{\partial}{\partial t_1} + (t_2 - 1)\frac{\partial}{\partial t_2} + \ldots + (t_n - 1)\frac{\partial}{\partial t_n} \right]^m \varphi(1, 1, \ldots, 1). \tag{9.1.11}$$

Also, let $q, \alpha+1, \alpha_0+1, K,$ and b be real positive numbers and

$$\left|\varphi(t_1, t_2, \ldots, t_n)\right| \leqslant K e^{b(|t_1-1|+|t_2-1|+\ldots+|t_n-1|)}. \tag{9.1.12}$$

Then,

$$\int_1^{1+i\infty} \cdots \int_1^{1+i\infty} \frac{e^{iq(t_1+t_2+\ldots t_n)}}{\sqrt{1-t_1}\sqrt{1-t_2}\ldots\sqrt{1-t_n}} \times \\ \times \frac{\varphi(t_1, t_2, \ldots, t_n) dt_1 dt_2 \ldots dt_n}{(t_1+\alpha_0)(t_2+1)(t_3+1)\ldots(t_{n-1}+1)(t_n+\alpha)} \sim \\ \sim \sum_{m=0}^{\infty} \frac{1}{m!} \int_1^{1+i\infty} \cdots \int_1^{1+i\infty} \frac{e^{iq(t_1+t_2+\ldots t_n)}}{\sqrt{1-t_1}\sqrt{1-t_2}\ldots\sqrt{1-t_n}} \times \\ \times \frac{b_m(t_1, t_2, \ldots, t_n) dt_1 dt_2 \ldots dt_n}{(t_1+\alpha_0)(t_2+1)(t_3+1)\ldots(t_{n-1}+1)(t_n+\alpha)} \tag{9.1.13}$$

where

$$b_m(t_1,t_2,\ldots,t_n)= \\ =\left[(t_1-1)\frac{\partial}{\partial t_1}+(t_2-1)\frac{\partial}{\partial t_2}+\ldots+(t_n-1)\frac{\partial}{\partial t_n}\right]^m \varphi(1,1,\ldots,1), \tag{9.1.14}$$

and the integral is evaluated on the left side of the branch cut, $1 \to 1+i\infty$, where $\arg\{(1-t)^{1/2}\} = -\pi/4$.

We also recall that the differential operator in the Taylor series is

$$\left(x_1\frac{\partial}{\partial x_1}+x_2\frac{\partial}{\partial x_2}+\ldots+x_n\frac{\partial}{\partial x_n}\right)^m = \\ = \sum_{p_1+p_2+\ldots+p_n=m}\frac{m!}{p_1!p_2!\ldots p_n!}x_1^{p_1}\frac{\partial^{p_1}}{\partial x_1^{p_1}}x_2^{p_2}\frac{\partial^{p_2}}{\partial x_2^{p_2}}\ldots x_n^{p_n}\frac{\partial^{p_n}}{\partial x_n^{p_n}}, \tag{9.1.15}$$

where each index p_l ($l = 1, 2, 3,\ldots,n$) spans all integers from zero to m ($p_l = 0, 1, 2, 3,\ldots,m$), but for each term on the right hand side of Eqn. (9.1.15), the sum of all p_l must be equal to m ($p_1+p_2+\ldots+p_n = m$). In addition, $\partial^0/\partial x^0 = 1$.

One should note that the lemmas presented above can be interpreted as a generalization of Watson's lemma [78] extended here for multiple integrals.

§9.2 Asymptotic Series for χ_n

Utilizing Lemma 2, it is possible to show that as $q \to \infty$, χ_n follows a Poincaré asymptotic series

$$\chi_2(z,\alpha) \sim \chi_1(2l,\alpha)\chi_1(z,1)+\sum_{m=1}^{\infty}\frac{\varphi_q^{(m)}(q,\alpha)}{\pi i^{m+1}}\int_1^{1+i\infty}\frac{e^{ikz(t-1)}}{\sqrt{1-t}}c_m(t)dt, \tag{9.2.1}$$

where $c_m(t)$ are the Taylor series coefficients

$$\frac{\sqrt{1+x}}{x+t}=\sum_{m=0}^{\infty}c_m(t)(x-1)^m; \tag{9.2.2}$$

$$\chi_n(z,\alpha)\sim\chi_1(2l,\alpha)\chi_1^{n-2}(2l,1)\chi_1(z,1)+\frac{\sqrt{1+\alpha}}{(i\pi)^n}\sum_{m=1}^{\infty}\frac{1}{m!}\int_1^{1+i\infty}\frac{e^{ikz(t_n-1)}}{\sqrt{1-t_n}}dt_n\times$$

$$\times\int_1^{1+i\infty}\cdots\int_1^{1+i\infty}e^{iq\sum_{s=1}^{n-1}(t_s-1)}\frac{u_m(t_1,t_2,\ldots,t_n)dt_1dt_2\ldots dt_{n-1}}{(t_1+\alpha)\prod_{p=1}^{n-1}\sqrt{1-t_p}\prod_{s=2}^{n-1}(1+t_s)}, \tag{9.2.3}$$

$$u_m(t_1,t_2,\ldots,t_n)=$$

$$=\left[(t_1-1)\frac{\partial}{\partial t_1}+(t_2-1)\frac{\partial}{\partial t_2}+\ldots+(t_{n-1}-1)\frac{\partial}{\partial t_{n-1}}\right]^m f_n(1,1,\ldots,1,t_n), \tag{9.2.4}$$

and

$$\underset{n=3,4\ldots}{f_n(t_1,t_2,\ldots,t_n)}=\frac{\sqrt{1+t_1}\prod_{s=2}^{n-1}(1+t_s)^{3/2}}{(t_1+t_2)(t_2+t_3)\ldots(t_{n-1}+t_n)}. \tag{9.2.5}$$

The right hand side of Eqn. (9.2.1) and each term in the series of Eqn. (9.2.3) is the sum of the products of single integrals of the type

$$\underset{m=1,2,3,\ldots}{\varphi_q^{(m)}(q,\alpha)}=\frac{\partial^m}{\partial q^m}\varphi(q,\alpha)=i^m\frac{\sqrt{1+\alpha}}{i\pi}\int_1^{1+i\infty}\frac{e^{iq(t-1)}}{\sqrt{1-t}}\frac{(t-1)^m}{t+\alpha}dt, \tag{9.2.6}$$

and

$$\hat{\varphi}_1(z) = \frac{\sqrt{2}}{i\pi} \int_1^{1+i\infty} \frac{e^{iz(t-1)}}{\sqrt{1-t}} \frac{dt}{(t+1)} = \varphi(z,1),$$

$$\underset{m=2,3,\ldots}{\hat{\varphi}_m(z)} = \frac{\sqrt{2}}{i\pi} \int_1^{1+i\infty} \frac{e^{iz(t-1)}}{\sqrt{1-t}} \frac{dt}{(t+1)^m} = ie^{-i2z} \int_\infty^z e^{i2s_1} \hat{\varphi}_{m-1}(s_1) ds_1 = \tag{9.2.7}$$

$$= i^{m-1} e^{-2iz} \int_\infty^z ds_1 \int_\infty^{s_1} ds_2 \ldots \int_\infty^{s_{m-3}} ds_{m-2} \int_\infty^{s_{m-2}} e^{2is_{m-1}} \varphi(s_{m-1}, 1) ds_{m-1},$$

where $\varphi(q, \alpha)$ is the function

$$\varphi(q,\alpha) = \frac{\sqrt{1+\alpha}}{i\pi} \int_1^{1+i\infty} \frac{e^{iq(t-1)}}{\sqrt{1-t}} \frac{dt}{t+\alpha} \equiv \chi_1(2l,\alpha), \tag{9.2.8}$$

which, as is shown in § 8.8, is described using a Fresnel integral, namely,

$$\varphi(q,\alpha) = \frac{2}{\sqrt{\pi}} e^{-i\frac{\pi}{4}} e^{-iq(1+\alpha)} \int_{\sqrt{q(1+\alpha)}}^{\infty} e^{it^2} dt. \tag{9.2.9}$$

It is important to recall that the integrals with limits from 1 to $1+i\infty$ are evaluated on the left side of the branch cut extending from $1 \to 1+i\infty$, where $\arg\{(1-t)^{1/2}\} = -\pi/4$.

Also, one should observe that $\varphi_q^{(m)}(q,\alpha)$ and $\hat{\varphi}_m(z)$ are algebraic functions of Fresnel integrals and are also power and exponential functions of q, α, and z. For instance,

$$\hat{\varphi}_2(z)=\frac{2}{\sqrt{\pi}}e^{i\frac{\pi}{4}}e^{-i2z}\left[z\int_{\sqrt{2z}}^{\infty}e^{it^2}dt+\frac{1}{\sqrt{2}}\int_{\infty}^{z}e^{i2x}x^{\frac{1}{2}}dx\right] \tag{9.2.10}$$

$$\begin{aligned}\hat{\varphi}_3(z)=&\frac{2}{\sqrt{\pi}}e^{i\frac{3\pi}{4}}e^{-i2z}\times\\ &\times\left[\frac{1}{2}z^2\int_{\sqrt{2z}}^{\infty}e^{it^2}dt+\frac{1}{\sqrt{2}}z\int_{\infty}^{z}e^{i2x}x^{\frac{1}{2}}dx-\frac{1}{2\sqrt{2}}\int_{\infty}^{z}e^{i2x}x^{\frac{3}{2}}dx\right]\end{aligned} \tag{9.2.11}$$

where

$$\int_{\infty}^{z}e^{i2x}x^{\frac{3}{2}}dx=\frac{1}{i2}\left[z^{\frac{3}{2}}e^{i2z}-\frac{3}{2}\int_{\infty}^{z}e^{i2x}x^{\frac{1}{2}}dx\right], \tag{9.2.12}$$

$$\int_{\infty}^{z}e^{i2x}x^{\frac{1}{2}}dx=\frac{1}{i2}\left[z^{\frac{1}{2}}e^{i2z}-\frac{1}{\sqrt{2}}\int_{\infty}^{\sqrt{2z}}e^{it^2}dt\right]. \tag{9.2.13}$$

We emphasize that the asymptotic expansion of Eqn. (9.2.3) involves the functions $\hat{\varphi}_m(kz)$ with the argument kz not just z.

Here, it is pertinent to explain the reason for the appearance of functions $\varphi_q^{(m)}(q,\alpha)$ and $\hat{\varphi}_m(z)$. The functions $\varphi_q^{(m)}(q,\alpha)$ are generated by terms $(t_s-1)^m$ in the Taylor expansion of Eqn. (9.2.4). Functions $\hat{\varphi}_m(z)$ result from differentiation of Eqn. (9.2.5) with respect to the variable t_{n-1}, which is coupled with the variable t_n in the factor $(t_{n-1}+t_n)^{-1}$. The derivative

$$\frac{\partial^m}{\partial t_{n-1}^m}\left(t_{n-1}+t_n\right)^{-1}=\left(-1\right)^m\frac{m!}{\left(t_{n-1}+t_n\right)^m}$$

at the point $t_{n-1} = 1$ generates the factor $(1+t_n)^{-m}$ and the function $\hat{\varphi}_m(z)$.

The physical meaning of these asymptotic series is of interest. The first terms in Eqns. (9.2.1) and (9.2.3) are expressed in terms of Fresnel integrals and show that edge waves of higher orders diffract according to a first-order approximation as plane waves. The remaining terms in these equations are described by derivatives and integrals of Fresnel integrals and incorporate the more complex character of edge waves.

§9.3 Estimates of $\varphi_q^{(m)}(q,\alpha)$, $\varphi(kz,1)$, and $\hat{\varphi}_m(kz)$

After substituting the variable t by $x = q(t-1)$ in Eqn. (9.2.6), it becomes clear that

$$\frac{\varphi_q^{(m)}(q,\alpha)}{\sqrt{1+\alpha}} = \begin{cases} O\left[q^{-m-1/2}/(1+\alpha)\right] & \text{as} \quad q\to\infty, \quad \alpha\neq-1, \\ O\left[q^{-m+1/2}\right] & \text{as} \quad q\to\infty, \quad \alpha=-1. \end{cases} \tag{9.3.1}$$

As a result, it is possible to assume that

$$\frac{\varphi_q^{(m)}(q,\alpha)}{\sqrt{1+\alpha}} = O\left[\frac{q^{-m+1/2}}{1+q(1+\alpha)}\right] \quad \text{as} \quad q\to\infty. \tag{9.3.2}$$

This relationship holds true if

$$q^{m-1/2}(1+q\epsilon)\int_1^{1+i\infty}\frac{e^{iq(t-1)}}{\sqrt{1-t}}\frac{(t-1)^m}{t+\epsilon-1}dt = \text{const} \quad \text{as} \quad q\to\infty, \tag{9.3.3}$$

where $\epsilon = 1+\alpha$. Setting $t-1 = \epsilon s$, $\lambda = q\epsilon$, we introduce the notation

$$\phi(\lambda) = \lambda^{m-1/2}(1+\lambda)\int_0^{i\infty} e^{i\lambda s}\frac{s^{m-1/2}}{1+s}ds. \tag{9.3.4}$$

One can show that $\phi(\lambda)$ = const as $\lambda \to \infty$ and $\lambda \to 0$. Really, from the obvious expressions

$$\phi(\lambda)=\frac{1+\lambda}{\lambda}\int_0^{i\infty} e^{it}\frac{t^{m-1/2}}{1+t/\lambda}dt, \tag{9.3.5}$$

and

$$\phi(\lambda)=(1+\lambda)\int_0^{i\infty} e^{it}\frac{t^{m-1/2}}{t+\lambda}dt, \tag{9.3.6}$$

it follows that

$$\lim_{\lambda\to\infty}\phi(\lambda)=\int_0^{i\infty} e^{it}t^{m-1/2}dt \tag{9.3.7}$$

and

$$\lim_{\lambda\to 0}\phi(\lambda)=\int_0^{i\infty} e^{it}t^{m-3/2}dt. \tag{9.3.8}$$

Since, $m = 1,2,3\ldots$, these integrals have finite values. Therefore,

$$\lim_{\lambda\to\infty}\phi(\lambda)=\text{const},\quad \lim_{\lambda\to 0}\phi(\lambda)=\text{const}. \tag{9.3.9}$$

These results prove Eqn. (9.3.3) and the estimate provided by Eqn. (9.3.2)

Using Eqn. (9.2.9) and the properties of the Fresnel integral, one can obtain the asymptotic estimates

$$\varphi(q,\alpha)=O\left[\frac{1}{1+\sqrt{q(1+\alpha)}}\right] \quad \text{as} \quad q\to\infty \tag{9.3.10}$$

and

$$\varphi(kz,1)=\begin{cases} O\left[1/\sqrt{kz}\right] & \text{as} \quad kz\to\infty, \\ 1+O\left[\sqrt{kz}\right] & \text{as} \quad kz\to 0. \end{cases} \tag{9.3.11}$$

The last expression can be written as

$$\varphi(kz,1)=O\left[\frac{1+\sqrt{kz}}{1+kz}\right] \quad \text{for} \quad 0\leqslant kz\leqslant\infty. \tag{9.3.12}$$

The asymptotic behavior of the function $\hat{\varphi}_m(kz)$, $m = 1, 2,\ldots$ follows from Eqn. (9.2.7). It is clear that

$$\hat{\varphi}_m(kz)=O(1) \quad \text{for} \quad kz\to 0. \tag{9.3.13}$$

By changing the variable of integration in Eqn. (9.2.7), we have

$$\underset{m=2,3,4,\ldots}{\hat{\varphi}_m(kz)}=\frac{\sqrt{2}}{\pi\sqrt{kz}}e^{-i\pi/4}\int_0^{i\infty}\frac{e^{is}}{\sqrt{s}}\frac{ds}{\left(2+s/kz\right)^m}. \tag{9.3.14}$$

From this it follows that

$$\hat{\varphi}_m(kz)=O\left[\frac{1}{\sqrt{kz}}\right] \quad \text{for} \quad kz\to\infty. \tag{9.3.15}$$

The estimates provided by Eqns. (9.3.13) and (9.3.15) can be combined as

$$\hat{\varphi}_m(kz) = O\left[\frac{1+\sqrt{kz}}{1+kz}\right] \quad \text{for} \quad 0 \leqslant kz \leqslant \infty. \tag{9.3.16}$$

A comparison of Eqns. (9.3.12) and (9.3.16) leads to the additional relationship

$$\hat{\varphi}_m(kz) = O[\varphi(kz,1)] \quad \text{for} \quad 0 \leqslant kz \leqslant \infty. \tag{9.3.17}$$

Utilization of these estimates will obtain some useful asymptotic estimates for χ_n in the next section.

§9.4 Asymptotic Representation for χ_n

Equations (9.2.1) and (9.2.3) allow the exact calculation of all terms in the asymptotic expansion for the function $\chi_n(z,\alpha)$. This section provides some useful approximations. First, we examine the accuracy of the first term in Eqn. (9.2.3). Its error is determined by the magnitude of the term with index $m = 1$ in the series given by Eqn. (9.2.3). This term equals.

$$\begin{aligned} T_{1\chi} &= \frac{\sqrt{1+\alpha}}{(i\pi)^n} \int_1^{1+i\infty} e^{ikz(t_n-1)} \frac{dt_n}{\sqrt{1-t_n}} \times \\ &\times \int_1^{1+i\infty} \cdots \int_1^{1+i\infty} e^{iq\sum_{s=1}^{n-1}(t_s-1)} \frac{u_1(t_1,t_2,\ldots,t_{n-1},t_n)\,dt_1 dt_2 \ldots dt_{n-1}}{(t_1+\alpha)\prod_{p=1}^{n-1}\sqrt{1-t_p}\prod_{s=2}^{n-1}(1+t_s)}, \end{aligned} \tag{9.4.1}$$

where $n = 3, 4, \ldots, q \gg 1$, and

$$\begin{aligned} &u_1(t_1,t_2,\ldots,t_{n-1},t_n) = \\ &= \left[(t_1-1)\frac{\partial}{\partial t_1} + (t_2-1)\frac{\partial}{\partial t_2} + \ldots + (t_{n-1}-1)\frac{\partial}{\partial t_{n-1}}\right] f_n(1,1,\ldots,1,t_n). \end{aligned} \tag{9.4.2}$$

The function f_n is given by Eqn. (9.2.5) and contains the factor $(t_{n-1}+t_n)^{-1}$ which couples t_{n-1} with t_n. An analysis of the function u_1 reveals that it contains the following terms

$$\frac{\partial}{\partial t_1} f_n(1,1,\ldots,1,t_n) = c_1 \frac{1}{1+t_n}, \tag{9.4.3}$$

$$\frac{\partial}{\partial t_s} f_n(1,1,\ldots,1,t_n) = c_2 \frac{1}{1+t_n} \quad \text{for} \quad s = 2,3,\ldots,n-2, \tag{9.4.4}$$

$$\frac{\partial}{\partial t_{n-1}} f_n(1,1,\ldots,1,t_n) = c_3 \frac{1}{1+t_n} + c_4 \frac{1}{(1+t_n)2}, \tag{9.4.5}$$

where $c_{1,2,3,4}$ are certain constants. After substitution of the function u_1 into Eqn (9.4.1) and referring to Eqns. (9.2.6) and (9.2.7), the function $T_{1\chi}$ becomes

$$\begin{aligned} T_{1\chi} &= \varphi(kz,1)\varphi^{n-3}(q,1)\left[d_1\varphi_q^{(1)}(q,\alpha)\varphi(q,1) + d_2(n-3)\varphi(q,\alpha)\varphi_q^{(1)}(q,1)\right] + \\ &+ \varphi(q,\alpha)\varphi_q^{(1)}(q,1)\varphi^{n-3}(q,1)\left[d_3\varphi(kz,1) + d_4\hat{\varphi}_2(kz)\right] \end{aligned} \tag{9.4.6}$$

with constants $d_{1,2,3,4}$. The $(n-3)$ identical derivatives of Eqn. (9.4.4) generated the $(n-3)$ identical terms in Eqn. (9.4.6).

Using the function $T_{1\chi}$, the function $\chi_n(z,\alpha)$ can be approximated as

$$\chi_n(z,\alpha) \approx \chi_1(2l,\alpha)\chi_1^{n-2}(2l,1)\chi_1(z,1) + T_{1\chi} \tag{9.4.7}$$

where

$$\chi_1(z,\alpha) = \varphi(kz,\alpha). \tag{9.4.8}$$

The approximation given by Eqn. (9.4.7) can also be written as

$$\chi_n(z,\alpha)=\chi_1(2l,\alpha)\chi_1^{n-2}(2l,1)\chi_1(z,1)\times$$
$$\times\left\{1+O\left[\frac{\varphi_q^{(1)}(q,\alpha)}{\varphi(q,\alpha)}\right]+O\left[(n-2)\frac{\varphi_q^{(1)}(q,1)}{\varphi(q,1)}\right]+\right. \qquad (9.4.9)$$
$$\left.+O\left[\frac{\hat{\varphi}_2(kz)}{\varphi(kz,1)}\frac{\varphi_q^{(1)}(q,1)}{\varphi(q,1)}\right]\right\}.$$

Referring to the estimates provided by Eqns. (9.3.2) and (9.3.13), this expression for $\chi_n(z,\alpha)$ can be further simplified as

$$\chi_n(z,\alpha)=\chi_1(2l,\alpha)\chi_1^{n-2}(2l,1)\chi_1(z,1)\left[1+O\left(\frac{n}{q}\right)\right] \qquad (9.4.10)$$

for $q \gg n, n = 2,3,4,\ldots,$ and $-1 \le \alpha \le 1$. Subsequently, for $q \gg n$

$$\chi_n(z,\alpha)-\chi_1(2l,\alpha)\chi_1^{n-2}(2l,1)\chi_1(z,1)=\frac{O[\varphi(kz,1)]}{q^{n/2}\left[1+\sqrt{q(1+\alpha)}\right]}. \qquad (9.4.11)$$

Utilizing Eqn. (9.4.10), it is not difficult to arrive at the following asymptotic relationship

$$\left|\chi_{n+m}(z,\alpha)-\frac{\chi_n(2l,\alpha)\chi_n(z,1)}{\chi_{n-1}(2l,1)}\left[\frac{\chi_n(2l,1)}{\chi_{n-1}(2l,1)}\right]^{m-1}\right|\leqslant$$
$$\leqslant\frac{|\varphi(kz,1)|}{1+\sqrt{q(1+\alpha)}}O\left(q^{-\frac{n+m}{2}}\right) \quad \text{for } q\gg n+m. \qquad (9.4.12)$$

This inequality plays an important role in the formulations of approximate expressions for the current.

§9.5 First-Order Approximation for the Current

Chapter 8 rigorously determined that the normalized current density was

$$J(z,\alpha_0)=j^{(0)}(z,\alpha_0)+\tilde{J}(z+l,\alpha_0)+\tilde{J}(l-z,-\alpha_0), \tag{9.5.1}$$

where

$$\tilde{J}(z,\alpha_0)=\sum_{n=1}^{\infty} j_n(z,\alpha_0), \tag{9.5.2}$$

$$\begin{gathered} j_n(z,\alpha_0)=-(-1)^n e^{-i\kappa\beta_n}\chi_n(z,-\beta_n)e^{iq(n-1)+ikz}, \\ \beta_n=(-1)^n\alpha_0, \quad q=2\kappa=2kl. \end{gathered} \tag{9.5.3}$$

Let us present $\tilde{J}(z,\alpha_0)$ in the form

$$\begin{aligned} \tilde{J}(z,\alpha_0)&=e^{i\kappa\alpha_0}\chi_1(z,\alpha_0)e^{ikz}-e^{-i\kappa\alpha_0}\sum_{n=2,4,\ldots}^{\infty}\chi_n(z,-\alpha_0)e^{iq(n-1)+ikz}+ \\ &+e^{i\kappa\alpha_0}\sum_{n=3,5,\ldots}^{\infty}\chi_n(z,\alpha_0)e^{iq(n-1)+ikz}. \end{aligned} \tag{9.5.4}$$

Replacing $\chi_n(z,\alpha)$ with the first terms of the asymptotic series of Eqns. (9.2.1) and (9.4.11), we can designate the resulting approximate expression for the current as $\tilde{J}_1$:

$$\begin{aligned} \tilde{J}_1(z,\alpha_0)&=e^{i\kappa\alpha_0}\chi_1(z,\alpha_0)e^{ikz}- \\ &-e^{-i\kappa\alpha_0}\chi_1(z,-\alpha_0)\chi_1(2l,1)e^{iq+ikz}\sum_{n=2,4,6\ldots}^{\infty}\left[\chi_1(2l,1)e^{iq}\right]^{n-2}+ \\ &+e^{i\kappa\alpha_0}\chi_1(z,\alpha_0)\chi_1(z,1)\chi_1(2l,1)e^{i2q+ikz}\sum_{n=3,5,7\ldots}^{\infty}\left[\chi_1(2l,1)e^{iq}\right]^{n-3}. \end{aligned}$$

Therefore,

$$\tilde{J}_1(z,\alpha_0)=e^{i\kappa\alpha_0}\chi_1(z,\alpha_0)e^{ikz}-$$
$$-\frac{e^{iq}}{\mathcal{D}_1}\left[\chi_1(2l,-\alpha_0)e^{-i\kappa\alpha_0}-\chi_1(2l,\alpha_0)\chi_1(2l,1)e^{iq+i\kappa\alpha_0}\right]\chi_1(z,1)e^{ikz}\,. \tag{9.5.5}$$

The denominator $\mathcal{D}_1$ characterizes the resonance in the strip and is given by

$$\mathcal{D}_1=1-\chi_1^2(2l,1)e^{i2q}\,. \tag{9.5.6}$$

It is interesting that if Eqn. (9.5.5) is substituted into Eqn. (9.5.1) that the result agrees exactly, other than in sign, with Eqn. (6.6.6), which was obtained using the parabolic equation method. The difference in sign is due to the fact that the component of the incident wave given by Eqn. (6.6.7) is opposite in sign to the analogous component corresponding to the incident wave in Eqn. (8.2.1). Thus, the derivation of Eqn. (9.5.5) with the estimate of its error can be considered as a rigorous substantiation of Eqn. (6.6.6).

It is not difficult to show that the approximation

$$J_1(z,\alpha_0)=j^{(0)}(z,\alpha_0)+\tilde{J}_1(z+l,\alpha_0)+\tilde{J}_1(l-z,-\alpha_0) \tag{9.5.7}$$

satisfies the edge and boundary conditions

$$J_1(\pm l,\alpha_0)=J_1(z,\pm 1)=0\,. \tag{9.5.8}$$

Further utilizing Eqn. (8.5.10) for $\chi_1(z,\alpha)$, we find that

$$J_1(z,\alpha_0)=O\left[\sqrt{k(l\pm z)}\right]\quad\text{for}\quad l\pm z\ll 1. \tag{9.5.9}$$

Let us present the current in the form

$$J(z,\alpha_0)=J_1(z,\alpha_0)+R_1(z,\alpha_0), \tag{9.5.10}$$

where

$$R_1(z,\alpha_0)=\tilde{R}_1(z+l,\alpha_0)+\tilde{R}_1(z-l,-\alpha_0). \qquad (9.5.11)$$

The value of R_1 characterizes the absolute error of the first-order approximation and is determined by the second term of the asymptotic expansion for $\chi_2(z,\pm\alpha_0)$. Using Eqn. (9.2.1) we find that

$$\tilde{R}_1(z,\alpha_0)=O\left[\varphi_q^{(1)}(q,-\alpha_0)\varphi(kz,1)\right] \text{ for } q\gg 1. \qquad (9.5.12)$$

Directing our attention to Eqns. (9.3.2) and (9.3.12), this expression may be rewritten more clearly as

$$\tilde{R}_1(z,\alpha_0)=O\left[\frac{1+\sqrt{kz}}{1+kz}\frac{q^{-1/2}}{1+q(1-\alpha_0)}\right] \qquad (9.5.13)$$

for $q\gg 1$ and $0\leqslant kz\leqslant\infty$.

It is also clear that on the basis of Eqns. (8.5.19), (8.5.20), and (9.5.8)

$$R_1(\pm l,\alpha_0)=R_1(z,\pm 1)=0, \qquad (9.5.14)$$

i.e., the absolute error of the first-order approximation is zero for two cases: 1) at the edges of the strip for any angle of incidence; 2) over the entire strip at grazing angles of incidence.

§9.6 N^th Order Approximation for the Current

9.6.1 Derivation of an Approximate Formula

Let us retain the first n terms in Eqn. (9.5.4), and use the asymptotic representation given by Eqn. (9.4.12) for the remainder. As a result, the following expression for the current

$$J(z,\alpha_0)=J_n(z,\alpha_0)+R_n(z,\alpha_0), \qquad (9.6.1)$$

is obtained, where

$$\left.\begin{aligned} J_n(z,\alpha_0) &= j^{(0)}(z,\alpha_0) + \tilde{J}_n(z+l,\alpha_0) + \tilde{J}_n(l-z,-\alpha_0), \\ R_n(z,\alpha_0) &= \tilde{R}_n(z+l,\alpha_0) + \tilde{R}_n(l-z,-\alpha_0). \end{aligned}\right\} \tag{9.6.2}$$

Also,

$$\tilde{J}_n(z,\alpha_0)e^{-ikz} = -\sum_{m=1}^{n}(-1)^m \chi_m(z,-\beta_m)e^{-i\kappa\beta_m + iq(m-1)} +$$

$$+(-1)^n \frac{\chi_n(z,\beta_n)\chi_n(z,1)}{\chi_{n-1}(2l,1)} e^{iqn+i\kappa\beta_n} \sum_{m=1,3,5\ldots}^{\infty} \left[\frac{\chi_n(2l,1)}{\chi_{n-1}(2l,1)}e^{iq}\right]^{m-1} -$$

$$-(-1)^n \frac{\chi_n(z,-\beta_n)\chi_n(z,1)\chi_n(2l,1)}{\chi_{n-1}^2(2l,1)} e^{iq(n+1)-i\kappa\beta_n} \sum_{m=2,4,6\ldots}^{\infty} \left[\frac{\chi_n(2l,1)}{\chi_{n-1}(2l,1)}e^{iq}\right]^{m-2},$$

from which

$$\begin{aligned} \tilde{J}_n(z,\alpha_0) &= -e^{ikz}\sum_{m=1}^{n}(-1)^m \chi_m(z,-\beta_m)e^{iq(m-1)-i\kappa\beta_m} + \\ &\quad + \frac{(-1)^n e^{iqn}}{\hat{D}_n} \frac{\chi_n(z,1)}{\chi_{n-1}(2l,1)} \times \\ &\quad \times \left[\chi_n(2l,\beta_n)e^{i\kappa\beta_n} - \chi_n(2l,-\beta_n)\frac{\chi_n(2l,1)}{\chi_{n-1}(2l,1)}e^{iq-i\kappa\beta_m}\right]e^{ikz}, \end{aligned} \tag{9.6.3}$$

where

$$\hat{D}_n = 1 - \left[\frac{\chi_n(2l,1)}{\chi_{n-1}(2l,1)}e^{iq}\right]^2, \quad n = 2,3,\ldots. \tag{9.6.4}$$

Equation (9.6.3) possesses clear physical meaning. It includes in *complete* form all edge waves of orders 1 to n. Edge waves of higher orders are calculated approximately and form a geometric progression, the sum of which leads to the emergence of the resonant denominator, $\hat{\mathcal{D}}_n$.

9.6.2 Verification of the Edge Conditions

To verify the edge conditions of Eqn. (8.5.19), note that the expression obtained for J_n is of the form

$$\begin{aligned} J_n(z,\alpha_0) = &-e^{-ikz\alpha_0} - \\ &-e^{ik(z+l)}\sum_{m=1}^{n}(-1)^m \chi_m(z+l,-\beta_m)e^{iq(m-1)-i\kappa\beta_m} - \\ &-e^{ik(l-z)}\sum_{m=1}^{n}(-1)^m \chi_m(l-z,\beta_m)e^{iq(m-1)+i\kappa\beta_m} + \\ &+A\chi_n(z+l,1)e^{ik(z+l)} + B\chi_n(l-z,1)e^{ik(l-z)}. \end{aligned} \tag{9.6.5}$$

Requiring that

$$J_n(\pm l,\alpha_0) = 0, \tag{9.6.6}$$

leads to a linear system of equations related by unknown coefficients A and B

$$\left.\begin{aligned} A + B\frac{\chi_n(2l,1)}{\chi_{n-1}(2l,1)}e^{iq} &= (-1)^n\frac{\chi_n(2l,\beta_n)}{\chi_{n-1}(2l,1)}e^{iqn+i\kappa\beta_n}, \\ A\frac{\chi_n(2l,1)}{\chi_{n-1}(2l,1)}e^{iq} + B &= (-1)^n\frac{\chi_n(2l,-\beta_n)}{\chi_{n-1}(2l,1)}e^{iqn-i\kappa\beta_n}. \end{aligned}\right\} \tag{9.6.7}$$

Thus,

$$A(\beta_n) = \frac{(-1)^n}{\hat{\mathcal{D}}_n} \frac{e^{iqn}}{\chi_{n-1}(2l,1)} \times$$

$$\times \left[\chi_n(2l,\beta_n) e^{i\kappa\beta_n} - \chi_n(2l,-\beta_n) \frac{\chi_n(2l,1)}{\chi_{n-1}(2l,1)} e^{iq - i\kappa\beta_n} \right], \tag{9.6.8}$$

$$B(\beta_n) = A(-\beta_n).$$

The expression for $A(\beta_n)$ completely agrees with the coefficient that precedes $\chi_n(z, 1)e^{ikz}$ in the resonant term of Eqn. (9.6.3). This means that the condition of Eqn. (9.6.6) is satisfied. Furthermore, Eqns. (9.4.10) and (8.5.10) lead to the relationship

$$J_n(z,\alpha_0) = O\left[\sqrt{k(l \pm z)}\right] \quad \text{for} \quad k(l \pm z) \ll 1, \tag{9.6.9}$$

characterizing the behavior of the current near the edge of the strip.

Equation (8.5.13) can also verify that J_n satisfies the condition that

$$J_n(z,\pm 1) = 0. \tag{9.6.10}$$

9.6.3 Estimate of the Error

The obtained approximation preserves the first n terms of Eqn. (9.5.4). Note that this series possesses a majorant in the form of a geometrical progression with the denominator given by Eqn. (8.6.11). Subsequently, the absolute error of the approximation for J_n is estimated by

$$\left|R_n(z,\alpha_0)\right| \leqslant \frac{\left[(2/\pi)\tan^{-1}\left(\sqrt{\pi/2q}\right)\right]^{n+1}}{1-(2/\pi)\tan^{-1}\left(\sqrt{\pi/2q}\right)}, \quad q = 2kl. \tag{9.6.11}$$

When $kl \gg n$ this estimate can be improved. Utilizing the asymptotic representation of Eqn. (9.4.12) with $m = 1$ and Eqn. (9.6.3), we determine that

$$\left|\tilde{R}_n(z,\alpha_0)\right| \leqslant \frac{\left|\varphi(kz,1)\right| O\left(q^{-\frac{n+1}{2}}\right)}{1+\sqrt{q\left[1+(-1)^n \alpha_0\right]}}, \quad \text{for} \quad q \gg n. \tag{9.6.12}$$

This expression can be written more clearly if we make use of Eqn. (9.3.12),

$$\left|\tilde{R}_n(z,\alpha_0)\right| \leqslant O\left\{\frac{1+\sqrt{kz}}{1+kz} \frac{q^{-\frac{n+1}{2}}}{1+\sqrt{q\left[1+(-1)^n \alpha_0\right]}}\right\} \tag{9.6.13}$$

for $q \gg n$ and $0 \leqslant kz \leqslant \infty$. In addition, from Eqns. (8.5.19), (8.5.20), (9.6.6), and (9.6.10) it follows that

$$R_n(\pm l,\alpha_0) = R_n(z,\pm 1) = 0\,. \tag{9.6.14}$$

Thus, the nth-order approximation of Eqn. (9.6.3) allows us to determine the current density on the strip with an accuracy of any predetermined degree of $(kl)^{-n/2}$ for $2kl \gg n$. In principle, according to Eqn. (9.6.11), the error of this approximation for a given q can be made as small as desired by increasing the value of n. However, this approximation is of practical value only for $2kl \gg n$, when calculations for χ_n can be conducted using asymptotic expressions, and when the values of χ_n decrease with increasing n sufficiently quickly.

10

Asymptotic Representation for the Scattering Pattern

The pattern of the scattered field in the far zone, in principle, can be determined by applying a Fourier transform to the asymptotic representations for the current. However, as § 10.5 will show, one should not do this since the expressions obtained using this approach do not satisfy the reciprocity theorem. Because of this, an asymptotic relationship will be sought for the field in the far zone, proceeding from the exact solutions of Eqns. (8.3.21) (8.3.22), and (8.3.23) of functional Eqn. (8.3.10).

§10.1 Exact Expressions for the Scattering Pattern and Some Properties of $\varphi_n(\alpha,\alpha_0)$

Utilizing Eqns. (8.4.6), (8.4.7), and (8.4.8), the scattered field in the far zone (for $y \geqslant 0$) can be written in the form

$$H_x = \Phi(\alpha,\alpha_0)\frac{e^{i(kr+\pi/4)}}{\sqrt{2\pi kr}}, \tag{10.1.1}$$

where

$$\alpha = \cos\vartheta, \quad \alpha_0 = \cos\vartheta_0, \tag{10.1.2}$$

$$\Phi(\alpha,\alpha_0)=\tilde{\Phi}(\alpha,\alpha_0)+\tilde{\Phi}(-\alpha,-\alpha_0). \tag{10.1.3}$$

The function $\tilde{\Phi}$ introduced here differs from Eqn. (8.4.8) by an exponential factor and is given by the series

$$\tilde{\Phi}(\alpha,\alpha_0)=e^{i\kappa\alpha}\sum_{n=1}^{\infty}(-1)^n\varphi_n(\alpha,-\beta_n)e^{iq(n-1)-i\kappa\beta_n} \tag{10.1.4}$$

that is a function of $\varphi_n(\alpha,\alpha_0)$, where $\beta_n=(-1)^n\alpha_0$,

$$\varphi_1(\alpha,\alpha_0)=\frac{\sqrt{1+\alpha}\sqrt{1+\alpha_0}}{\alpha+\alpha_0}, \tag{10.1.5}$$

$$\varphi_2(\alpha,\alpha_0)e^{iq}=\frac{\sqrt{1+\alpha}\sqrt{1+\alpha_0}}{i\pi}\int_1^{1+i\infty}\frac{h(t)dt}{(t+\alpha)(t+\alpha_0)}, \tag{10.1.6}$$

$$\underset{n=3,4,5\ldots}{\varphi_n(\alpha,\alpha_0)e^{iq(n-1)}}=\frac{\sqrt{1+\alpha}\sqrt{1+\alpha_0}}{(i\pi)^{n-1}}\times$$

$$\times\int_1^{1+i\infty}\cdots\int_1^{1+i\infty}\frac{h(t_1)h(t_2)\ldots h(t_{n-1})dt_1dt_2\ldots dt_{n-1}}{(t_1+\alpha_0)(t_2+t_1)(t_3+t_2)\ldots(t_{n-1}+t_{n-2})(t_{n-1}+\alpha)}, \tag{10.1.7}$$

$$h=\sqrt{\frac{1+t}{1-t}}e^{iqt},\quad q=2\kappa=2kl,$$

and the integral is evaluated on the left side of the branch cut $1\rightarrow 1+i\infty$, where $\arg\{(1-t)^{1/2}\}=-\pi/4$.

Turning our attention to the properties of φ_n, we first note that

$$\varphi_n(\alpha,\alpha_0)=\varphi_n(\alpha_0,\alpha). \tag{10.1.8}$$

Let us further determine $\varphi_n(-1,\alpha_0)$. To do this, let us present φ_n as

$$\varphi_n(\alpha,\alpha_0)e^{iq(n-1)} = \frac{\sqrt{1+\alpha}\sqrt{1+\alpha_0}}{(i\pi)^{n-1}}\times$$

$$\times\int\limits_1^{1+i\infty}\cdots\int\limits_1^{1+i\infty}\frac{h(t_1)h(t_2)\ldots h(t_{n-2})dt_1dt_2\ldots dt_{n-2}}{(t_1+\alpha_0)(t_2+t_1)(t_3+t_2)\ldots(t_{n-3}+t_{n-4})(t_{n-2}+t_{n-3})}I(t_{n-2},\alpha), \tag{10.1.9}$$

where

$$I(t_{n-2},\alpha) = -\frac{1}{2}\int\limits_{-\infty}^{\infty}\sqrt{\frac{1+t_{n-1}}{1-t_{n-1}}}\frac{e^{iqt_{n-1}}dt_{n-1}}{(t_{n-1}+\widehat{t_{n-2}})(t_{n-1}+\widehat{\alpha})} =$$

$$= \int\limits_1^{1+i\delta}\frac{h(t_{n-1})dt_{n-1}}{(t_{n-1}+t_{n-2})(t_{n-1}+\alpha)} + \int\limits_{1+i\delta}^{1+i\infty}\frac{h(t_{n-1})dt_{n-1}}{(t_{n-1}+t_{n-2})(t_{n-1}+\alpha)},\quad \delta\geqslant 0. \tag{10.1.10}$$

It is evident that

$$\lim_{\alpha\to-1}\sqrt{1+\alpha}\,\mathrm{I}(t_{n-2},\alpha) = \lim_{\alpha\to-1}\sqrt{1+\alpha}\int\limits_1^{1+i\delta}\frac{h(t_{n-1})dt_{n-1}}{(t_{n-1}+t_{n-2})(t_{n-1}+\alpha)} =$$

$$= \lim_{\alpha\to-1}\frac{\sqrt{1+\alpha}\sqrt{2}e^{iq}}{1+t_{n-2}}\int\limits_1^{1+i\infty}\frac{dt_{n-1}}{\sqrt{1-t_{n-1}}(t_{n-1}+\alpha)}.$$

Therefore, it follows that

$$\lim_{\alpha\to-1}\varphi_n(\alpha,\alpha_0)e^{iq(n-1)} = -\lim_{\alpha\to-1}\frac{\sqrt{1+\alpha}\sqrt{1+\alpha_0}}{2(i\pi)^{n-1}}\times \tag{10.1.11}$$

$$\times\int_1^{1+i\infty}\cdots\int_1^{1+i\infty}\frac{h(t_1)h(t_2)\ldots h(t_{n-2})dt_1dt_2\ldots dt_{n-2}}{(t_1+\alpha_0)(t_2+t_1)(t_3+t_2)\ldots(t_{n-3}+t_{n-4})(t_{n-2}+t_{n-3})}\times$$

$$\times\int_{-\infty}^{\infty}\frac{\sqrt{2}e^{iq}}{\sqrt{1-t_{n-1}}}\frac{dt_{n-1}}{(t_{n-1}+\widehat{\alpha})(t_{n-2}+1)}.$$

Applying the Cauchy integral theorem to the third line of Eqn. (10.1.11), we obtain the following relationships

$$\left.\begin{aligned}\varphi_n(-1,\alpha_0)&=\varphi_{n-1}(1,\alpha_0),\\ \varphi_n(\alpha,-1)&=\varphi_{n-1}(\alpha,1),\end{aligned}\right\} \tag{10.1.12}$$

and in addition,

$$\varphi_1(-1,\alpha_0)=0,\quad \varphi_2(-1,\alpha_0)=\varphi_1(1,\alpha_0)=\sqrt{\frac{2}{1+\alpha_0}}. \tag{10.1.13}$$

Utilizing Eqns. (10.1.8) and (10.1.12), it is possible to prove that

$$\Phi(\alpha,\alpha_0)=\Phi(\alpha_0,\alpha) \tag{10.1.14}$$

and

$$\Phi(\pm1,\alpha_0)=0,\quad \Phi(\alpha,\pm1)=0. \tag{10.1.15}$$

Equation (10.1.14) is a formulation of the reciprocity theorem. The first part of Eqn. (10.1.15) demonstrates the continuity of the field in the $y = 0$ plane for $|z| > l$, and the second part of this equation shows that at grazing angles of incidence no scattered field is created.

Using transformations analogous to those which we have performed with χ_n in § 8.6, it is possible to show that

$$|\varphi_n(\alpha,\alpha_0)| < \sqrt{1+\frac{\pi}{q}}\left(\frac{2}{\pi}\tan^{-1}\sqrt{\frac{\pi}{2q}}\right)^{n-3}. \tag{10.1.16}$$

Thus, the series of Eqn. (10.1.4), as does the series of Eqn. (8.5.1), possesses a majorant geometrical progression with a ratio given by Eqn. (8.6.11), and consequently, converges absolutely and uniformly for any value of $q > 0$.

§10.2 Asymptotic Representations for $\varphi_n(\alpha,\alpha_0)$

10.2.1 Asymptotic Series for $\varphi_n(\alpha,\alpha_0)$

Applying Lemma II in § 9.1 to $\varphi_n(\alpha, \alpha_0)$, it is possible to obtain the following asymptotic expansion as $q \to \infty$:

$$\varphi_2(\alpha,\alpha_0) \sim \frac{1}{\alpha-\alpha_0}\sum_{m=0}^{\infty}\frac{a_m}{i^m}\left[\sqrt{1+\alpha}\,\varphi_q^{(m)}(q,\alpha_o) - \sqrt{1+\alpha_0}\,\varphi_q^{(m)}(q,-\alpha)\right], \tag{10.2.1}$$

where a_m is a Taylor series coefficient

$$\sqrt{1+t} = \sum_{m=0}^{\infty} a_m (t-1)^m, \tag{10.2.2}$$

$$\varphi_n(\alpha,\alpha_0) \sim \varphi(q,\alpha)\varphi(q,\alpha_o)\varphi^{n-3}(q,1) + \frac{\sqrt{1+\alpha}\sqrt{1+\alpha_0}}{(i\pi)^{n-1}}\times \tag{10.2.3}$$

$$\times\sum_{m=1}^{\infty}\frac{1}{m!}\int_{1}^{1+i\infty}\cdots\int_{1}^{1+i\infty}\frac{e^{iq\sum_{s=1}^{n-1}(t_s-1)}\hat{u}_m(t_1,t_2,\ldots,t_{n-1})dt_1dt_2\ldots dt_{n-1}}{(t_1+\alpha)(t_{n-1}+\alpha_0)\prod_{p=1}^{n-1}\sqrt{1-t_p}\prod_{s=2}^{n-2}(1+t_s)}, \qquad (10.2.3, \text{cont.})$$

where

$$\hat{u}_m(t_1,t_2,\ldots,t_{n-1})= \qquad (10.2.4)$$

$$=\left[(t_1-1)\frac{\partial}{\partial t_1}+(t_2-1)\frac{\partial}{\partial t_2}+\ldots+(t_{n-1}-1)\frac{\partial}{\partial t_{n-1}}\right]^m \hat{f}_n(1,1,\ldots,1),$$

$$\underset{n=4,5\ldots}{\hat{f}_n(t_1,t_2,\ldots,t_{n-1})}=\frac{\sqrt{1+t_1}\sqrt{1+t_{n-1}}}{(t_2+t_1)(t_3+t_2)\ldots(t_{n-1}+t_{n-2})}\prod_{s=2}^{n-2}(1+t_s)^{3/2}, \qquad (10.2.5)$$

$$\hat{f}_3(t_1,t_2)=\frac{\sqrt{1+t_1}\sqrt{1+t_2}}{t_1+t_2}. \qquad (10.2.6)$$

The first term in Eqn. (10.2.3) shows that in a first-order approximation, the diffraction of edge waves is identical to the diffraction of plane waves. The remaining terms exhibit the more complicated structure of edge waves and are expressed as derivatives of Fresnel integrals $\varphi_q^{(m)}(q,\alpha)$, determined using Eqns. (9.2.6) and (9.2.9).

Equation (10.2.3) represents the exact total asymptotic expansion for the function $\varphi_n(\alpha,\alpha_0)$. Now our objective is to derive simple approximations for it similar to those given by Eqns. (9.4.10) and (9.4.12) for the function $\chi_n(z,\alpha)$. Retaining only the term with $m = 1$ in the series of Eqn. (10.2.3), one obtains

$$\varphi_n(\alpha,\alpha_0)\approx\varphi(q,\alpha)\varphi(q,\alpha_0)\varphi^{n-3}(q,1)+T_{1\varphi}, \qquad (10.2.7)$$

where

$$T_{1\varphi} = \frac{\sqrt{1+\alpha}\sqrt{1+\alpha_0}}{(i\pi)^{n-1}} \times$$

$$\times \int_1^{1+i\infty} \cdots \int_1^{1+i\infty} \frac{e^{iq\sum_{s=1}^{n-1}(t_s-1)} \hat{u}_1(t_1,t_2,\ldots,t_{n-1})dt_1 dt_2 \ldots dt_{n-1}}{(t_1+\alpha)(t_{n-1}+\alpha_0)\prod_{p=1}^{n-1}\sqrt{1-t_p}\prod_{s=2}^{n-2}(1+t_s)}, \quad (10.2.8)$$

$$\hat{u}_1 = \left[(t_1-1)\frac{\partial}{\partial t_1} + (t_2-1)\frac{\partial}{\partial t_2} + \ldots + (t_{n-1}-1)\frac{\partial}{\partial t_{n-1}}\right]\hat{f}_n(1,1,\ldots,1), \quad (10.2.9)$$

and the function $\hat{f}_n$ is given by Eqn. (10.2.5). It is clear that at the point $t_1 = t_2 \ldots = t_{n-1} = 1$,

$$\frac{\partial \hat{f}_n}{\partial t_2} = \frac{\partial \hat{f}_n}{\partial t_3} = \ldots = \frac{\partial \hat{f}_n}{\partial t_{n-2}}. \quad (10.2.10)$$

The $(n-3)$ identical derivatives of Eqn. (10.2.10) generate $(n-3)$ identical integrals in the function $T_{1\varphi}$. After substituting Eqn. (10.2.9) into Eqn. (10.2.8), it follows that

$$T_{1\varphi} = \varphi^{n-3}(q,1)\Big\{O\big[\varphi_q^{(1)}(q,\alpha)\varphi(q,\alpha_0)\big] + O\big[\varphi_q^{(1)}(q,\alpha_0)\varphi(q,\alpha)\big] + \\ +(n-3)\varphi^{n-4}(q,1)O\big[\varphi_q^{(1)}(q,1)\varphi(q,\alpha)\varphi(q,\alpha_0)\big]\Big\}. \quad (10.2.11)$$

Subsequent substitution of $T_{1\varphi}$ into Eqn. (10.2.7) leads to the approximation

$$\varphi_n(\alpha,\alpha_0) = \varphi(q,\alpha)\varphi(q,\alpha_0)\varphi^{n-3}(q,1) \times \\ \times\left\{1 + O\left[\frac{\varphi_q^{(1)}(q,\alpha)}{\varphi(q,\alpha)}\right] + O\left[\frac{\varphi_q^{(1)}(q,\alpha_0)}{\varphi(q,\alpha_0)}\right] + (n-3)O\left[\frac{\varphi_q^{(1)}(q,1)}{\varphi(q,1)}\right]\right\}. \quad (10.2.12)$$

It follows from Eqns. (9.3.2) and (9.3.10) that

$$\frac{\varphi_q^{(1)}(q,\alpha)}{\varphi(q,\alpha)}=\frac{\sqrt{1+\alpha}}{\sqrt{q}}O\left[\frac{1+\sqrt{q(1+\alpha)}}{1+q(1+\alpha)}\right]\leqslant O\left[\frac{1}{q}\right],\quad \frac{\varphi_q^{(1)}(q,1)}{\varphi(q,1)}=O\left[\frac{1}{q}\right] \tag{10.2.13}$$

for $q \gg 1$ and $-1 \leq \alpha \leq 1$. This results in the asymptotic expression

$$\varphi_n(\alpha,\alpha_0)=\varphi(q,\alpha)\varphi(q,\alpha_0)\varphi^{n-3}(q,1)\left\{1+O\left[\frac{n}{q}\right]\right\}, \tag{10.2.14}$$

which is valid for $n = 3, 4, 5, \ldots, q \gg n$ and $-1 \leq \alpha \leq 1$.

Introducing the function

$$U_{s,n}(\alpha,\alpha_0)=\varphi_s(\alpha,\alpha_0)-\frac{\varphi_n(1,\alpha)\varphi_n(1,\alpha_0)}{\varphi_{n-1}(1,1)}\left[\frac{\varphi_n(1,1)}{\varphi_{n-1}(1,1)}\right]^{s-n-1}, \tag{10.2.15}$$

it is beneficial to study some of its asymptotic representations as $q \to \infty$, which will prove useful for further research.

10.2.2 Estimate of $U_{n,2}(\alpha,\alpha_0)$

Since $\varphi_1(1, 1) = 1$, then

$$U_{n,2}(\alpha,\alpha_0)=\varphi_n(\alpha,\alpha_0)-\varphi_2(1,\alpha)\varphi_2(1,\alpha_0)\varphi_2^{n-3}(1,1), \tag{10.2.16}$$

where

$$\left.\begin{aligned}\varphi_2(1,\alpha)&=\frac{\sqrt{2(1+\alpha)}}{i\pi}\int_1^{1+i\infty}\frac{e^{iq(t-1)}}{\sqrt{1-t}}\frac{f_2(t)dt}{t+\alpha},\\ f_2(t)&=\frac{1}{\sqrt{1+t}},\end{aligned}\right\} \tag{10.2.17}$$

$$\varphi_n(\alpha,\alpha_0)=\frac{\sqrt{1+\alpha}\sqrt{1+\alpha_0}}{(i\pi)^{n-1}}\times$$

$$\times\int\limits_1^{1+i\infty}\cdots\int\limits_1^{1+i\infty}\frac{e^{iq\sum_{s=1}^{n-1}(t_s-1)}\hat{\mathrm{f}}_n(t_1,t_2,\ldots,t_{n-1})dt_1dt_2\ldots dt_{n-1}}{(t_1+\alpha)(t_{n-1}+\alpha_0)\prod_{p=1}^{n-1}\sqrt{1-t_p}\prod_{s=2}^{n-2}(1+t_s)}. \tag{10.2.18}$$

The function $U_{n,2}$ may be written in the form

$$U_{n,2}(\alpha,\alpha_0)=\frac{\sqrt{1+\alpha}\sqrt{1+\alpha_0}}{(i\pi)^{n-1}}\times$$

$$\times\int\limits_1^{1+i\infty}\cdots\int\limits_1^{1+i\infty}\mathcal{K}_{n,2}(t_1,t_2,\ldots,t_{n-1})\left[F_{n,2}(t_1,t_2,\ldots,t_{n-1})-1\right]dt_1dt_{2\ldots}dt_{n-1}, \tag{10.2.19}$$

where

$$\mathcal{K}_{n,2}(t_1,t_2,\ldots,t_{n-1})=\frac{2^{n-2}\prod_{s=1}^{n-1}f_2(t_s)e^{iq\sum_{s=1}^{n-1}(t_s-1)}}{(t_1+\alpha)(t_{n-1}+\alpha_0)\prod_{p=1}^{n-1}\sqrt{1-t_p}\prod_{s=2}^{n-2}(1+t_s)}, \tag{10.2.20}$$

$$F_{n,2}(t_1,t_2,\ldots,t_{n-1})=\frac{(1+t_1)(1+t_{n-1})}{2^{n-2}(t_1+t_2)(t_2+t_3)\ldots(t_{n-2}+t_{n-1})}\prod_{s=2}^{n-2}(1+t_s)^2, \tag{10.2.21}$$

$n=4,5\ldots$

and

$$F_{n,2}(1,1,\ldots,1)=1,\quad F_{3,2}(t_1,t_2)=\frac{(1+t_1)(1+t_2)}{2(t_1+t_2)}. \tag{10.2.22}$$

It is possible to show that in the neighborhood of $t_1 = t_2 = \ldots = t_{n-1} = 1$ the expansion of $F_{n,2} - 1$ in a Taylor series begins with quadratic terms and is of the form

$$F_{n,2}(t_1,t_2,\ldots,t_{n-1}) - 1 = \frac{1}{4}\sum_{m=1}^{n-2}(t_m - 1)(t_{m+1} - 1) + \ldots . \quad n = 3,4\ldots \tag{10.2.23}$$

Therefore, applying Lemma II of § 9.1 to $U_{n,2}$, as $q \to \infty$, we obtain

$$\left.\begin{aligned} U_{3,2}(\alpha,\alpha_0) &\sim -\frac{1}{4}\varphi_q^{(1)}(q,\alpha)\varphi_q^{(1)}(q,\alpha_0), \\ U_{4,2}(\alpha,\alpha_0) &\sim -\frac{1}{4}\varphi_q^{(1)}(q,1)\left[\varphi(q,\alpha)\varphi_q^{(1)}(q,\alpha_0) + \varphi_q^{(1)}(q,\alpha)\varphi(q,\alpha_0)\right], \end{aligned}\right\} \tag{10.2.24}$$

$$\begin{aligned} \underset{n=5,6,7\ldots}{U_{n,2}(\alpha,\alpha_0)} \sim -\frac{1}{4}\varphi^{n-5}(q,1)\varphi_q^{(1)}(q,1)\Big[&\varphi(q,\alpha)\varphi_q^{(1)}(q,\alpha_0)\varphi(q,1) + \\ &+\varphi_q^{(1)}(q,\alpha)\varphi(q,\alpha_0)\varphi(q,1) + (n-4)\varphi(q,\alpha)\varphi(q,\alpha_0)\varphi_q^{(1)}(q,1)\Big]. \end{aligned} \tag{10.2.25}$$

10.2.3 Asymptotic Representation for $\varphi_{m+n}(\alpha,\alpha_0)$

According to Eqn. (10.2.15), $U_{n+m,n}$ is equal to

$$U_{n+m,n}(\alpha,\alpha_0) = \varphi_{n+m}(\alpha,\alpha_0) - \frac{\varphi_n(1,\alpha)\varphi_n(1,\alpha_0)}{\varphi_{n-1}(1,1)}\left[\frac{\varphi_n(1,1)}{\varphi_{n-1}(1,1)}\right]^{m-1}. \tag{10.2.26}$$

We are interested in the asymptotic representation of this function for $q \gg m+n$. This is done by substituting Eqn. (10.2.14) into Eqn. (10.2.26)

$$\begin{aligned} U_{n+m,n}(\alpha,\alpha_0) = {}&\varphi(q,\alpha)\varphi(q,\alpha_0)\varphi^{m+n-3}(q,1)(1+\rho_{m+n}) - \\ &-\frac{\varphi(q,\alpha)\varphi(q,\alpha_0)\varphi^{2n-4}(q,1)(1+\rho_n)^2}{\varphi^{(n-2)m}(q,1)(1+\rho_{n-1})^m}\varphi^{(n-1)(m-1)}(q,1)(1+\rho_n)^{m-1}, \end{aligned}$$

where $q \gg m+n$ and

$$\rho_n = O\left(\frac{n}{q}\right).$$

As a result

$$\left|U_{n+m,n}(\alpha,\alpha_0)\right| \leqslant O\left[\frac{m+n}{q}\left|\varphi(q,\alpha)\varphi(q,\alpha_0)\varphi^{m+n-3}(q,1)\right|\right], \tag{10.2.27}$$

and therefore,

$$\begin{aligned}\left|\varphi_{n+m}(\alpha,\alpha_0) - \frac{\varphi_n(1,\alpha)\varphi_n(1,\alpha_0)}{\varphi_{n-1}(1,1)}\left[\frac{\varphi_n(1,1)}{\varphi_{n-1}(1,1)}\right]^{m-1}\right| &\leqslant \\ \leqslant \frac{O\left[q^{-\frac{n+m-1}{2}}\right]}{\left[1+\sqrt{q(1+\alpha)}\right]\left[1+\sqrt{q(1+\alpha_0)}\right]},\end{aligned} \tag{10.2.28}$$

for $q \gg m+n$. This inequality plays a fundamental role in deriving approximate expressions for the scattering pattern. Examining Eqns. (10.2.25) and (10.2.28) in detail for $n = 2$ shows that Eqn. (10.2.28) gives an excessive value for the difference on the left hand side by the factor $q = 2kl$.

§10.3 First-Order Approximation for the Scattering Pattern

It follows from Eqn. (10.2.1) that

$$\begin{aligned}\varphi_2(\alpha,\alpha_0) \sim &\frac{\sqrt{2}}{\alpha-\alpha_0}\left[\sqrt{1+\alpha}\varphi(q,\alpha_0) - \sqrt{1+\alpha_0}\varphi(q,\alpha)\right] + \\ &+ \frac{1}{\alpha-\alpha_0}O\left[\sqrt{1+\alpha}\varphi_q^{(1)}(q,\alpha_0) - \sqrt{1+\alpha_0}\varphi_q^{(1)}(q,\alpha)\right]\end{aligned} \tag{10.3.1}$$

for $q \gg 1$. Transforming the first term on the right hand side, we obtain

$$\sqrt{1+\alpha}\varphi(q,\alpha_0)-\sqrt{1+\alpha_0}\varphi(q,\alpha)=$$

$$=\sqrt{1+\alpha}\left[\varphi(q,\alpha_0)-\sqrt{\frac{1+\alpha_0}{2}}\varphi(q,1)\right]- \tag{10.3.2}$$

$$-\sqrt{1+\alpha_0}\left[\varphi(q,\alpha)-\sqrt{\frac{1+\alpha}{2}}\varphi(q,1)\right].$$

However,

$$\varphi(q,\alpha)-\sqrt{\frac{1+\alpha}{2}}\varphi(q,1)=(1-\alpha)\frac{\sqrt{1+\alpha}}{i\pi}\int_1^{1+i\infty}\frac{e^{iq(t-1)}}{\sqrt{1-t}}\frac{dt}{(t+\alpha)(t+1)}=$$

$$=\frac{1-\alpha}{2}\varphi(q,\alpha)+(1-\alpha)O\left[\varphi_q^{(1)}(q,\alpha)\right]\quad \text{for} \quad q\gg 1. \tag{10.3.3}$$

Therefore, substituting Eqn. (10.3.3) into (10.3.2), we obtain the following asymptotic expression for $\varphi_2(\alpha,\alpha_0)$:

$$\varphi_2(\alpha,\alpha_0)\sim\frac{1}{\alpha-\alpha_0}\left[(1-\alpha_0)\sqrt{\frac{1+\alpha}{2}}\varphi(q,\alpha_0)-(1-\alpha)\sqrt{\frac{1+\alpha_0}{2}}\varphi(q,\alpha)\right]+ \tag{10.3.4}$$

$$+\frac{1}{\alpha-\alpha_0}O\left[\sqrt{1+\alpha}\varphi_q^{(1)}(q,\alpha_0)-\sqrt{1+\alpha_0}\varphi_q^{(1)}(q,\alpha)\right] \text{ for } q\gg 1.$$

This expression possesses the same error as Eqn. (10.3.1); however, it is more suitable for calculating the scattering pattern.

Substituting Eqn. (10.3.4) into Eqns. (10.1.3) and (10.1.4) and using the first term of the asymptotic expansion of Eqn. (10.2.7) for $\varphi_n(\alpha,\alpha_0)$ with $n=3,4,\ldots,$we find that

$$\Phi(\alpha,\alpha_0)=\Phi_1(\alpha,\alpha_0)+P_1(\alpha,\alpha_0), \tag{10.3.5}$$

where

$$\left.\begin{aligned}\Phi_1(\alpha,\alpha_0)&=\tilde{\Phi}_1(\alpha,\alpha_0)+\tilde{\Phi}_1(-\alpha,-\alpha_0),\\ P_1(\alpha,\alpha_0)&=\tilde{P}_1(\alpha,\alpha_0)+\tilde{P}_1(-\alpha,-\alpha_0).\end{aligned}\right\} \tag{10.3.6}$$

Here,

$$\tilde{\Phi}_1(\alpha,\alpha_0)=-\frac{\sqrt{1+\alpha}\sqrt{1+\alpha_0}}{\alpha+\alpha_0}e^{i\kappa(\alpha+\alpha_0)}+$$

$$+\frac{e^{iq}e^{i\kappa(\alpha-\alpha_0)}}{\alpha+\alpha_0}\left[(1+\alpha_0)\sqrt{\frac{1+\alpha}{2}}\varphi(q,-\alpha_0)-(1-\alpha)\sqrt{\frac{1-\alpha_0}{2}}\varphi(q,\alpha)\right]- \tag{10.3.7}$$

$$-\frac{e^{2iq}e^{i\kappa\alpha}}{\mathcal{D}_1}\varphi(q,\alpha)\left[\varphi(q,\alpha_0)e^{i\kappa\alpha_0}-\varphi(q,-\alpha_0)\varphi(q,1)e^{iq-i\kappa\alpha_0}\right],$$

and the resonant denominator is given by,

$$\mathcal{D}_1=1-\varphi^2(q,1)e^{i2q}, \tag{10.3.8}$$

agrees with the analogous value given by Eqn. (9.5.6) for the current.

Note that $\varphi(q,\alpha)$ is a Fresnel integral given by Eqn. (9.2.9) and that $\varphi(q,-1)=1$. Taking into account the property of $\varphi(q,\alpha)$, it is not difficult to show that

$$\Phi_1(\pm1,\alpha_0)=\Phi_1(\alpha,\pm1)=0. \tag{10.3.9}$$

Thus, this approximation satisfies the rigorous nature of Eqn. (10.1.15). Moreover, it satisfies the reciprocity theorem, since

$$\Phi_1(\alpha,\alpha_0)=\Phi_1(\alpha_0,\alpha). \tag{10.3.10}$$

The absolute error of the first-order approximation $\tilde{P}_1(\alpha,\alpha_0)$, according to Eqns. (9.3.2) and (10.3.4), is equal to

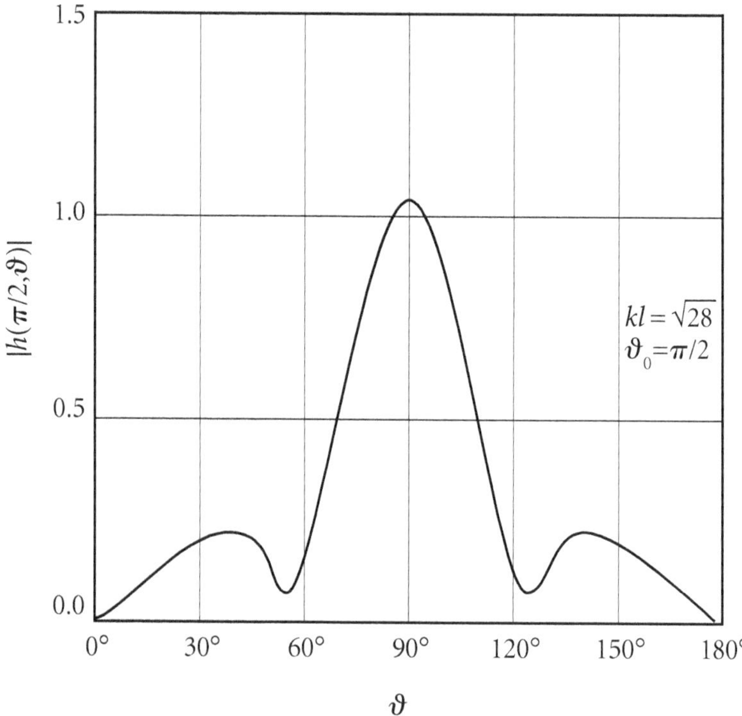

Figure 10.3.1: Scattering pattern of a strip for a plane wave at normal incidence.

$$\tilde{P}_1(\alpha,\alpha_0)=\frac{\sqrt{1+\alpha}\,\sqrt{1+\alpha_0}\,O\left(\sqrt{q}\right)}{\left[1+q(1+\alpha)\right]\left[1+q(1+\alpha_0)\right]} \quad \text{for} \quad q\gg 1. \tag{10.3.11}$$

Moreover, Eqns. (10.3.9) and (10.3.5) also show that

$$P_1(\alpha,\pm 1)=P_1(\pm 1,\alpha_0)=0. \tag{10.3.12}$$

Figures 10.3.2 and 10.3.1 depict the behavior of

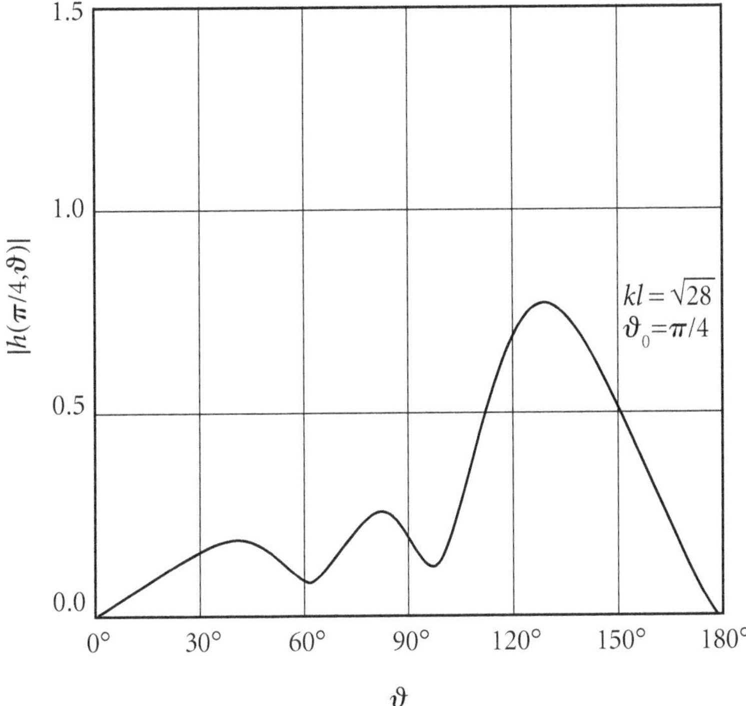

Figure 10.3.2: Scattering pattern of a strip illuminated by a plane wave as a function of the observation angle.

$$h(\vartheta_0, \vartheta) = \frac{1}{ly}\Phi_1(\alpha, \alpha_0) \tag{10.3.13}$$

for $\vartheta_0 = \pi/4$ and $\vartheta_0 = \pi/2$. They show that for $kl = \sqrt{28}$ the curves, within the limits of graphical accuracy, coincide with the exact curves for h which were calculated from a rigorous series obtained performing a separation of variables in elliptic coordinates.

It is interesting to note that the expression obtained for $\Phi_1(\alpha,\alpha_0)$ completely agrees with Eqn. (7.8.10), which was previously obtained using physical considerations. Therefore, the estimate provided by Eqn. (10.3.11) provides a rigorous substantiation of the approximation given by Eqn. (7.8.10).

Equation (10.3.7) does not differ in its degree of complexity from the expression of Khaskind and Vainshtein [122], but it does possess a greater accuracy. In particular, the total scattering cross section

$$S(\alpha_0) = -\frac{2}{k}\Im\mathrm{m}\{\Phi(-\alpha_0,\alpha_0)\} \tag{10.3.14}$$

produced by both Eqns. (10.3.7) and (7.8.18) at normal incidence ($\alpha_0 = 0$) is equal to

$$S(0) = 4l\left[1 - 2\sqrt{\frac{2}{\pi}}\frac{\cos(q-\pi/4)}{q^{3/2}} + \frac{2}{\pi}\frac{\cos(2q)}{q^2} + O\left(q^{-5/2}\right)\right]. \tag{10.3.15}$$

Whereas from the results of [122] it follows that

$$S(0) = 4l\left[1 + \frac{\sin 2q}{2\pi q} - 2\sqrt{\frac{2}{\pi}}\frac{\cos(q-\pi/4)}{q^{3/2}} + \frac{7}{8\pi}\frac{\cos(2q)}{q^2} + O\left(q^{-5/2}\right)\right]. \tag{10.3.16}$$

As was to be expected, Eqn. (10.3.15) agrees with the well known result given by Eqn. (31.5) in reference [121]. However, Eqn. (10.3.16) provided by Khaskind and Vainshtein correctly describes only the first term of the asymptotic expansion, which does not depend on kl, i.e., the term corresponding to approximate geometrical optics. Only when $\sin 2q = 0$ does it also give the first correction term on the order of $(kl)^{-3/2}$. In comparison with Eqn. (10.3.16), it is also interesting to estimate the error of our previous expression, published in references [5,13]. It is possible to show that according to this formula

$$S(0) = 4l\left[1 + \frac{\sin 2q}{\pi q} - \frac{3}{\sqrt{\pi}}\frac{\cos(q-\pi/4)}{q^{3/2}} + \frac{\cos(2q)}{\pi q^2} + O\left(q^{-5/2}\right)\right]. \tag{10.3.17}$$

Comparing Eqns. (10.3.16) and (10.3.17) with the exact result given by Eqn. (10.3.15), we see that Eqn. (10.3.17), derived using intuitive arguments, and Eqn. (10.3.16) obtained by Khaskind and Vainshtein using a significantly more complicated method, possess identical degrees of error.

§10.4 Nth-Order Approximation for the Scattering Pattern

10.4.1 Derivation of an Approximate Formula

Let us retain the first n terms of Eqn. (10.1.4) and represent the remaining terms using the asymptotic representation of Eqn. (10.2.28). In so doing, Eqn. (10.1.3) becomes

$$\Phi(\alpha,\alpha_0)=\Phi_n(\alpha,\alpha_0)+P_n(\alpha,\alpha_0), \tag{10.4.1}$$

where

$$\left.\begin{aligned}\Phi_n(\alpha,\alpha_0)&=\tilde{\Phi}_n(\alpha,\alpha_0)+\tilde{\Phi}_n(-\alpha,-\alpha_0),\\ \mathrm{P}_n(\alpha,\alpha_0)&=\tilde{\mathrm{P}}_n(\alpha,\alpha_0)+\tilde{\mathrm{P}}_n(-\alpha,-\alpha_0).\end{aligned}\right\} \tag{10.4.2}$$

Here,

$$\tilde{\Phi}_n(\alpha,\alpha_0)=e^{i\kappa\alpha}\sum_{m=1}^{n}(-1)^m\varphi_m(\alpha,-\beta_m)e^{iq(m-1)-i\kappa\beta_m}-$$

$$-(-1)^n e^{i\kappa\alpha}\frac{\varphi_n(1,\alpha)\varphi_n(1,\beta_n)}{\varphi_{n-1}(1,1)}\sum_{m=1,3,5\ldots}^{\infty}\left[\frac{\varphi_n(1,1)}{\varphi_{n-1}(1,1)}e^{iq}\right]^{m-1}e^{iqn+i\kappa\beta_n}+$$

$$+(-1)^n e^{i\kappa\alpha}\frac{\varphi_n(1,\alpha)\varphi_n(1,-\beta_n)\varphi_n(1,1)}{\varphi_{n-1}^2(1,1)}\sum_{m=2,4,6\ldots}^{\infty}\left[\frac{\varphi_n(1,1)}{\varphi_{n-1}(1,1)}e^{iq}\right]^{m-2}e^{iq(n+1)-i\kappa\beta_n}$$

or

$$\tilde{\Phi}_n(\alpha,\alpha_0)=e^{i\kappa\alpha}\sum_{m=1}^{n}(-1)^m\varphi_m(\alpha,-\beta_m)e^{iq(m-1)-i\kappa\beta_m}-(-1)^n\frac{e^{iqn}}{\mathcal{D}_n}\times$$

$$\times\left[\varphi_n(1,\beta_n)e^{i\kappa\beta_n}-\varphi_n(1,-\beta_n)\frac{\varphi_n(1,1)}{\varphi_{n-1}(1,1)}e^{iq-i\kappa\beta_n}\right]\frac{\varphi_n(1,\alpha)}{\varphi_{n-1}(1,1)}e^{i\kappa\alpha}, \tag{10.4.3}$$

where $\beta_n = (-1)^n\alpha_0$,

$$\mathcal{D}_n=1-\left[\frac{\varphi_n(1,1)}{\varphi_{n-1}(1,1)}e^{iq}\right]^2, \tag{10.4.4}$$

and $n = 2, 3, 4\ldots$. It is possible to show that Eqn. (10.4.2) for $\tilde{\Phi}_n$ satisfies the reciprocity theorem,

$$\Phi_n(\alpha,\alpha_0)=\Phi_n(\alpha_0,\alpha)\ . \tag{10.4.5}$$

10.4.2 Verification of the Boundary Conditions

Equation (10.4.3) shows that the scattered field can be written as

$$\begin{aligned}\Phi_n(\alpha,\alpha_0)&=e^{i\kappa\alpha}\sum_{m=1}^{n}(-1)^m\varphi_m(\alpha,-\beta_m)e^{iq(m-1)-i\kappa\beta_m}+\\&+e^{-i\kappa\alpha}\sum_{m=1}^{n}(-1)^m\varphi_m(-\alpha,\beta_m)e^{iq(m-1)+i\kappa\beta_m}+\\&+A\varphi_n(1,\alpha)e^{i\kappa\alpha}+B\varphi_n(1,-\alpha)e^{-i\kappa\alpha}\ .\end{aligned} \tag{10.4.6}$$

Imposing the requirement that

$$\Phi_n(\pm 1,\alpha_0)=0 \tag{10.4.7}$$

on $\Phi_n(\alpha, \alpha_0)$, we obtain the following linear equations

$$\left.\begin{aligned} A+B\frac{\varphi_n(1,1)}{\varphi_{n-1}(1,1)}e^{iq} &= -(-1)^n\frac{\varphi_n(1,\beta_n)}{\varphi_{n-1}(1,1)}e^{iqn+i\kappa\beta_n}, \\ A\frac{\varphi_n(1,1)}{\varphi_{n-1}(1,1)}e^{iq}+B &= -(-1)^n\frac{\varphi_n(1,-\beta_n)}{\varphi_{n-1}(1,1)}e^{iqn-i\kappa\beta_n}. \end{aligned}\right\} \tag{10.4.8}$$

Thus,

$$\begin{aligned} A(\beta_n) &= \frac{-(-1)^n}{\varphi_{n-1}(1,1)}\frac{e^{iqn}}{\mathcal{D}_n}\left[\varphi_n(1,\beta_n)e^{i\kappa\beta_n}-\varphi_n(1,-\beta_n)\frac{\varphi_n(1,1)}{\varphi_{n-1}(1,1)}e^{iq-i\kappa\beta_n}\right], \\ B(\beta_n) &= A(-\beta_n). \end{aligned} \tag{10.4.9}$$

Substituting the values for A and B into Eqn. (10.4.6), we see that it coincides with Eqn. (10.4.2) where $\tilde{\Phi}(\alpha,\alpha_0)$ is determined using Eqn. (10.4.3). Thus, the approximation derived for $\tilde{\Phi}_n(\alpha,\alpha_0)$ satisfies the strict relationship that

$$\Phi_n(\pm 1,\alpha_0)=\Phi_n(\alpha,\pm 1)=0. \tag{10.4.10}$$

10.4.3 Estimate of the Error

The absolute error of the nth-order approximation is determined mainly by the error in the approximation for the $(n+1)^{\text{th}}$ term, i.e., by the value of $U_{n+1,n}(\alpha,\alpha_0)$. Therefore, according to Eqns. (10.2.27) and (10.2.28), it is estimated to be

$$\begin{aligned} \left|\tilde{P}_n(\alpha,\alpha_0)\right| &= O\left[U_{n+1,n}(a,a_0)\right] \leqslant \\ &\leqslant \frac{O\left[q^{-\frac{n}{2}}\right]}{\left[1+\sqrt{q(1+\alpha)}\right]\left[1+\sqrt{q\left(1+(-1)^n\alpha_0\right)}\right]}, \end{aligned} \tag{10.4.11}$$

where $n = 2, 3\ldots$ and $q \gg n$. For $n = 2$ this estimate can be improved referencing the asymptotics of Eqns. (10.2.24) and (9.3.2). It turns out that

$$\left|\tilde{P}_2(\alpha,\alpha_0)\right| = O\left[U_{3,2}(\alpha,\alpha_0)\right] \leqslant$$
$$\leqslant \frac{\sqrt{1+\alpha_0}\sqrt{1+\alpha}\, O(q^{-1})}{\left[1+q(1+\alpha)\right]\left[1+q(1+\alpha_0)\right]} \quad \text{for} \quad q \gg 1. \tag{10.4.12}$$

In addition, Eqns. (10.1.15), (10.4.1), and (10.4.10) yield

$$P_n(\alpha,\pm 1) = P_n(\pm 1,\alpha_0) = 0\,. \tag{10.4.13}$$

This equality and Eqn. (10.4.12) show that Eqn. (10.4.11) provides an excessive value for the approximation error.

Thus, the nth-order approximation given by Eqn. (10.4.3) can determine the scattering pattern in the far zone to within an accuracy to any order of $(kl)^{-n/2}$ for $2kl \gg n$. According to Eqns. (10.1.16) and (10.4.11), the error of this approximation for any given value of q can, in principle, be made arbitrarily small by increasing the value of n. However, these expressions possess practical value only when $2kl \gg n$, when φ_n can be represented asymptotically and decreases with increasing n sufficiently quickly.

10.4.4 Total Scattering Cross Section

This section will conclude with a verification of these results. One can improve upon Eqn. (10.3.15), by using the second-order approximation to calculate more complete asymptotic expansions of the total scattering cross section given by Eqn. (10.3.14). Only the case of a normally incident plane wave is considered. According to Eqns. (10.4.2) and (10.4.3)

$$\Phi_2(\alpha,\alpha) = \tilde{\Phi}_2(\alpha,\alpha) + \tilde{\Phi}_2(-\alpha,-\alpha), \tag{10.4.14}$$

where

$$\tilde{\Phi}_2(\alpha,\alpha) = -\varphi_1(\alpha,\alpha)e^{iq\alpha} + \varphi_2(\alpha,-\alpha)e^{iq} -$$
$$-\frac{e^{i2q}}{\mathcal{D}_2}\left[\varphi_2^2(1,\alpha)e^{iq\alpha} - \varphi_2(1,\alpha)\varphi_2(1,-\alpha)\varphi_2(1,1)e^{iq}\right], \tag{10.4.15}$$

$$\mathcal{D}_2 = 1 - \left[\frac{\varphi_2(1,1)}{\varphi_1(1,1)} e^{iq} \right]^2, \quad \varphi_1(\alpha,\alpha) = \frac{1+\alpha}{2\alpha}, \quad \varphi_1(1,1) = 1. \tag{10.4.16}$$

Setting $\alpha = 0$ in Eqn. (10.4.14), we obtain

$$\Phi_2(0,0) = -1 - iq + 2\varphi_2(0,0)e^{iq} - 2\frac{\varphi_2^2(1,0)e^{i2q}}{1+\varphi_2(1,1)e^{iq}}, \tag{10.4.17}$$

and

$$\left.\begin{aligned} \varphi_2(1,0) &= \frac{\sqrt{2}}{i\pi} \int_1^{1+i\infty} \frac{e^{iq(t-1)}}{\sqrt{1-t}} \frac{dt}{t\sqrt{1+t}}, \\ \varphi_2(0,0) &= \frac{1}{i\pi} \int_1^{1+i\infty} \sqrt{\frac{1+t}{1-t}} \frac{e^{iq(t-1)}}{t^2} dt, \\ \varphi_2(1,1) &= \frac{2}{i\pi} \int_1^{1+i\infty} \frac{e^{iq(t-1)}}{\sqrt{1-t}} \frac{dt}{(1+t)^{3/2}}. \end{aligned}\right\} \tag{10.4.18}$$

For $q \gg 1$ the asymptotic expressions for these functions can be written as

$$\left.\begin{aligned} \varphi_2(1,0) &= \frac{e^{i\pi/4}}{\sqrt{\pi}} q^{-1/2} + \frac{5}{8}\frac{e^{-i\pi/4}}{\sqrt{\pi}} q^{-3/2} + O\left(q^{-5/2}\right), \\ \varphi_2(0,0) &= \sqrt{\frac{2}{\pi}} e^{-i\pi/4} q^{-1/2} + \frac{7}{4}\frac{e^{-i\pi/4}}{\sqrt{2\pi}} q^{-3/2} + O\left(q^{-5/2}\right), \\ \varphi_2(1,1) &= \frac{e^{i\pi/4}}{\sqrt{2\pi}} q^{-1/2} + \frac{3}{8}\frac{e^{-i\pi/4}}{\sqrt{2\pi}} q^{-3/2} + O\left(q^{-5/2}\right). \end{aligned}\right\} \tag{10.4.19}$$

Utilizing these equations to determine an expression for $\Phi_2(0,0)$ and substituting it into Eqn. (10.3.14), the asymptotic expansion of the total scattering cross section is found to be

$$S(0)=4l\left[1-2\sqrt{\frac{2}{\pi}}\frac{\cos(q-\pi/4)}{q^{3/2}}+\frac{2}{\pi}\frac{\cos(2q)}{q^{2}}-\right.$$

$$-\frac{1}{\pi}\sqrt{\frac{2}{\pi}}\frac{\cos(3q+\pi/4)-(7\pi/4)\cos(q+\pi/4)}{q^{5/2}}- \tag{10.4.20}$$

$$\left.-\frac{1}{\pi^{2}}\frac{\sin 4q-(5\pi/2)\sin 2q}{q^{3}}+O\left(q^{-7/2}\right)\right].$$

As should be expected, this expression for S completely agrees with Eqn. (31.5) of reference [121].

§10.5 Relationship Between Approximations for the Current and the Scattering Pattern

Previously, we used an independent method to find an approximate expression for the current density excited on a strip and the far field scattering pattern. We will now explain how these two expressions are related.

From Eqns. (10.1.1), (10.1.3), and (8.4.7), it follows that

$$\tilde{\Phi}(\alpha,\alpha_0)=2\pi ik\sqrt{1-\alpha^2}\sum_{n=1}^{\infty}F_n(k\alpha,k\alpha_0)e^{i\kappa\alpha}. \tag{10.5.1}$$

However, referring to Eqns. (8.3.1) and (8.3.2) this expression can be rewritten as

$$\tilde{\Phi}(\alpha,\alpha_0)=ik\sqrt{1-\alpha^2}e^{i\kappa\alpha}\int_{-\infty}^{\infty}\tilde{J}(z,\alpha_0)e^{-ikz\alpha}dz. \tag{10.5.2}$$

The quantity of $\tilde{\Phi}(\alpha,\alpha_0)$ is a function of $\varphi_n(\alpha,\alpha_0)$, but $\tilde{J}(z,\alpha_0)$ is a function of $\chi_n(z,\alpha_0)$. The connection between φ_n and χ_n is determined according to Eqns. (8.5.7) and (10.1.7) by

$$\chi_n(z,\alpha_0)e^{ikz} = -\frac{1}{2\pi i}\int_{-\infty}^{\infty} e^{ikzt}\frac{\varphi_n(t,\alpha_0)}{\sqrt{1-t^2}}dt, \tag{10.5.3}$$

$$\varphi_n(\alpha,\alpha_0) = -ik\sqrt{1-\alpha^2}\int_{-\infty}^{\infty} e^{ik(1-\alpha)z}\chi_n(z,\alpha_0)dz \quad . \tag{10.5.4}$$

Substituting Eqn. (9.6.3) into Eqn. (10.5.2) and utilizing Eqn. (10.5.4) we obtain

$$\begin{aligned}\tilde{\Phi}'_n(\alpha,\alpha_0) &= e^{i\kappa\alpha}\sum_{m=1}^{n}(-1)^m\varphi_m(\alpha,-\beta_m)e^{iq(m-1)-i\kappa\beta_m} - (-1)^n\frac{e^{iqn}}{\hat{D}_n}\times \\ &\times\left[\chi_n(2l,\beta_n)e^{i\kappa\beta_n} - \chi_n(2l,-\beta_n)\frac{\chi_n(2l,1)}{\chi_{n-1}(2l,1)}e^{iq-i\kappa\beta_n}\right]\frac{\varphi_n(1,\alpha)e^{i\kappa\alpha}}{\chi_{n-1}(2l,1)}.\end{aligned} \tag{10.5.5}$$

Since $\varphi_n(1,\alpha) \neq \chi_n(2l,\alpha)$, it is evident that

$$\tilde{\Phi}'_n(\alpha,\alpha_0) \neq \tilde{\Phi}_n(\alpha,\alpha_0), \tag{10.5.6}$$

where $\tilde{\Phi}_n(\alpha,\alpha_0)$ is determined by Eqn. (10.4.3), and in addition,

$$\Phi'_n(\alpha,\alpha_0) \neq \Phi'_n(\alpha_0,\alpha), \tag{10.5.7}$$

where

$$\Phi'_n(\alpha,\alpha_0) = \tilde{\Phi}'_n(\alpha,\alpha_0) + \tilde{\Phi}'_n(-\alpha,-\alpha_0). \tag{10.5.8}$$

It is also clear that $\Phi_n'(\alpha, \alpha_0)$ also fails to satisfy the boundary condition given by Eqn. (10.1.15). Thus, direct integration of the approximate expression for the current, the error of which can be made arbitrarily small, leads to an equation for the scattering pattern that fails to satisfy both the reciprocity theorem and the boundary conditions.

This result is rather unexpected and therefore unusual. It is well known that the integration of the current $j^{(0)}(z, \alpha_0)$ used in physical optics gives an expression for the scattered field that fails to satisfy both the reciprocity theorem and the boundary conditions (see, for instance, [13]). An analogous fact has been observed for thin dipoles [14]. It was expected, however, that if the current were calculated with sufficient accuracy, its integration would produce a scattering pattern that would satisfy both the reciprocity theorem and the boundary conditions. We now see that such expectations are not justified.

Because of this, an important role will be played by the development of methods that allow the use of approximations for the current to determine a scattered field that satisfies the reciprocity theorem and the boundary conditions. An example of such a method was the development of the theory of radiation in Chapter 7.

We will once again return to Eqn. (10.5.5) and estimate the difference

$$\tilde{\Phi}_n'(\alpha,\alpha_0)-\tilde{\Phi}_n(\alpha,\alpha_0)=(-1)^n\varphi_n(1,\alpha)e^{iqn+i\kappa\alpha}\times$$
$$\times\left\{\frac{1}{\varphi_{n-1}(1,1)\mathcal{D}_n}\left[\varphi_n(1,\beta_n)e^{i\kappa\beta_n}-\varphi_n(1,-\beta_n)\frac{\varphi_n(1,1)}{\varphi_{n-1}(1,1)}e^{iq-i\kappa\beta_n}\right]-\right. \tag{10.5.9}$$
$$\left.-\frac{1}{\chi_{n-1}(2l,1)\hat{\mathcal{D}}_n}\left[\chi_n(2l,\beta_n)e^{i\kappa\beta_n}-\chi_n(2l,-\beta_n)\frac{\chi_n(2l,1)}{\chi_{n-1}(2l,1)}e^{iq-i\kappa\beta_n}\right]\right\}.$$

In accordance with Eqns. (9.4.8), (9.4.10), and (10.2.14)

$$\chi_n(2l,\alpha)=\varphi(q,\alpha)\varphi^{n-1}(q,1)\left[1+O\left(\frac{n}{q}\right)\right], \tag{10.5.10}$$

$$\varphi_n(\alpha,1)=\varphi(q,\alpha)\varphi^{n-2}(q,1)\left[1+O\left(\frac{n}{q}\right)\right], \tag{10.5.10, cont.}$$

for $q \gg n$. With the use of these relationships, it can be shown that if $q \gg n$

$$\left|\frac{\chi_n(2l,\beta_n)}{\chi_{n-1}(2l,1)}-\frac{\varphi_n(1,\beta_n)}{\varphi_{n-1}(1,1)}\right|\leqslant\frac{O(q^{-1})}{1+\sqrt{q(1+\beta_n)}}, \tag{10.5.11}$$

$$\left|\mathcal{D}_n-\hat{\mathcal{D}}_n\right|\leqslant O(q^{-2}), \tag{10.5.12}$$

$$\left|\frac{\chi_n(2l,-\beta_n)\chi_n(2l,1)}{\chi_{n-1}^2(2l,1)}-\frac{\varphi_n(1,-\beta_n)\varphi_n(1,1)}{\varphi_{n-1}^2(1,1)}\right|\leqslant\frac{O(q^{-3/2})}{1+\sqrt{q(1-\beta_n)}}. \tag{10.5.13}$$

It follows that

$$\left|\tilde{\Phi}_n'(\alpha,\alpha_0)-\tilde{\Phi}_n(\alpha,\alpha_0)\right|\leqslant\frac{O(q^{-n/2})}{\left[1+\sqrt{q(1+\alpha)}\right]\left[1+\sqrt{q(1+(-1)^n\alpha_0)}\right]}, \tag{10.5.14}$$

for $q \gg n$.

Comparing this inequality with Eqn. (10.4.11), we see that $\tilde{\Phi}_n'$ agrees with $\tilde{\Phi}_n$ within an accuracy on the order of $\tilde{P}_n(\alpha,\alpha_0)$, which is the order of error with which $\tilde{\Phi}_n(\alpha,\alpha_0)$ is determined.

The approximate expression for the far field scattering pattern will now be used to calculate the current on a strip. From Eqn. (10.5.2), we have

$$\tilde{J}(z,\alpha_0)=\frac{1}{2\pi i}\int_{-\infty}^{\infty}e^{ik(z-l)\alpha}\frac{\tilde{\Phi}(\alpha,\alpha_0)}{\sqrt{1-\alpha^2}}d\alpha. \tag{10.5.15}$$

Substituting the approximation for $\tilde{\Phi}_n(\alpha,\alpha_0)$ of Eqn. (10.4.3), we find that

$$\tilde{J}_n'(z,\alpha_0)=-e^{ikz}\sum_{m=1}^{n}(-1)^m\chi_m(z,-\beta_m)e^{iq(m-1)-i\kappa\beta_m}+(-1)^n\frac{e^{iqn}}{\mathcal{D}_n}\times$$
$$\times\left[\varphi_n(1,\beta_n)e^{i\kappa\beta_n}-\varphi_n(1,-\beta_n)\frac{\varphi_n(1,1)}{\varphi_{n-1}(1,1)}e^{iq-i\kappa\beta_n}\right]\frac{\chi_n(z,1)}{\chi_{n-1}(2l,1)}e^{ikz}. \tag{10.5.16}$$

Comparing this formula with Eqn (9.6.3), we see that they differ. It follows that the current

$$J_n'(z,\alpha_0)=j^{(0)}(z,\alpha_0)+\tilde{J}_n'(z+l,\alpha_0)+\tilde{J}_n'(l-z,-\alpha_0), \tag{10.5.17}$$

calculated from an arbitrarily accurate expression for the scattering pattern does not equal zero at the edges of the strip and fails to approach zero when the illuminating plane wave reaches grazing angles of incidence:

$$J_n'(\pm l,\alpha_0)\neq 0,\quad J_n'(z,\pm 1)\neq 0. \tag{10.5.18}$$

That is, $J_n'(z,\alpha_0)$ does not satisfy the edge and limiting conditions given by Eqns. (8.5.19) and (8.5.20).

Using Eqns. (10.5.11), (10.5.12), and (10.5.13), it is easily shown that

$$\tilde{J}_n'(z,\alpha_0)-\tilde{J}_n(z,\alpha_0)=O\left[\tilde{R}_n(z,\alpha_0)\right] \tag{10.5.19}$$

for $q \gg n$ and $0 \leqslant z \leqslant \infty$, where $\tilde{R}_n(z,\alpha_0)$ is determined using Eqn. (9.6.13). It goes without saying that $\tilde{J}_n'(z,\alpha_0)$ calculated using $\tilde{\Phi}_n(\alpha,\alpha_0)$ differs from $\tilde{J}(z,\alpha_0)$ by a value on the order of its error.

§10.6 Additional Comments

- Some results of this chapter were published in references [18, 22, 24].
- The first approximation given by Eqn. (10.3.6) was used in [67] for calculation of the scattered field. As was shown in reference [68], the data in Fig. 8 of [67] related to Eqn. (10.3.6) are the results of incorrect numerical calculations in [67].

11

Plane Wave Diffraction at a Strip Oriented in the Direction of Polarization (Dirichlet Boundary Problem)

Previously, Chapters 8–10 examined the asymptotic solution to the Neumann boundary problem. This chapter will present an analogous but abbreviated examination of the Dirichlet boundary problem.

§11.1 Formulation and Solution of the Functional Equations

Let there be a plane wave

$$E_x = e^{-ik\left(z\alpha_0 + y\sqrt{1-\alpha_0^2}\right) - i\omega t}, \quad H_x = 0, \quad \alpha_0 = \cos\vartheta_0 \tag{11.1.1}$$

incident on a perfectly conducting strip bounded by the coordinates $y = 0$, $-\infty \leqslant x \leqslant \infty$, and $-l \leqslant z \leqslant l$. As was done in § 8.2, it is possible to show that edge waves of a scattered field exhibit symmetry. Therefore, the incident wave of Eqn. (11.1.1) excites a current on a strip given by

$$j_x(z,\alpha_0) = \frac{c}{2\pi} J(z,\alpha_0), \tag{11.1.2}$$

where

$$J(z,\alpha_0) = j^{(0)}(z,\alpha_0) + \tilde{J}(z+l,\alpha_0) + \tilde{J}(l-z,-\alpha_0). \tag{11.1.3}$$

Here,

$$j^{(0)}(z,\alpha_0) = \sqrt{1-\alpha_0^2}\, e^{-ikz\alpha_0} \tag{11.1.4}$$

represents the current excited on an infinite plane ($-\infty \leqslant z \leqslant \infty, y = 0$) and

$$\tilde{J}(z,\alpha_0) = \sum_{n=1}^{\infty} j_n(z,\alpha_0) \tag{11.1.5}$$

describes the current waves, excited due to the diffraction occurring at the ends of the strip.

Each term of the edge wave will be presented as a Fourier transform

$$j_n(z,\alpha_0) = \int_{-\infty}^{\infty} G_n(t,k\alpha_0)e^{izt}\,dt \tag{11.1.6}$$

and $\mathbf{E}_n(z, y, \alpha_0)$ will be the electric vector of the field radiated by this current. According to Eqn. (1.2.2) the tangential component E_x is equal to

$$E_x = ikA_x, \tag{11.1.7}$$

where

$$A_x = \frac{i}{2}\int_{-\infty}^{\infty} j_n(\zeta,\alpha_0) H_0^{(1)}\left(k\sqrt{y^2+(z-\zeta)^2}\right) d\zeta. \tag{11.1.8}$$

Replacing the Hankel function with Eqn. (8.3.5), we obtain

$$A_x = i\int_{-\infty}^{\infty} G_n(w,k\alpha_0)\frac{e^{i(vy+wz)}}{v}dw\,, \quad y \geqslant 0. \tag{11.1.9}$$

Thus,

$$E_{nx}(z,0,\alpha_o) = -k\int_{-\infty}^{\infty} G_n(t,k\alpha_0)\frac{e^{izt}}{\sqrt{k^2-t^2}}dt\,. \tag{11.1.10}$$

As a result, the conditions

$$J(z,\alpha_0)=0 \quad \text{for} \quad |z|>l \tag{11.1.11}$$

$$E_{nx}(\zeta,0,\alpha_0)=0 \quad \text{for} \quad \zeta>0 \tag{11.1.12}$$

lead to the recursive system of functional equations.

$$\left.\begin{aligned}
&\int_{-\infty}^{\infty} G_1(t,k\alpha_0)e^{i\zeta t}dt = -j^{(0)}(\zeta-l,\alpha_0), \quad \zeta<0\\
&\int_{-\infty}^{\infty} \left[G_n(t,k\alpha_0)+G_{n-1}(-t,-k\alpha_0)e^{-i2lt}\right]e^{i\zeta t}dt = 0, \quad \zeta<0, \; n-2,3,\ldots\\
&\int_{-\infty}^{\infty} G_n(t,k\alpha_0)\frac{e^{i\zeta t}}{\sqrt{k^2-t^2}}dt = 0, \quad \zeta>0, \quad n=1,2,\ldots
\end{aligned}\right\} \tag{11.1.13}$$

where the contour of integration traverses along the real axis and skirts above the branch point at $t=-k$ and skirts below the branch point at $t=k$.

Applying the factorization method to this system of equations, it is found that

$$G_1(k\alpha, k\alpha_0) = \frac{1}{2\pi ik}\frac{\sqrt{1-\alpha}\sqrt{1-\alpha_0}}{\alpha+\widehat{\alpha_0}}e^{ik\alpha_0}, \tag{11.1.14}$$

$$\underset{n=2,3,\ldots}{G_n(t, k\alpha_0)} = -\frac{\sqrt{k-t}}{2\pi i}\int_{-\infty}^{\infty}\frac{e^{-i2lu}}{\sqrt{k-u}(u-\underset{\smile}{t})}G_{n-1}(-u, -k\alpha_0)du. \tag{11.1.15}$$

Here, as before, the $\frown$ and $\smile$ symbols indicate how the contour of integration skirts around the corresponding pole.

§11.2 Scattering Pattern and the Edge Wave Equation

Equations (11.1.7) and (11.1.9) show that the surface currents $j_n(z+l, \alpha_0)$ and $j_n(l-z, -\alpha_0)$ respectively radiate the fields

$$\left.\begin{aligned} E_{nx} &= -k\int_{-\infty}^{\infty}G_n(w, k\alpha_0)\frac{e^{ivy}}{v}e^{iw(z+l)}dw, \\ E_{nx} &= -k\int_{-\infty}^{\infty}G_n(-w, -k\alpha_0)\frac{e^{ivy}}{v}e^{iw(z-l)}dw \end{aligned}\right\} y \geqslant 0. \tag{11.2.1}$$

where $v = (k^2-w^2)^{1/2}$ and $\Im\mathfrak{m}\{v\} \geq 0$. Thus, the total scattered field for $y \geqslant 0$ is

$$E_x = -e^{ik\left(-z\alpha_0 + y\sqrt{1-\alpha_0^2}\right)} - \\ -k\int_{-\infty}^{\infty}\sum_{n=1}^{\infty}\left[G_n(w, k\alpha_0)e^{iwl} + G_n(-w, -k\alpha_0)e^{-iwl}\right]\frac{e^{i(vy+wz)}}{v}dw. \tag{11.2.2}$$

Making the substitutions

$$w = k\cos\tau, \quad v = k\sin\tau, \quad z = r\cos\vartheta, \quad y = r\sin\vartheta$$

and applying the method of steepest descent [78], scattered far field is

$$E_x = 2\pi ik\sum_{n=1}^{\infty}\left[G_n(k\alpha, k\alpha_0)e^{i\kappa\alpha} + G_n(-k\alpha, -k\alpha_0)e^{-i\kappa\alpha}\right]\frac{e^{i(kr+\pi/4)}}{\sqrt{2\pi kr}}, \qquad (11.2.3)$$

where $\alpha = \cos\vartheta$ and $\alpha_0 = \cos\vartheta_0$.

Let us introduce the function

$$G(\alpha, \alpha_0) = \sum_{n=1}^{\infty} G_n(k\alpha, k\alpha_0). \qquad (11.2.4)$$

According to Eqn. (11.1.15) the terms of this series are determined by the recursive formula

$$\underset{n=2,3,\ldots}{G_n(k\alpha, k\alpha_0)} = \frac{\sqrt{1-\alpha}}{2\pi i}\int_{-\infty}^{\infty}\frac{e^{iqu}}{\sqrt{1+u}\,(u+\widehat{\alpha})}G_{n-1}(ku, -k\alpha_0)\,du, \qquad (11.2.5)$$

where $q = 2\kappa = 2kl$ and the contour of integration skirts above the branch point at $u = -1$, above the pole at $u = -\alpha$, and skirts below the branch point at $u = 1$. It is evident that the series given by Eqn. (11.2.4) is a Neumann series of the integral equation

$$G(\alpha, \alpha_0) = \tilde{G}_1(k\alpha, k\alpha_0) + \frac{\sqrt{1-\alpha}}{2\pi i}\int_{-\infty}^{\infty}\frac{e^{iqu}}{\sqrt{1+u}\,(u+\widehat{\alpha})}G(u, -\alpha_0)\,du, \qquad (11.2.6)$$

where

$$\tilde{G}_1(k\alpha, k\alpha_0) = \frac{\sqrt{1-\alpha}\sqrt{1-\alpha_0}}{\alpha+\alpha_0} e^{i\kappa\alpha_0} . \qquad (11.2.7)$$

This integral equation was first obtained (for $G(-\alpha, \alpha_0)$) by reference [124] using a different approach and was utilized in references [122,125, 134] for an asymptotic solution to this problem. Our derivation provides a physical meaning: the nth term in the Neumann series of this equation describes the scattering pattern of the nth-order edge wave. Therefore, either Eqn. (11.2.6) or Eqn. (8.4.10) can be called the *edge wave equation.*

§11.3 Infinite Series and the Integral Equation for the Current

11.3.1 Series of Functions $\xi_n(z, \alpha_0)$ and Some of Their Properties

Utilizing Eqns. (11.1.5), (11.1.6), and (11.1.14), (11.1.15) it is possible to obtain the following expression for the current

$$\tilde{J}(z, \alpha_0) = e^{ikz} \sum_{m=1}^{\infty} (-1)^m \xi_m(z, -\beta_m) e^{iq(m-1)-i\kappa\beta_m} , \qquad (11.3.1)$$

where, as before, $\beta_m = (-1)^m \alpha_0$ and

$$\xi_1(z, \alpha) = -\frac{1}{2\pi i} \int_{-\infty}^{\infty} e^{ikz(t-1)} \frac{\sqrt{1-t}}{t+\widehat{\alpha}} dt , \qquad (11.3.2)$$

$$\xi_2(z, \alpha) e^{iq+ikz} = -\frac{1}{2(\pi i)^2} \int_1^{1+i\infty} \frac{g(t_1) dt_1}{t_1+\alpha} \int_{-\infty}^{\infty} e^{ikzt_2} \frac{\sqrt{1-t_2}}{t_2+\widehat{t_1}} dt_2 , \qquad (11.3.3)$$

$$\underset{n=3,4,\ldots}{\xi_n(z,\alpha)e^{iq(n-1)+ikz}} = -\frac{1}{2(\pi i)^n}\times$$

$$\times\int_1^{1+i\infty}\cdots\int_1^{1+i\infty}\frac{g(t_1)g(t_2)\ldots g(t_{n-1})dt_1dt_2\ldots dt_{n-1}}{(t_1+\alpha)(t_2+t_1)(t_3+t_2)\ldots(t_{n-1}+t_{n-2})}\times \quad (11.3.4)$$

$$\times\int_{-\infty}^{\infty} e^{ikzt_n}\frac{\sqrt{1-t_n}}{t_n+t_{n-1}}dt_n\,,$$

where

$$g(t)=\sqrt{\frac{1-t}{1+t}}e^{iqt}\,,\quad q=2\kappa,\quad \kappa=kl\,. \qquad (11.3.5)$$

For positive values of z, we can write these equations as

$$\xi_1(z,\alpha)=\sqrt{2}\psi(kz,\alpha),\quad \psi(kz,\alpha)=\frac{i}{\sqrt{2(1+\alpha)}}\varphi_{kz}^{(1)}(kz,\alpha) \qquad (11.3.6)$$

$$\xi_2(z,\alpha)e^{iq+ikz}=\frac{1}{(\pi i)^2}\int_1^{1+i\infty}\frac{g(t_1)dt_1}{t_1+\alpha}\int_1^{1+i\infty}e^{ikzt_2}\frac{\sqrt{1-t_2}}{t_2+t_1}dt_2\,, \qquad (11.3.7)$$

$$\underset{n=3,4,\ldots}{\xi_n(z,\alpha)e^{iq(n-1)+ikz}} = \frac{1}{(\pi i)^n}\times$$

$$\times\int_1^{1+i\infty}\cdots\int_1^{1+i\infty}\frac{g(t_1)g(t_2)\ldots g(t_{n-1})dt_1dt_2\ldots dt_{n-1}}{(t_1+\alpha)(t_2+t_1)(t_3+t_2)\ldots(t_{n-1}+t_{n-2})}\times \qquad (11.3.8)$$

$$\times \int_{1}^{1+i\infty} e^{ikzt_n} \frac{\sqrt{1-t_n}}{t_n + t_{n-1}} dt_n, \tag{11.3.8, cont.}$$

where the integral is evaluated on the left side of the branch cut extending from $1 \to 1+i\infty$, where $\arg\{(1-t)^{1/2}\} = -\pi/4$, and $\varphi_q^{(1)}(q, \alpha)$ is the derivative of the Fresnel integral (see, Eqns. (9.2.6) and (9.2.9)).

From these formulas it follows that for $z > 0$ and $z \to 0$

$$\xi_n(z,\alpha) = O\left(\frac{1}{\sqrt{kz}}\right) \quad \text{for} \quad n = 1, 2, \ldots, \tag{11.3.9}$$

i.e., near the edge of the strip each partial edge current wave possesses the same singularity as a Sommerfeld first-order wave.

Moreover, for $z < 0$ Eqns. (11.3.2), (11.3.3), and (11.3.4) indicate that

$$\left.\begin{aligned} \xi_1(z,\alpha) &= \sqrt{1+\alpha}\, e^{-ikz(1+\alpha)}, \\ \xi_n(z,\alpha) &= \xi_{n-1}(2l - z, \alpha) e^{-i2kz}, \end{aligned}\right\} z < 0 \tag{11.3.10}$$

which easily verify that $J(z, \alpha_0) = 0$ for $|z| > l$.

11.3.2 Integral Equation for the Current

Multiplying both parts of Eqn. (11.1.15) by e^{izt} and integrating with respect to t from $-\infty$ to ∞, we obtain a recursive formula for the current

$$j_n(z,\alpha_0) = \frac{1}{\pi} \int_{-\infty}^{\infty} K(s - 2l, z) j_{n-1}(s, -\alpha_0) ds, \tag{11.3.11}$$

where

$$K(x,z) = -\frac{1}{4\pi i}\int_{-\infty}^{\infty} e^{izt}\sqrt{k-t}\,dt \int_{-\infty}^{\infty} \frac{e^{ixu}}{\sqrt{k-u}}\frac{du}{u-t}, \tag{11.3.12}$$

from which

$$K(x,z) = 0 \quad \text{for} \quad x < 0, \tag{11.3.13}$$

$$K(x,z) = -\frac{1}{2}\int_{-\infty}^{\infty} e^{i(x+z)u}\,du = -\pi\delta(x+z) \quad \text{for} \quad z < 0. \tag{11.3.14}$$

It follows that Eqn. (11.3.11) can be rewritten as

$$j_n(z,\alpha_0) = \frac{1}{\pi}\int_0^{\infty} K(x,z)\,j_{n-1}(2l+x,-\alpha_0)\,dx, \tag{11.3.15}$$

yielding, in particular, that

$$j_n(z,\alpha_0) = -j_{n-1}(2l-z,-\alpha_0) \quad \text{for} \quad z < 0. \tag{11.3.16}$$

It is evident that the series

$$j(z,\alpha_0) = \sum_{n=1}^{\infty} j_n(z,\alpha_0) \tag{11.3.17}$$

is the Neumann series for an integral equation of the second kind,

$$j(z,\alpha_0) = j_1(z,\alpha_0) + \frac{1}{\pi}\int_0^{\infty} K(x,z)\,j(2l+x,-\alpha_0)\,dx, \tag{11.3.18}$$

where

$$j_1(z,\alpha_0) = -\sqrt{1-\alpha_0}\,\xi_1(z,\alpha_0)e^{ikz+i\kappa\alpha_0}. \tag{11.3.19}$$

Applying Babinet's principle to the results of Schwarzschild [139], it is possible to show that the edge wave current given by Eqn. (11.3.17) satisfies Eqn. (11.3.18) when using a kernel given by

$$\hat{K}(x,z) = \sqrt{\frac{x}{z}}\frac{e^{ik(x+z)}}{x+z} \quad \text{for} \quad x>0 \quad \text{and} \quad z>0. \tag{11.3.20}$$

Calculations similar to those performed in § 8.7 ensure the equivalence of kernels given by Eqns. (11.3.12) and (11.3.20). Thus, as was the case with the Neumann boundary problem, the solution given by Eqn. (11.3.1) is equivalent to Schwarzschild's solution and differs only in the form written. Schwarzschild demonstrated the uniform convergence of his solution. Therefore, there is no need to dwell on this issue other than cite the proof provided earlier in Chapter 8.

We have now obtained a formal solution to the functional equations for the Dirichlet boundary problem and established a connection of this solution with independent well-established research. The next section will present an asymptotic representation of this solution.

§11.4 Asymptotic Representation for $\xi_n(z,\alpha)$

Applying Lemma 2 of § 9.1 to $\xi_n(z,\alpha)$ for $z \geq 0$ one can obtain the following asymptotic relationship as $q \to \infty$:

$$\xi_2(z,\alpha) \sim \sqrt{2}\psi(q,\alpha)\psi(kz,1) + \\ + \frac{1}{i\pi}\sum_{m=1}^{\infty}\frac{\sqrt{2}}{i^m}\frac{\partial^m \psi(q,\alpha)}{\partial q^m}\int_1^{1+i\infty} e^{ikz(t-1)}\sqrt{1-t}\,d_m(t)dt \tag{11.4.1}$$

where $d_m(t)$ is the Taylor series coefficient

$$(1+x)^{-1/2}(x+t)^{-1}=\sum_{m=0}^{\infty}d_m(t)(x-1)^m, \tag{11.4.2}$$

and

$$\xi_n(z,\alpha)\sim\sqrt{2}\psi(q,\alpha)\psi^{n-2}(q,1)\psi(kz,1)+$$

$$+\frac{1}{(i\pi)^n}\sum_{m=1}^{\infty}\frac{1}{m!}\int_1^{1+i\infty}e^{ikz(t_n-1)}\sqrt{1-t_n}dt_n\times \tag{11.4.3}$$

$$\times\int_1^{1+i\infty}\ldots\int_1^{1+i\infty}e^{iq\sum_{s=1}^{n-1}(t_s-1)}\frac{\prod_{s=1}^{n-1}\sqrt{1-t_s}\,v_m(t_1,\ldots t_n)dt_1\ldots dt_{n-1}}{(t_1+\alpha)\prod_{s=2}^{n-1}(1+t_s)},$$

where

$$v_m(t_1,t_2,\ldots t_n)=$$

$$=\left[(t_1-1)\frac{\partial}{\partial t_1}+(t_2-1)\frac{\partial}{\partial t_2}+\ldots+(t_{n-1}-1)\frac{\partial}{\partial t_{n-1}}\right]^m g_n(1,1,\ldots,1,t_n) \tag{11.4.4}$$

$$\underset{n=3,4,\ldots}{g_n(t_1,t_2,\ldots,t_n)}=\frac{(1+t_1)^{-1/2}\prod_{s=2}^{n-1}\sqrt{1+t_s}}{(t_1+t_2)(t_2+t_3)\ldots(t_{n-1}+t_n)}. \tag{11.4.5}$$

According to Eqn. (11.3.6),

$$\psi(q,\alpha)=\frac{1}{i\pi\sqrt{2}}\int_1^{1+i\infty}e^{iq(t-1)}\frac{\sqrt{1-t}}{t+\alpha}dt=\frac{i}{\sqrt{2(1+\alpha)}}\varphi_q^{(1)}(q,\alpha). \tag{11.4.6}$$

The right hand side of Eqn. (11.4.1), as well as each term in the series of Eqn. (11.4.3), is a sum of products of single integrals of the type $\varphi_q^{(m)}(q,\alpha)$ and

$$\frac{\sqrt{2}}{i\pi}\int_1^{1+i\infty}\frac{e^{iz(t-1)}}{\sqrt{1-t}}\frac{t-1}{(1+t)^m}dt=\frac{1}{i}\frac{d}{dz}\hat{\varphi}_m(z), \tag{11.4.7}$$

where $\varphi_q^{(m)}(q,\alpha)$ and $\hat{\varphi}_m(z)$ are determined using Eqns. (9.2.6) and (9.2.7). One should emphasize the following. Though these functions look quite complicated, they can be expressed only through Fresnel integral and power and exponential functions of q, α, and z.

The nature of the origination of functions $\varphi_q^{(m)}(q,\alpha)$ and $\hat{\varphi}_m(z)$ was explained in § 9.2. Functions $\varphi_q^{(m)}(q,\alpha)$ are generated by terms $(t_s-1)^m$ in the Taylor expansion given by Eqn. (11.4.4). Functions $\hat{\varphi}_m(z)$ result from the differentiation of Eqn. (11.4.5) with respect to variable t_{n-1} which is coupled with variable t_n in the factor $(t_{n-1}+t_n)^{-1}$. The mth order derivative of this factor at the point $t_{n-1}=1$ generates the quantity $(1+t_n)^{-m}$ and subsequently the function $\hat{\varphi}_m(z)$.

Asymptotic properties of functions $\varphi_q^{(m)}(q,\alpha)$ are described by Eqn. (9.3.2) while the asymptotic behavior of functions $d\hat{\varphi}_m(z)/dz$ follow directly from Eqn. (11.4.7). Introducing a new variable, $x=z(t-1)$, one immediately sees that

$$\frac{d}{dz}\hat{\varphi}_m(z)=O\left[z^{-3/2}\right] \quad \text{for} \quad z\to\infty. \tag{11.4.8}$$

Setting $z=0$ in Eqn. (11.4.7) shows that

$$\frac{d}{dz}\hat{\varphi}_m(z)=O[1] \quad \text{for} \quad m=2,3,4\ldots \quad \text{as} \quad z\to 0. \tag{11.4.9}$$

According to Eqns. (9.2.7) and (9.3.11)

$$\frac{d}{dz}\hat{\varphi}_1(z)=\frac{d}{dz}\varphi(z,1)=O\left[z^{-1/2}\right] \quad \text{for} \quad z\to 0. \tag{11.4.10}$$

In this chapter, the function $\psi(kz,\alpha)$ plays a key role. Its behavior is determined by Eqn. (11.4.6) and by the asymptotic properties of the function $\varphi(kz,\alpha)$ described by Eqns. (9.3.1), (9.3.2) and (9.3.11). In accordance with these expressions,

$$\psi(q,\alpha)=O\left[\frac{q^{-1/2}}{1+q(1+\alpha)}\right],$$
$$\psi_q^{(1)}(q,\alpha)=\frac{\partial}{\partial q}\psi(q,\alpha)=O\left[\frac{q^{-3/2}}{1+q(1+\alpha)}\right] \tag{11.4.11}$$

for $q \gg 1$ and $-1 \le \alpha \le 1$. From a comparison of functions $\psi(kz,1)$ and $d\hat{\varphi}_2(kz)/dkz$, it also follows that

$$d\hat{\varphi}_2(kz)/dkz = O[\psi(kz,1)] \quad \text{for} \quad kz\to\infty, \tag{11.4.12}$$

$$\left[\frac{d\hat{\varphi}_2(kz)}{dkz}\Big/\psi(kz,1)\right]\to 0 \quad \text{for} \quad kz\to 0. \tag{11.4.13}$$

These two asymptotics can be combined in the following relationship

$$\left|\frac{d}{dkz}\hat{\varphi}_2(kz)\right| \leqslant O[\psi(kz,1)] \quad \text{for} \quad 0\leqslant kz\leqslant\infty. \tag{11.4.14}$$

Using these asymptotic expressions for functions $\varphi_q^{(m)}(q,\alpha)$, $\hat{\varphi}_m(kz)$, and $\psi(q,\alpha)$, one can derive key approximations for the function $\xi_n(z,\alpha)$ similar to those in Chapter 9. Let us first examine the following approximation

$$\xi_n(z,\alpha)\approx\sqrt{2}\psi(q,\alpha)\psi^{n-2}(q,1)\psi(kz,1)+T_{1\xi} \tag{11.4.15}$$

where

$$T_{1\xi} = \frac{1}{(i\pi)^n} \int\limits_{1}^{1+i\infty} \sqrt{1-t_n}\, e^{ikz(t_n-1)} dt_n \times$$

$$\times \int\limits_{1}^{1+i\infty} \cdots \int\limits_{1}^{1+i\infty} e^{iq\sum\limits_{s=1}^{n-1}(t_s-1)} \frac{\prod\limits_{s=1}^{n-1}\sqrt{1-t_s}\, v_1(t_1,t_2,\ldots,t_n) dt_1 dt_2 \ldots dt_{n-1}}{(t_1+\alpha)\prod\limits_{s=2}^{n-1}(1+t_s)}, \tag{11.4.16}$$

$$v_1(t_1,t_2,\ldots,t_{n-1},t_n) =$$

$$= \left[(t_1-1)\frac{\partial}{\partial t_1} + (t_2-1)\frac{\partial}{\partial t_2} + \ldots + (t_{n-1}-1)\frac{\partial}{\partial t_{n-1}}\right] g_n(1,1,\ldots,1,t_n), \tag{11.4.17}$$

The function g_n is defined by Eqn. (11.4.5). It is necessary only to note that it contains the factor $(t_{n-1}+t_n)^{-1}$ which couples t_{n-1} with t_n and generates the factor $(1+t_n)^{-2}$ in function v_1 and the factor $d\hat{\varphi}_2(kz)/dkz$ in the function $T_{1\xi}$. Omitting other details of calculation that were previously explained in Chapter 9, we provide the final asymptotic estimation for $T_{1\xi}$

$$T_{1\xi} = \psi(kz,1)\psi^{n-3}(q,1)\times$$

$$\times\left\{O\left[\psi_q^{(1)}(q,\alpha)\psi(q,1)\right] + (n-2)O\left[\psi(q,\alpha)\psi_q^{(1)}(q,1)\right] + \right. \tag{11.4.18}$$

$$\left. +\psi(q,\alpha)\psi_q^{(1)}(q,1)O\left[\frac{d\hat{\varphi}_2(kz)/dkz}{\psi(kz,1)}\right]\right\}.$$

After substituting this result into Eqn. (11.4.15), the latter takes the form

$$\xi_n(z,\alpha) \approx \sqrt{2}\psi(q,\alpha)\psi^{n-2}(q,1)\psi(kz,1)\times$$
$$\times\left\{1+O\left[\frac{\psi_q^{(1)}(q,\alpha)}{\psi(q,\alpha)}\right]+(n-2)O\left[\frac{\psi_q^{(1)}(q,1)}{\psi(q,1)}\right]+\right. \tag{11.4.19}$$
$$\left.+\frac{\psi_q^{(1)}(q,1)}{\psi(q,1)}O\left[\frac{d\hat{\varphi}_2(kz)/dkz}{\psi(kz,1)}\right]\right\}.$$

The asymptotic estimations shown in Eqns. (11.4.11) and (11.4.14) allow one to simplify this expression as

$$\xi_n(z,\alpha) \approx \sqrt{2}\psi(q,\alpha)\psi^{n-2}(q,1)\psi(kz,1)\left\{1+O\left[\frac{n}{q}\right]\right\} \tag{11.4.20}$$

with $q \gg n, 0 \le z \le \infty$, and $-1 \le \alpha \le 1$.

Utilizing these formulas, it is not difficult to show that

$$\left|\xi_{n+m}(z,\alpha)-\frac{\xi_n(z,1)\xi_n(2l,\alpha)}{\xi_{n-1}(2l,1)}\left[\frac{\xi_n(2l,1)}{\xi_{n-1}(2l,1)}\right]^{m-1}\right| \leqslant$$
$$\leqslant \left|\psi(q,\alpha)\psi(kz,1)\psi^{m+n-2}(q,1)\right| O\left(\frac{n+m}{q}\right) \tag{11.4.21}$$

or

$$\left|\xi_{n+m}(z,\alpha)-\frac{\xi_n(z,1)\xi_n(2l,\alpha)}{\xi_{n-1}(2l,1)}\left[\frac{\xi_n(2l,1)}{\xi_{n-1}(2l,1)}\right]^{m-1}\right| \leqslant$$
$$\leqslant O\left[\frac{q^{-3(m+n-1)/2}}{\sqrt{kz}(1+kz)\left[1+q(1+\alpha)\right]}\right] \tag{11.4.22}$$

for $q \gg n+m$ and $0 \leqslant z \leqslant \infty$. The asymptotic representations given by Eqns. (11.4.20) and (11.4.22) play an important role in the development of approximate expressions for the current.

§11.5 First-Order Approximation for the Current

Let us present the current of Eqn. (11.3.1) in the form

$$\begin{aligned}\tilde{J}(z,\alpha_0) &= -\sqrt{1-\alpha_0}\, e^{i\kappa\alpha_0}\xi_1(z,\alpha_0)e^{ikz} + \\ &+ e^{ikz}\sqrt{1+\alpha_0}\, e^{-i\kappa\alpha_0}\sum_{m=2,4,\ldots}^{\infty}\xi_m(z,-\alpha_0)e^{iq(m-1)} - \\ &- e^{ikz}\sqrt{1-\alpha_0}\, e^{i\kappa\alpha_0}\sum_{m=3,5,\ldots}^{\infty}\xi_m(z,\alpha_0)e^{iq(m-1)} .\end{aligned} \tag{11.5.1}$$

Substituting the first terms of the asymptotic representations of Eqns. (11.4.1) and (11.4.20) into Eqn. (11.5.1), we obtain

$$\tilde{J}(z,\alpha_0) = \tilde{J}_1(z,\alpha_0) + \tilde{R}_1(z,\alpha_0), \tag{11.5.2}$$

where

$$\begin{aligned}\tilde{J}_1(z,\alpha_0) &= -\sqrt{2(1-\alpha_0)}\,\psi(kz,\alpha_0)e^{i\kappa\alpha_0+ikz} + \frac{\sqrt{2}}{\mathcal{D}_1}\psi(kz,1)e^{iq}\times \\ &\times\left[\sqrt{1+\alpha_0}\,\psi(q,-\alpha_0)e^{-i\kappa\alpha_0} - \sqrt{1-\alpha_0}\,\psi(q,\alpha_0)\psi(q,1)e^{i\kappa\alpha_0+iq}\right]e^{ikz},\end{aligned} \tag{11.5.3}$$

$$\mathcal{D}_1 = 1-\psi^2(q,1)e^{i2q}, \tag{11.5.4}$$

and the absolute error is given by

$$\tilde{R}_1(z,\alpha_0) = O\left\{\frac{q^{-3/2}}{\sqrt{kz}(1+kz)\left[1+q(1-\alpha_0)\right]}\right\} \tag{11.5.5}$$

for $q \gg 1$ and $0 \leqslant z \leqslant \infty$. Using this approximation, the total normalized current density of Eqn. (11.1.3) is of the form

$$J(z,\alpha_0) = J_1(z,\alpha_0) + R_1(z,\alpha_0), \tag{11.5.6}$$

where

$$J_1(z,\alpha_0) = \sqrt{1-\alpha_0^2}\, e^{-ikz\alpha_0} + \tilde{J}_1(z+l,\alpha_0) + \tilde{J}_1(l-z,-\alpha_0), \tag{11.5.7}$$

$$R_1(z,\alpha_0) = \tilde{R}_1(z+l,\alpha_0) + \tilde{R}_1(l-z,-\alpha_0). \tag{11.5.8}$$

It is interesting to compare these expressions with the results of reference [125], which calculated some of the first terms of the asymptotic expansion for $q \gg 1$. Substituting z in Eqn. (11.5.3) with $l-z$, α_0 with $-\alpha_0$, and assuming $k(l-z) \gg 1$ and $q(1 \pm \alpha_0) \gg 1$, we find the following value for the wave of current traveling from the end $z = l$:

$$\begin{aligned} \tilde{J}_1(\zeta,\alpha_0) = {} & \frac{e^{i\frac{3\pi}{4}}}{2\sqrt{\pi}} \frac{\sqrt{1-\alpha_0}}{1+\alpha_0} e^{ikl\alpha_0} \left[1 - i\frac{3}{2k\zeta(1+\alpha_0)}\right] \frac{e^{ik\zeta}}{(k\zeta)^{3/2}} - \\ & - \frac{i}{32\pi} \frac{\sqrt{1+\alpha_0}}{1-\alpha_0} \frac{e^{ikl(2-\alpha_0)+ik\zeta}}{(kl)^{3/2}(k\zeta)^{3/2}}, \quad \zeta = l - z. \end{aligned} \tag{11.5.9}$$

This is identical to Eqn. (26) in reference [125] if the typographical errors are rectified. Namely, Eqns. (26–2) and (26–3) should read

$$\hat{\psi}_2(x|\alpha_0) = 3\frac{e^{i\frac{3\pi}{4}}}{4\sqrt{\pi}} \frac{\sqrt{1-\alpha_0}}{(1+\alpha_0)^2} \frac{e^{i\epsilon(\alpha_0-x+1)}}{(1-x)^{5/2}},$$

$$\hat{\psi}_3(x|\alpha_0) = \frac{1}{32\pi} \frac{\sqrt{1+\alpha_0}}{1-\alpha_0} \frac{e^{-i\epsilon(\alpha_0+x-3)}}{(1-x)^{3/2}}.$$

§11.6 Nth-Order Approximation for the Current

Let us retain the first n terms of the series of Eqn. (11.3.1) and let the remainder of the terms be replaced by the asymptotic representation provided by Eqn. (11.4.22). As a result, the total current can be written as

$$J(z,\alpha_0) = J_n(z,\alpha_0) + R_n(z,\alpha_0), \tag{11.6.1}$$

where

$$J_n(z,\alpha_0) = \sqrt{1-\alpha_0^2}\,e^{-ikz\alpha_0} + \tilde{J}_n(z+l,\alpha_0) + \tilde{J}_n(l-z,-\alpha_0) \tag{11.6.2}$$

$$R_n(z,\alpha_0) = \tilde{R}_n(z+l,\alpha_0) + \tilde{R}_n(l-z,-\alpha_0), \tag{11.6.3}$$

$$\begin{aligned}\tilde{J}_n(z,\alpha_0) = {} & e^{ikz}\sum_{m=1}^{n}(-1)^m \xi_m(z,-\beta_m)e^{iq(m-1)-i\kappa\beta_m} - \\ & -(-1)^n \frac{e^{iqn}}{\hat{\mathcal{D}}_n}\Bigg[\sqrt{1-\beta_n}\,\xi_n(2l,\beta_n)e^{i\kappa\beta_n} - \\ & -\sqrt{1+\beta_n}\,\xi_n(2l,-\beta_n)\frac{\xi_n(2l,1)}{\xi_{n-1}(2l,1)}e^{iq-i\kappa\beta_m}\Bigg]\frac{\xi_n(z,1)}{\xi_{n-1}(2l,1)}e^{ikz},\end{aligned} \tag{11.6.4}$$

$$\beta_n = (-1)^n \alpha_0,$$

$$\hat{\mathcal{D}}_n = 1 - \left[\frac{\xi_n(2l,1)}{\xi_{n-1}(2l,1)}e^{iq}\right]^2, \quad \text{for} \quad n = 2,3,\ldots. \tag{11.6.5}$$

For $q \gg n$ and $0 \leqslant z \leqslant \infty$ the absolute error $\tilde{R}_n(z,\alpha_0)$ is estimated to be

$$\left|\tilde{R}_n(z,\alpha_0)\right| \leqslant O\left\{\frac{q^{-3n/2}}{\sqrt{kz}(1+kz)\left[1+q\left[1+(-1)^n\alpha_0\right]\right]}\right\}. \tag{11.6.6}$$

§11.7 Scattering Pattern Represented by a Series of Functions $\psi_n(\alpha, \alpha_0)$

According to Eqn. (11.2.3), the scattered field can be represented in the far zone as

$$E_x = -\Psi(\alpha,\alpha_0)\frac{e^{i(kr+\pi/4)}}{\sqrt{2\pi kr}}, \tag{11.7.1}$$

where $\alpha = \cos\vartheta$ and $\alpha_0 = \cos\vartheta_0$,

$$\Psi(\alpha,\alpha_0) = \tilde{\Psi}(\alpha,\alpha_0) + \tilde{\Psi}(-\alpha,-\alpha_0), \tag{11.7.2}$$

$$\tilde{\Psi}(\alpha,\alpha_0) = \sqrt{1-\alpha}e^{i\kappa\alpha}\sum_{m=1}^{\infty}(-1)^m\sqrt{1+\beta_m}\psi_m(\alpha,-\beta_m)e^{iq(m-1)-i\kappa\beta_m}. \tag{11.7.3}$$

The function $\psi_n(\alpha, \alpha_0)$ is of the following form

$$\psi_1(\alpha,\alpha_0) = \frac{1}{\alpha+\alpha_0}, \quad \psi_2(\alpha,\alpha_0)e^{iq} = \frac{1}{\pi i}\int_1^{1+i\infty}\frac{g(t)dt}{(t+\alpha)(t+\alpha_0)}, \tag{11.7.4}$$

$$\underset{n=3,4,\ldots}{\psi_n(\alpha,\alpha_0)e^{iq(n-1)}} = \frac{1}{(\pi i)^{n-1}}\times$$

$$\times\int_1^{1+i\infty}\cdots\int_1^{1+i\infty}\frac{g(t_1)g(t_2)\ldots g(t_{n-1})dt_1dt_2\ldots dt_{n-1}}{(t_1+\alpha_0)(t_2+t_1)(t_3+t_2)\ldots(t_{n-1}+t_{n-2})(t_{n-1}+\alpha)}, \tag{11.7.5}$$

where $g(t)$ is given by Eqn. (11.3.5), and the integrals are evaluated along the left side of the branch cut from $1 \to 1+i\infty$, where $(1-t)^{1/2} = |(1-t)^{1/2}|e^{-i\pi/4}$. It is evident that $\psi_n(\alpha, \alpha_0)$ is symmetrical

$$\psi_n(\alpha,\alpha_0)=\psi_n(\alpha_0,\alpha), \tag{11.7.6}$$

and it follows that the scattering pattern satisfies the reciprocity theorem, i.e.,

$$\Psi(\alpha,\alpha_0)=\Psi(\alpha_0,\alpha). \tag{11.7.7}$$

According to Schwarzschild [139], the series of Eqn. (11.7.3) converges absolutely and uniformly for any value of $q > 0$.

Further, one should note that $\psi_n(\alpha, \alpha_0)$ satisfies the relationship

$$\lim_{\alpha\to-1}\sqrt{1-\alpha^2}\,\frac{\partial}{\partial\alpha}\psi_n(\alpha,\alpha_0)=\frac{1}{2}\psi_{n-1}(1,\alpha_0), \tag{11.7.8}$$

the utilization of which can be used to prove that

$$\lim_{\alpha\to\pm1}\sqrt{1-\alpha^2}\,\frac{\partial}{\partial\alpha}\Psi(\alpha,\alpha_0)=0. \tag{11.7.9}$$

This equation demonstrates the continuity of the scattered field for $y = 0$ and $|z| > l$. Indeed, H_z of the scattered field is proportional to the derivative

$$\frac{\partial}{\partial\vartheta}E_x=-\sqrt{1-\alpha^2}\,\frac{\partial}{\partial\alpha}E_x\,.$$

The requirement that the field remains continuous at $y = 0$, $|z| > l$ is equivalent to the condition that $H_z = 0$, i.e., to Eqn. (11.7.9).

§11.8 Asymptotic Representation for $\psi_n(\alpha,\alpha_0)$

Applying Lemma 2 from § 9.1 to $\psi_n(\alpha, \alpha_0)$ one can obtain the following (Poincaré) asymptotic expansion as $q \to \infty$:

$$\psi_2(\alpha,\alpha_0) \sim \frac{1}{\alpha-\alpha_0}\sum_{m=0}^{\infty}\frac{b_m}{i^{m+1}}\left[\frac{\varphi_q^{(m+1)}(q,\alpha)}{\sqrt{1+\alpha}} - \frac{\varphi_q^{(m+1)}(q,\alpha_0)}{\sqrt{1+\alpha_0}}\right], \tag{11.8.1}$$

where b_m are the Taylor series coefficients

$$\frac{1}{\sqrt{1+t}} = \sum_{m=0}^{\infty} b_m (t-1)^m ; \tag{11.8.2}$$

$$\psi_{\substack{n \\ n=3,4,\ldots}}(\alpha,\alpha_0) \sim \frac{1}{2}\psi(q,\alpha)\psi(q,\alpha_0)\psi^{n-3}(q,1) + \frac{1}{(i\pi)^{n-1}} \times$$

$$\times \sum_{m=1}^{\infty}\frac{1}{m!}\int_1^{1+i\infty}\cdots\int_1^{1+i\infty} e^{iq\sum_{s=1}^{n-1}(t_s-1)} \frac{\prod_{s=1}^{n-1}\sqrt{1-t_s}\,\hat{v}_m(t_1,\ldots t_{n-1})dt_1\ldots dt_{n-1}}{(t_1+\alpha_0)(t_{n-1}+\alpha)\prod_{s=2}^{n-2}(1+t_s)}, \tag{11.8.3}$$

where

$$\hat{v}_m(t_1,t_2,\ldots t_{n-1}) =$$

$$= \left[(t_1-1)\frac{\partial}{\partial t_1} + (t_2-1)\frac{\partial}{\partial t_2} + \ldots + (t_{n-1}-1)\frac{\partial}{\partial t_{n-1}}\right]^m \hat{g}_n(1,1,\ldots,1) \tag{11.8.4}$$

$$\hat{g}_3(t_1,t_2) = \frac{1}{\sqrt{1+t_1}\sqrt{1+t_2}(t_1+t_2)}, \tag{11.8.5}$$

$$\hat{g}_{\substack{n \\ n=4,5,6,\ldots}}(t_1,t_2,\ldots,t_{n-1}) = \frac{(1+t_1)^{-1/2}(1+t_{n-1})^{-1/2}\prod_{s=2}^{n-2}\sqrt{1+t_s}}{(t_1+t_2)(t_2+t_3)\ldots(t_{n-2}+t_{n-1})}, \tag{11.8.6}$$

and $\psi(q,\alpha)$ is determined by Eqn. (11.3.6). It is evident that each term in Eqn. (11.8.3) is the sum of the products of single integrals of the type $\varphi_q^{(m)}(q,\alpha)$, as given by Eqn. (9.2.6).

We omit the details of the asymptotic examination of the first term with $m = 1$ in the series of Eqn. (11.8.3). This examination is similar to those explained in detail in §§ 9.4, 10.2, and 11.4. Using the results of this examination, one obtains

$$\psi_n(\alpha,\alpha_0)=\frac{1}{2}\psi(q,\alpha)\psi(q,\alpha_0)\psi^{n-3}(q,1)\times$$
$$\times\left\{1+O\left[\frac{\psi_q^{(1)}(q,\alpha)}{\psi(q,\alpha)}+\frac{\psi_q^{(1)}(q,\alpha_0)}{\psi(q,\alpha_0)}\right]+(n-3)O\left[\frac{\psi_q^{(1)}(q,1)}{\psi(q,1)}\right]\right\}. \tag{11.8.7}$$

Referring to Eqns. (11.3.6) and (11.4.11), we find that

$$\psi_n(\alpha,\alpha_0)=\frac{1}{2}\psi(q,\alpha)\psi(q,\alpha_0)\psi^{n-3}(q,1)\{1+O[n/q]\} \quad \text{for} \quad q\gg n. \tag{11.8.8}$$

Finally, utilizing this equation, it is not difficult to obtain another important asymptotic relationship

$$\left|\psi_{n+m}(\alpha,\alpha_0)-\frac{\psi_n(1,\alpha)\psi_n(1,\alpha_0)}{\psi_{n-1}(1,1)}\left[\frac{\psi_n(1,1)}{\psi_{n-1}(1,1)}\right]^{m-1}\right|\leqslant$$
$$\leqslant\frac{qO\left[q^{-3(m+n-1)/2}\right]}{[1+q(1+\alpha)][1+q(1+\alpha_0)]} \quad \text{for } q\gg n+m. \tag{11.8.9}$$

This inequality may be improved for $n = 2$. This requires investigating the asymptotic behavior of the function

$$V_{n,2}(\alpha,\alpha_0)=\psi_n(\alpha,\alpha_0)\psi_1^{n-2}(1,1)-\psi_2(1,\alpha)\psi_2(1,\alpha_0)\psi_2^{n-3}(1,1). \tag{11.8.10}$$

Utilizing the integrals given by Eqns. (11.7.4) and (11.7.5), Eqn. (11.8.10) may be rewritten as

$$\mathcal{V}_{n,2}(\alpha,\alpha_0)=\frac{1}{(\pi i)^{n-1}}\times$$

$$\times\int_1^{1+i\infty}\cdots\int_1^{1+i\infty}\mathcal{K}_{n,2}(t_1,t_2,\ldots t_{n-1})\left[F_{n,2}(t_1,t_2,\ldots t_{n-1})-1\right]dt_1dt_2\ldots dt_{n-1}, \tag{11.8.11}$$

where

$$\mathcal{K}_{n,2}(t_1,t_2,\ldots t_{n-1})=\frac{e^{iq\sum\limits_{s=1}^{n-1}(t_s-1)}\prod\limits_{s=1}^{n-1}\sqrt{1-t_s}}{(t_1+\alpha_0)(t_{n-1}+\alpha)\prod\limits_{s=1}^{n-1}(1+t_s)^{3/2}\prod\limits_{s=2}^{n-2}(1+t_s)}, \tag{11.8.12}$$

and the function $F_{n,2}(t_1, t_2,\ldots t_n)$ is determined by Eqns. (10.2.21) and (10.2.22). An expansion of $F_{n,2}-1$ into a Taylor series about the point $t_1=t_2=\ldots=t_{n-1}=1$ possesses the form of Eqn. (10.2.23). Considering these remarks and applying Lemma 2 from § 9.1 to $\mathcal{V}_{n,2}$ for $q\gg n$, we find that

$$\left.\begin{aligned}
&\mathcal{V}_{3,2}(\alpha,\alpha_0)\sim-\frac{1}{16}\psi_q^{(1)}(q,\alpha)\psi_q^{(1)}(q,\alpha_0),\\
&\mathcal{V}_{4,2}(\alpha,\alpha_0)\sim-\frac{1}{32}\psi_q^{(1)}(q,1)\left[\psi(q,\alpha)\psi_q^{(1)}(q,\alpha_0)+\psi_q^{(1)}(q,\alpha)\psi(q,\alpha_0)\right],
\end{aligned}\right\} \tag{11.8.13}$$

$$\begin{aligned}
\mathcal{V}_{\substack{n,2\\ n=4,5}}(\alpha,\alpha_0)\sim&-\frac{1}{2^{n+1}}\psi_q^{(1)}(q,1)\psi^{n-5}(q,1)\times\\
&\times\Big\{(n-4)\psi(q,\alpha)\psi(q,\alpha_0)\psi_q^{(1)}(q,1)+\\
&+\psi(q,1)\left[\psi(q,\alpha)\psi_q^{(1)}(q,\alpha_0)+\psi_q^{(1)}(q,\alpha)\psi(q,\alpha_0)\right]\Big\}.
\end{aligned} \tag{11.8.14}$$

From here,

$$\psi_3(\alpha,\alpha_0)-\frac{\psi_2(1,\alpha)\psi_2(1,\alpha_0)}{\psi_1(1,1)}\sim-\frac{1}{8}\psi_q^{(1)}(q,\alpha)\psi_q^{(1)}(q,\alpha_0), \tag{11.8.15}$$

$$\begin{aligned}\psi_n(\alpha,\alpha_0)-\frac{\psi_2(1,\alpha)\psi_2(1,\alpha_0)}{\psi_1(1,1)}\left[\frac{\psi_2(1,1)}{\psi_1(1,1)}\right]^{n-3}\sim\\ \sim-\frac{1}{8}\psi_q^{(1)}(q,1)\psi^{n-5}(q,1)\{(n-4)\psi(q,\alpha)\psi(q,\alpha_0)\psi_q^{(1)}(q,1)+\\ \psi(q,1)[\psi(q,\alpha)\psi_q^{(1)}(q,\alpha_0)+\psi_q^{(1)}(q,\alpha)\psi(q,\alpha_0)]\}.\end{aligned} \tag{11.8.16}$$

Further utilizing Eqns. (9.3.2), (11.3.6), and (11.4.11), we obtain

$$\begin{aligned}\psi_n(\alpha,\alpha_0)-\frac{\psi_2(1,\alpha)\psi_2(1,\alpha_0)}{\psi_1(1,1)}\left[\frac{\psi_2(1,1)}{\psi_1(1,1)}\right]^{n-3}=\\ =\frac{O\left[q^{-3(n-1)/2}\right]}{[1+q(1+\alpha)][1+q(1+\alpha_0)]}\quad\text{for } n=3,4,\ldots\text{ and } q\gg n.\end{aligned} \tag{11.8.17}$$

This asymptotic approximation permits the construction of approximate expressions for the scattered field.

§11.9 First-Order Approximation for the Scattering Pattern

Limiting the asymptotic expansion of Eqn. (11.8.1) to the first term yields

$$\begin{aligned}\psi_2(\alpha,\alpha_0)=-\frac{\psi(q,\alpha)-\psi(q,\alpha_0)}{\alpha-\alpha_0}+\\ +O\left[\frac{\varphi_q^{(2)}(q,\alpha)}{\sqrt{1+\alpha}}-\frac{\varphi_q^{(2)}(q,\alpha_0)}{\sqrt{1+\alpha_0}}\right]\quad\text{for } q\gg 1.\end{aligned} \tag{11.9.1}$$

Our goal is to construct such an expression for the total scattered field, which would satisfy the boundary condition given by Eqn. (11.7.9). With this purpose in mind, let us add the zero quantity $\psi(q,1) - \psi(q,1)$ to the right hand side of Eqn. (11.9.1) and consider the combinations $\psi(q,\alpha) - \psi(q,1)$ and $\psi(q,\alpha_0) - \psi(q,1)$. One can show that

$$\psi(q,\alpha)-\psi(q,1)=\frac{1-\alpha}{2}\psi(q,\alpha)+(1-\alpha)O\left[\frac{\varphi_q^{(2)}(q,\alpha)}{\sqrt{1+\alpha}}\right] \quad \text{for} \quad q \gg 1.$$

This approximation allows one to transform Eqn. (11.9.1) [in the limit of its error bound] into the form

$$\begin{aligned}\psi_2(\alpha,\alpha_0)=&-\frac{1}{\alpha-\alpha_0}\left[\frac{1-\alpha}{2}\psi(q,\alpha)-\frac{1-\alpha_0}{2}\psi(q,\alpha_0)\right]+\\&+O\left[\frac{\varphi_q^{(2)}(q,\alpha)}{\sqrt{1+\alpha}}-\frac{\varphi_q^{(2)}(q,\alpha_0)}{\sqrt{1+\alpha_0}}\right].\end{aligned} \tag{11.9.2}$$

Substituting Eqn. (11.9.2) into Eqn. (11.7.3) and using the asymptotic representation given by Eqn. (11.8.8) for the remainder of ψ_n, we find that

$$\begin{aligned}\tilde{\Psi}_1(\alpha,\alpha_0)=&-\frac{\sqrt{1-\alpha}\sqrt{1-\alpha_0}}{\alpha+\alpha_0}e^{i\kappa(\alpha+\alpha_0)}-\\&-\frac{\sqrt{1-\alpha}\sqrt{1+\alpha_0}}{\alpha+\alpha_0}\left[\frac{1-\alpha}{2}\psi(q,\alpha)-\frac{1+\alpha_0}{2}\psi(q,-\alpha_0)\right]e^{iq+i\kappa(\alpha-\alpha_0)}-\\&-\frac{\sqrt{1-\alpha}}{2\mathcal{D}_1}e^{iq}\left[\sqrt{1-\alpha_0}\psi(q,\alpha_0)e^{i\kappa\alpha_0}-\right.\\&\left.-\sqrt{1+\alpha_0}\psi(q,-\alpha_0)\psi(q,1)e^{iq-i\kappa\alpha_0}\right]\psi(q,\alpha)e^{i\kappa\alpha},\end{aligned} \tag{11.9.3}$$

where

$$\mathcal{D}_1 = 1 - \psi^2(q,1)e^{i2q}\,. \tag{11.9.4}$$

Observing that

$$\left.\begin{aligned} &\lim_{\alpha\to-1}\sqrt{1-\alpha^2}\,\frac{\partial}{\partial\alpha}\psi(q,\alpha) = \frac{1}{2}, \\ &\lim_{\alpha\to1}\sqrt{1-\alpha^2}\,\frac{\partial}{\partial\alpha}\psi(q,\alpha) = 0, \end{aligned}\right\} \tag{11.9.5}$$

it is not difficult to prove the that

$$\lim_{\alpha\to\pm1}\sqrt{1-\alpha^2}\,\frac{\partial}{\partial\alpha}\Psi_1(\alpha,\alpha_0) = 0\,, \tag{11.9.6}$$

where

$$\Psi_1(\alpha,\alpha_0) = \tilde{\Psi}_1(\alpha,\alpha_0) + \tilde{\Psi}_1(-\alpha,-\alpha_0). \tag{11.9.7}$$

The absolute error of this approximation is determined by the formula

$$P_1(\alpha,\alpha_0) = \tilde{P}_1(\alpha,\alpha_0) + \tilde{P}_1(-\alpha,-\alpha_0), \tag{11.9.8}$$

where according to Eqn. (11.9.2)

$$\tilde{P}_1(\alpha,\alpha_0) = \frac{O\left[q^{-1/2}\right]}{\left[1+q(1+\alpha)\right]\left[1+q(1-\alpha_0)\right]} \quad \text{for } q \gg 1. \tag{11.9.9}$$

From Eqn. (11.9.6) it also follows that

$$\lim_{\alpha\to\pm1}\sqrt{1-\alpha^2}\,\frac{\partial}{\partial\alpha}P_1(\alpha,\alpha_0) = 0\,. \tag{11.9.10}$$

To verify these results, we will calculate the total scattering cross section

$$S=-\frac{2}{k}\Im\mathfrak{m}\{\Psi(-\alpha_0,\alpha_0)\} \tag{11.9.11}$$

for normal incidence ($\alpha_0 = 0$). Utilizing Eqn. (11.9.3), we have

$$\Psi_1(0,0)=1-iq+\left[\psi(q,0)-\frac{\partial\psi(q,0)}{\partial\alpha}\right]e^{iq}-\frac{\psi^2(q,0)}{1+\psi(q,1)e^{iq}}e^{i2q}. \tag{11.9.12}$$

Substituting the asymptotics

$$\left.\begin{aligned}\psi(q,0)&=\frac{e^{-i\pi/4}}{2\sqrt{2\pi}}q^{-3/2}+O\left(q^{-5/2}\right),\\ \frac{\partial\psi(q,0)}{\partial\alpha}&=-\frac{e^{-i\pi/4}}{2\sqrt{2\pi}}q^{-3/2}+O\left(q^{-5/2}\right),\end{aligned}\right\} \tag{11.9.13}$$

we find the total scattering cross section is equal to

$$S=4l\left[1-\frac{\sin(q-\pi/4)}{\sqrt{2\pi}q^{5/2}}+O\left(q^{-7/2}\right)\right], \tag{11.9.14}$$

which agrees with the two first terms of the asymptotic representation given by Eqn. (31.4) in reference [121].

The first-order approximation given by Eqn. (11.9.3) possesses the same complexity as the expressions found in reference [5] and [134] but is comparatively more exact. Effectively, it is possible to show that reference [5] yields

$$S=4l\left[1-\frac{\sin(q-\pi/4)}{2\sqrt{\pi}q^{5/2}}+O\left(q^{-3}\right)\right], \tag{11.9.15}$$

while reference[134] provides

$$S=4l\left[1-\frac{203}{128}\frac{\sin(q-\pi/4)}{\sqrt{2\pi}q^{5/2}}+O\left(q^{-7/2}\right)\right]. \tag{11.9.16}$$

Notice that according to the definition

$$S/4l = -\Im\mathrm{m}\left\{\Psi(-\alpha_0, \alpha_0)/q\right\} \tag{11.9.17}$$

the normalized total cross section is less than the directivity pattern by the factor $1/q$. Thus, the expression for the scattering pattern, obtained earlier in [5] and [134] possess idential orders of absolute error equal to $(kl)^{-3/2}$. The error of the first-order approximation given by Eqn. (11.9.3) according to Eqn. (11.9.9) is on the order of $(kl)^{-5/2}$ for both $q(1\pm\alpha) \gg 1$ and $q(1\pm\alpha_0) \gg 1$.

§11.10 Nth-Order Approximation for the Scattering Pattern

Retaining the first n terms of the series of Eqn. (11.7.3) and representing the remainder with the asymptotic expression of Eqn. (11.8.9), we obtain

$$\begin{aligned}\tilde{\Psi}_n(\alpha,\alpha_0) &= \sqrt{1-\alpha}\, e^{i\kappa\alpha} \sum_{m=1}^{n} (-1)^m \sqrt{1+\beta_m}\, \psi_m(\alpha,-\beta_m) e^{iq(m-1)-i\kappa\beta_m} - \\ &-(-1)^n \frac{\sqrt{1-\alpha}}{\mathcal{D}_n} \frac{\psi_n(1,\alpha)}{\psi_{n-1}(1,1)} e^{i(qn+\kappa\alpha)} \times \\ &\times \left[\sqrt{1-\beta_n}\,\psi_n(1,\beta_n) e^{i\kappa\beta_n} - \sqrt{1+\beta_n}\,\psi_n(1,-\beta_n) \frac{\psi_n(1,1)}{\psi_{n-1}(1,1)} e^{iq-i\kappa\beta_n}\right],\end{aligned} \tag{11.10.1}$$

where

$$\mathcal{D}_n = 1 - \left[\frac{\psi_n(1,1)}{\psi_{n-1}(1,1)} e^{iq}\right]^2, \quad n = 2,3,\ldots. \tag{11.10.2}$$

The scattering pattern is determined by

$$\Psi(\alpha,\alpha_0) = \Psi_n(\alpha,\alpha_0) + P_n(\alpha,\alpha_0), \tag{11.10.3}$$

where

$$\Psi_n(\alpha,\alpha_0)=\tilde{\Psi}_n(\alpha,\alpha_0)+\tilde{\Psi}_n(-\alpha,-\alpha_0), \tag{11.10.4}$$

$$P(\alpha,\alpha_0)=\tilde{P}_n(\alpha,\alpha_0)+\tilde{P}_n(-\alpha,-\alpha_0). \tag{11.10.5}$$

The value of P_n characterizes the absolute error of this approximation and according to Eqn. (11.8.9) is approximately

$$\left|\tilde{P}_n(\alpha,\alpha_0)\right|\leqslant\frac{O\left(q^{-(3n-2)/2}\right)}{\left[1+q(1+\alpha)\right]\left[1+q\left[1+(-1)^n\alpha_0\right]\right]}, \quad n=2,3,\ldots \tag{11.10.6}$$

for $q \gg n$. For $n = 2$ this estimate can be improved. Referring to Eqns. (11.8.15) and (11.4.11), we can determine that

$$\left|\tilde{P}_2(\alpha,\alpha_0)\right|\leqslant\frac{O\left(q^{-3}\right)}{\left[1+q(1+\alpha)\right]\left[1+q(1+\alpha_0)\right]} \quad \text{for} \quad q\gg 1. \tag{11.10.7}$$

It should be noted that $\Psi_n(\alpha, \alpha_0)$ satisfies the reciprocity theorem and the limiting condition that

$$\lim_{\alpha\to\pm1}\sqrt{1-\alpha^2}\,\frac{\partial}{\partial\alpha}\Psi_n(\alpha,\alpha_0)=0, \tag{11.10.8}$$

which is not difficult to verify with the use of Eqn. (11.7.8). Therefore, $P_n(\alpha, \alpha_0)$ also satisfies the condition that

$$\lim_{\alpha\to\pm1}\sqrt{1-\alpha^2}\,\frac{\partial}{\partial\alpha}P_n(\alpha,\alpha_0)=0. \tag{11.10.9}$$

This equality and Eqn. (11.10.7) show that the error estimate provided by Eqn. (11.10.6) is excessive.

To verify these results, it is interesting to compare them with the literature. This may be accomplished by calculating the second-order approximation of the total scattering cross section for a plane wave at both normal and grazing incidence. It is possible to show that

$$\Psi_2(0,0)=1-iq+2\psi_2(0,0)e^{iq}-4\frac{\psi_2^2(1,0)}{1+2\psi_2(1,1)e^{iq}}e^{i2q}. \tag{11.10.10}$$

Substituting the expressions

$$\psi_2(0,0)=\frac{e^{-i\pi/4}}{2\sqrt{2\pi}}q^{-3/2}-\frac{27}{16}\frac{e^{i\pi/4}}{\sqrt{2\pi}}q^{-5/2}+O\left(q^{-7/2}\right), \tag{11.10.11}$$

$$\psi_2(1,0)=\frac{e^{-i\pi/4}}{4\sqrt{2\pi}}q^{-3/2}-\frac{21}{32}\frac{e^{i\pi/4}}{\sqrt{2\pi}}q^{-5/2}+O\left(q^{-7/2}\right), \tag{11.10.12}$$

$$\psi_2(1,1)=\frac{e^{-i\pi/4}}{8\sqrt{2\pi}}q^{-3/2}-\frac{15}{64}\frac{e^{i\pi/4}}{\sqrt{2\pi}}q^{-5/2}+O\left(q^{-7/2}\right), \tag{11.10.13}$$

Eqn. (11.9.11) becomes

$$S=4l\left[1-\frac{\sin(q-\pi/4)}{\sqrt{2\pi}}q^{-5/2}+\right.$$
$$\left.+\frac{27}{8}\frac{\sin(q+\pi/4)}{\sqrt{2\pi}}q^{-7/2}-\frac{\cos 2q}{8\pi}q^{-4}+O\left(q^{-9/2}\right)\right] \tag{11.10.14}$$

for a plane wave at normal incidence. According to Eqn.(11.10.7), all the terms written here are correct.

For a plane wave at grazing incidence

$$S=-\frac{2}{k}\Im\mathrm{m}\left\{\Psi_2(-1,1)\right\}, \tag{11.10.15}$$

where

$$\Psi_2(-1,1)=\lim_{\alpha\to-1}\left\{\left[-\sqrt{2(1+\alpha)}\psi_1(-\alpha,-1)e^{-i\kappa(\alpha+1)}+\right.\right.$$
$$\left.\left.+\sqrt{2(1-\alpha)}\psi_2(\alpha,-1)e^{iq+i\kappa(\alpha-1)}\right]+\frac{8}{\mathcal{D}_2}\psi_2(1,1)\psi_2^2(1,-1)e^{i2q}\right\}. \tag{11.10.16}$$

Note that for $q\gg 1$

$$\sigma(\alpha)=-\sqrt{2(1+\alpha)}\psi_1(-\alpha,-1)e^{-i\kappa(\alpha+1)}+$$
$$+\sqrt{2(1-\alpha)}\psi_2(\alpha,-1)e^{iq+i\kappa(\alpha-1)}=$$
$$=\sqrt{\frac{2}{1+\alpha}}e^{-i\kappa(\alpha+1)}-\sqrt{\frac{1-\alpha}{1+\alpha}}\varphi(q,\alpha)e^{i\kappa(\alpha+1)}- \tag{11.10.17}$$
$$-\frac{\sqrt{2(1-\alpha)}}{i\pi}\int_1^{1+i\infty}\frac{e^{iq(t-1)}f(t)dt}{\sqrt{1-t}(t+\alpha)},$$

where

$$f(t)=-\frac{t-1}{4\sqrt{2}}+\frac{3}{32\sqrt{2}}(t-1)^2-$$
$$-\frac{5}{128\sqrt{2}}(t-1)^3+\frac{35}{2^{11}\sqrt{2}}(t-1)^4+O\left[(t-1)^5\right]. \tag{11.10.18}$$

Applying the limit as $\alpha\to-1$ and noting that

$$\int_1^{1+i\infty}\frac{e^{iq(t-1)}}{\sqrt{1-t}}(t-1)^m dt=\frac{i^{m+3/2}}{q^{m+1/2}}\Gamma\left(m+\frac{1}{2}\right), \tag{11.10.19}$$

we obtain

$$\sigma(-1)=\frac{2\sqrt{2}e^{-i\frac{\pi}{4}}}{\sqrt{\pi}}q^{1/2}+\frac{\Gamma(1/2)e^{i\frac{\pi}{4}}}{2\pi\sqrt{2}}q^{-1/2}-\frac{3\Gamma(3/2)e^{i\frac{3\pi}{4}}}{2^4\pi\sqrt{2}}q^{-3/2}-$$
$$-\frac{5\Gamma(5/2)e^{i\frac{\pi}{4}}}{2^6\sqrt{2\pi}}q^{-5/2}+\frac{35\Gamma(7/2)e^{i\frac{3\pi}{4}}}{2^{10}\sqrt{2\pi}}q^{-7/2}+O\left(q^{-9/2}\right). \quad (11.10.20)$$

The asymptotic expansion for $\psi_2(1, 1)$ is provided by Eqn. (11.10.13) and for $\psi_2(1, -1)$ is equal to

$$\psi_2(1,-1)=-\frac{e^{i\pi/4}}{2\sqrt{2\pi}}q^{-1/2}-\frac{3}{2^4}\frac{e^{-i\pi/4}}{\sqrt{2\pi}}q^{-3/2}+$$
$$+\frac{45}{2^8}\frac{e^{i\pi/4}}{\sqrt{2\pi}}q^{-5/2}+O\left[q^{-7/2}\right]. \quad (11.10.21)$$

Inserting Eqns. (11.10.13), (11.10.19), and (11.10.21) into Eqn. (11.10.16), and employing Eqn. (11.10.15), we find that

$$S=4l\left\{\frac{2}{\sqrt{\pi}}q^{-1/2}-\frac{1}{4\sqrt{\pi}}q^{-3/2}+\frac{3}{64\sqrt{\pi}}q^{-5/2}+\right.$$
$$+\frac{1}{\sqrt{\pi}}\left[\frac{15}{512}-\frac{\sin(2q+\pi/4)}{8\sqrt{2\pi}}\right]q^{-7/2}- \quad (11.10.22)$$
$$\left.-\frac{1}{\sqrt{\pi}}\left[\frac{525}{16384}-\frac{21\cos(2q+\pi/4)}{64\sqrt{2\pi}}\right]q^{-9/2}+O\left(q^{-11/2}\right)\right\}.$$

According to Eqn. (11.10.7) all of the terms outlined here are correct. As was to be expected, Eqns. (11.10.14) and (11.10.22) completely agree with corresponding expressions (31.4) and (31.7) of reference [121].

§11.11 Relationship Between the Approximations for the Current and the Scattering Pattern

Utilizing Eqns. (11.2.3) and (11.7.1) we may write

$$\tilde{\Psi}(\alpha,\alpha_0) = -2\pi ik\sum_{n=1}^{\infty} G_n(k\alpha, k\alpha_0)e^{i\kappa\alpha}. \tag{11.11.1}$$

Equations (11.1.5) and (11.1.6) can be used to rewrite this expression as

$$\tilde{\Psi}(\alpha,\alpha_0) = -ike^{i\kappa\alpha}\int_{-\infty}^{\infty}\tilde{J}(z,\alpha_0)e^{-ikz\alpha}dz. \tag{11.11.2}$$

Substituting Eqn. (11.6.4) and referring to the relationships

$$\xi_n(z,\alpha_0)e^{ikz} = -\frac{1}{2\pi i}\int_{-\infty}^{\infty} e^{ikzt}\sqrt{1-t}\,\psi_n(t,\alpha_0)dt, \tag{11.11.3}$$

$$\psi_n(\alpha,\alpha_0) = -\frac{ik}{\sqrt{1-\alpha}}\int_{-\infty}^{\infty} e^{ikz(1-\alpha)}\xi_n(z,\alpha_0)dz, \tag{11.11.4}$$

we obtain

$$\tilde{\Psi}_n'(\alpha,\alpha_0) = \sqrt{1-\alpha}\sum_{m=1}^{n}(-1)^m\sqrt{1+\beta_m}\,\psi_m(\alpha,-\beta_m)e^{iq(m-1)-i\kappa\beta_m} -$$

$$-(-1)^n\frac{\sqrt{1-\alpha}}{\hat{\mathcal{D}}_n}\frac{\psi_n(1,\alpha)}{\xi_{n-1}(2l,1)}e^{i(qn+\kappa\alpha)}\times \tag{11.11.5}$$

$$\times\left[\sqrt{1-\beta_n}\,\xi_n(2l,\beta_n)e^{i\kappa\beta_n}-\right.$$

$$\left.-\sqrt{1+\beta_n}\,\xi_n(2l,-\beta_n)\frac{\xi_n(2l,1)}{\xi_{n-1}(2l,1)}e^{iq-i\kappa\beta_n}\right], \quad (11.11.5,\text{cont.})$$

where $\hat{D}_n$ is determined by Eqn. (11.6.5). A comparison of Eqn. (11.11.5) with Eqn. (11.10.1) shows that the function $\tilde{\Psi}_n'(\alpha,\alpha_0)$ is different from $\tilde{\Psi}_n(\alpha,\alpha_0)$. It is evident that

$$\Psi_n'(\alpha,\alpha_0)=\tilde{\Psi}_n'(\alpha,\alpha_0)+\tilde{\Psi}_n'(-\alpha,-\alpha_0) \quad (11.11.6)$$

fails to satisfy both the reciprocity theorem and the boundary conditions given by Eqn. (11.10.8). Thus, direct integration of the approximate expression for the current yields a scattering pattern which fails to satisfy both the reciprocity theorem and the boundary conditions.

Taking into account Eqn. (11.8.8) and

$$\xi_n(2l,\alpha)=\sqrt{2}\psi(q,\alpha)\psi^{n-1}(q,1)\left[1+O\left(\frac{n}{q}\right)\right], \quad (11.11.7)$$

it is not difficult to show that for $q \gg n$

$$\left|\frac{\xi_n(2l,\alpha)}{\xi_{n-1}(2l,1)}-\frac{\psi_n(1,\alpha)}{\psi_{n-1}(1,1)}\right|\leqslant\frac{O\left(q^{-3/2}\right)}{1+q(1+\alpha)}, \quad (11.11.8)$$

$$\left|D_n-\hat{D}_n\right|\leqslant O\left(q^{-4}\right). \quad (11.11.9)$$

Therefore,

$$\left|\tilde{\Psi}_n'(\alpha,\alpha_0)-\tilde{\Psi}_n(\alpha,\alpha_0)\right| \leqslant \frac{O\left(q^{-(3n-2)/2}\right)}{\left[1+q(1+\alpha)\right]\left[1+q\left[1+(-1)^n\alpha_0\right]\right]} \quad . \tag{11.11.10}$$

Comparing this with Eqn. (11.10.6), we see that $\tilde{\Psi}_n'(\alpha, \alpha_0)$ differs from $\tilde{\Psi}_n(\alpha, \alpha_0)$ in its error bound.

From Eqn. (11.11.2) it follows that

$$\tilde{J}(z,\alpha_0) = -\frac{1}{2\pi i}\int_{-\infty}^{\infty} \tilde{\Psi}(\alpha,\alpha_0)e^{-ik(l-z)\alpha}d\alpha . \tag{11.11.11}$$

Substituting the approximate expression for the pattern given by $\tilde{\Psi}_n(\alpha, \alpha_0)$, and using Eqn. (11.11.3), we find that

$$\begin{aligned}\tilde{J}_n'(z,\alpha_0) &= e^{ikz}\sum_{m=1}^{n}(-1)^m\sqrt{1+\beta_m}\,\xi_m(z,-\beta_m)e^{iq(m-1)-i\kappa\beta_m} - \\ &-(-1)^n\frac{1}{\mathcal{D}_n}\frac{\xi_n(z,1)}{\psi_{n-1}(1,1)}e^{i(qn+kz)}\times \\ &\times\left[\sqrt{1-\beta_n}\psi_n(1,\beta_n)e^{i\kappa\beta_n}-\sqrt{1+\beta_n}\psi_n(1,-\beta_n)\frac{\psi_n(1,1)}{\psi_{n-1}(1,1)}e^{iq-i\kappa\beta_n}\right].\end{aligned} \tag{11.11.12}$$

This expression differs from $\tilde{J}_n(z, \alpha_0)$, determined by Eqn. (11.6.4) by the error bound, given by Eqn. (11.6.6).

§11.12 Fundamental Results of the Mathematical Theory of Edge Diffraction

- This theory is based on the solution of functional equations. Exact and asymptotic series are presented for the current on a strip and for the scattered field in the far zone.
- The terms in the exact series are represented through new special functions, containing multiple integrals. These series differ from the classical series with Mathieu functions, in that they converge faster at shorter wavelengths.
- It is shown that these exact series are equivalent to the Schwarzschild's solution [139] and to Neumann series for integral equations formulated by Westpfahl [124].
- Uniform asymptotic expansions are constructed for the currents and scattering patterns. They are valid for arbitrary position of the observation point on the strip and in the far zone. They are applicable for arbitrary directions of the incident wave, including grazing directions. The error of these formulas does not exceed a value on the order of $(kl)^{-n/2}$ for $kl \gg n$ where n is any positive integer.
- It is shown that the direct integration of the approximate expressions for the current (scattering pattern) leads to expressions for the scattering pattern (current) which fail to satisfy both the reciprocity theorem and the boundary conditions.
- The results of this research provide a rigorous foundation of approximate solutions, found in the previous sections with the aid of parabolic equations and the physical theory of diffraction. Some results of this research were published in [23].
- The mathematical theory developed here can be applied to other diffraction problems. In the next chapter, it is extended for investigation of diffraction at an open resonator.

12

Edge Diffraction at Open-Ended Parallel Plate Resonator

This chapter will examine edge waves excited by a plane wave incident upon a parallel plate waveguide of a finite size. The principal characteristic of this structure is that it exhibits resonant properties, i.e., the ability at some frequencies to store strong electromagnetic fields. The parallel plate waveguide is a simple case of an open resonator, which is utilized in quasioptical generators. The characteristic frequencies of such a resonator were first researched by Fox and Li [142]. These authors used Huygen's principle to formulate an integral equation and numerically determine some of its characteristic functions and characteristic numbers. Subsequently, Vainshtein [112] used physical considerations to conduct an analytical investigation of the characteristic oscillations of a two dimensional planar resonator. Note also that single-mode parallel plate resonators with geometrical parameters $kb < 3\pi/2$ and $kl \gg 1$ were considered by Jones [143] and Williams [144]. Here, $2b$ is the distance between the plates and $2l$ is their width. The solution of the problem for the case $kb \ll 1, kl \ll 1$ was given by Pimenov [145].

The plane wave excitation of such a resonator is less well understood. It was examined by Fialkovskii [147], who derived an approximate expression for the resonant component of the field between the plates. His results, serve principally as the foundation of the characteristic functions constructed by Vainshtein [112].

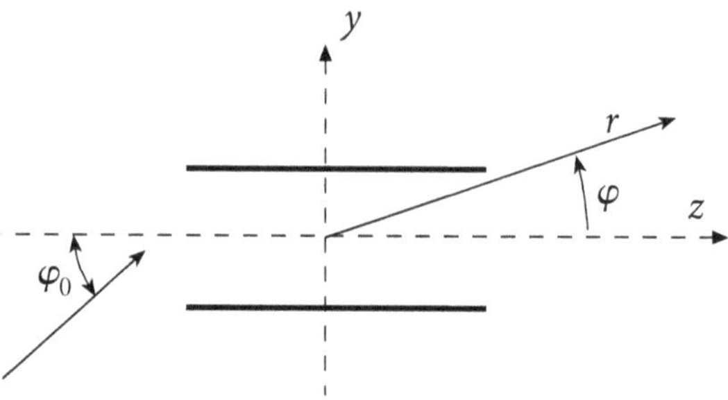

Figure 12.1.1: Diffraction of a plane wave from a parallel plate waveguide.

Here, we are interested in a physical description of the formation of the diffracted field created by multiple edge wave diffraction. From this point of view, the theory developed in Chapters 8–11 is appropriate for investigating the properties of the scattered field in the far zone and inside the resonator.

§12.1 Derivation of the Fundamental Functional Equations

Let us examine the diffraction of a plane wave

$$H_{0x} = e^{ik(z\cos\varphi_0 + y\sin\varphi_0)}, \quad 0 < \varphi_0 < \pi, \tag{12.1.1}$$

from two parallel (perfectly conducting) plates, the orientation of which is described by

$$-\infty \leqslant x \leqslant \infty, \quad -l \leqslant z \leqslant l, \quad y = \pm b, \tag{12.1.2}$$

as is shown by Fig. 12.1.1. Rewriting Eqn. (12.1.1) in the form

$$H_x = H_{0x}^{(1)} + H_{0x}^{(2)}$$

will allow us to examine this problem in two parts:

A) excitation of an open resonator by

$$H_{0x}^{(1)} = i\sin\left(ky\sin\varphi_0\right)e^{ikz\cos\varphi_0}, \tag{12.1.3}$$

and

B) excitation of an open resonator by

$$H_{0x}^{(2)} = \cos\left(ky\sin\varphi_0\right)e^{ikz\cos\varphi_0}. \tag{12.1.4}$$

Problem A corresponds to the excitation of a system by an even electric field (that is an even function of y),

$$E_z^0 = \frac{2\pi}{ck}g(z) \quad \text{for} \quad y = \pm b, \tag{12.1.5}$$

where

$$\begin{aligned} &g(z) = G(u)e^{iuz}, \qquad && G(u) = \frac{c}{2\pi}v^0\cos v^0 b, \\ &v^0 = \sqrt{k^2 - u^2}, && u = k\cos\varphi_0. \end{aligned} \tag{12.1.6}$$

And problem B represents the excitation of the system by an odd electric field (an odd function of y),

$$E_z^0 = \pm\frac{2\pi}{ck}g(z) \quad \text{for} \quad y = \pm b, \tag{12.1.7}$$

where

$$\begin{aligned} &g(z) = G(u)e^{iuz}, \qquad && G(u) = \frac{ic}{2\pi}v^0\sin v^0 b, \\ &u = k\cos\varphi_0, && v^0 = \sqrt{k^2 - u^2}. \end{aligned} \tag{12.1.8}$$

The terms even and odd excitations (in relationship to the electric field) will be used frequently from here on. Also note that in accordance with the above representation of the incident field, the diffracted field can be represented by $H_x = H_x^{(1)} + H_x^{(2)}$. Here, $H_x^{(1)}$ is excited by the incident field $H_{0x}^{(1)}$, and $H_x^{(2)}$ is excited by the incident field $H_{0x}^{(2)}$. Sometimes to simplify expressions, we omit the indices 1 and 2.

For the case of even excitation, the scattered field is of the form

$$\left.\begin{aligned} H_x &= -\frac{4\pi}{c}\int_{-\infty}^{\infty} F(w,u)e^{i(wz+vy)}\cos vb\,dw \quad \text{for} \quad y>b, \\ H_x &= -i\frac{4\pi}{c}\int_{-\infty}^{\infty} F(w,u)e^{i(wz+vb)}\sin vy\,dw \quad \text{for} \quad -b<y<b, \\ H_x &= \frac{4\pi}{c}\int_{-\infty}^{\infty} F(w,u)e^{i(wz-vy)}\cos vb\,dw \quad \text{for} \quad y<-b, \end{aligned}\right\} \tag{12.1.9}$$

where

$$v=\sqrt{k^2-w^2}, \quad \Im\mathrm{m}\{v\}\geqslant 0. \tag{12.1.10}$$

The contour of integration proceeds along the real axis and skirts above the branch cut $-k \rightarrow -k-i\infty$ and below the branch cut $k \rightarrow k+i\infty$. For v we choose that branch for which $v = k$ for $w = 0$. The field expressed by Eqn. (12.1.9) corresponds to a surface current density of

$$j_z = \int_{-\infty}^{\infty} F(w,u)e^{iwz}\,dw, \quad y=\pm b \tag{12.1.11}$$

and an electric field given by

$$E_z = -\frac{2\pi}{ck} \int_{-\infty}^{\infty} L(w) F(w,u) e^{iwz}\, dw, \quad y = \pm b \tag{12.1.12}$$

where

$$L(w) = v\left(1 + e^{2ivb}\right) = 2v \cos vb e^{ivb}. \tag{12.1.13}$$

For the case of odd excitation, the scattered field is of the form

$$\left.\begin{aligned}
H_x &= i\frac{4\pi}{c} \int_{-\infty}^{\infty} F(w,u) e^{i(wz+vy)} \sin vb\, dw \quad \text{for} \quad y > b, \\
H_x &= \frac{4\pi}{c} \int_{-\infty}^{\infty} F(w,u) e^{i(wz+vb)} \cos vy\, dw \quad \text{for} \quad -b < y < b, \\
H_x &= i\frac{4\pi}{c} \int_{-\infty}^{\infty} F(w,u) e^{i(wz-vy)} \sin vb\, dw \quad \text{for} \quad y < -b.
\end{aligned}\right\} \tag{12.1.14}$$

At $y = \pm b$ this scattered field corresponds to a surface current density of

$$j_z = \pm \int_{-\infty}^{\infty} F(w,u) e^{iwz}\, dw. \tag{12.1.15}$$

Thus, its electric field is equal to

$$E_z = \mp \frac{2\pi}{ck} \int_{-\infty}^{\infty} L(w) F(w,u) e^{iwz}\, dw, \tag{12.1.16}$$

where

$$L(w) = v\left(1 - e^{i2vb}\right) = -i2v\sin(vb)e^{ivb}. \tag{12.1.17}$$

The roots of the equation,

$$L(w) = 0, \tag{12.1.18}$$

determine the propagation constants of the waves in an infinite waveguide with plates at $y = \pm b$. In the case of an even excitation, these roots are equal to

$$w_m = \sqrt{k^2 - \left(\frac{\pi}{2b}\right)^2 (2m-1)^2}, \quad m = 1,2,3\ldots \tag{12.1.19}$$

and in the case of odd excitation,

$$w_m = \sqrt{k^2 - \left(\frac{\pi}{2b}\right)^2 (2m)^2}, \quad m = 0,1,2,\ldots \tag{12.1.20}$$

For square roots of negative argument, w_m is defined as $w_m = i|w_m|$, and in the remaining cases, $\Im m\{w_m\} = 0$, $\Re e\{w_m\} > 0$. In addition to these roots, Eqn. (12.1.18) also possesses roots $w = -w_m$. The root $w_0 = k$ occurring as a result of an odd excitation corresponds to a purely transverse wave.

Enforcing the boundary conditions

$$j_z = 0 \quad \text{at} \quad y = \pm b \quad \text{and} \quad |z| > l, \tag{12.1.21}$$

$$E_z + E_z^0 = 0 \quad \text{at} \quad y = \pm b \quad \text{and} \quad |z| < l, \tag{12.1.22}$$

problems A and B are reduced to a solution of the functional equations

$$\left.\begin{aligned}
&\int_{-\infty}^{\infty} F(w,u)e^{iwz}\,dw = 0, \quad |z|>l, \\
&\int_{-\infty}^{\infty} L(w)F(w,u)e^{iwz}\,dw = g(z), \quad |z|<l,
\end{aligned}\right\} \tag{12.1.23}$$

The solution to these equations is studied in the following section.

§12.2 Formulation and Solution of the Functional Equations for Edge Waves

For an even excitation, the incident field is given by Eqn. (12.1.3) which can be written as

$$H_{0x}^{(1)} = \frac{1}{2}\left[e^{ik(z\cos\varphi_0 + y\sin\varphi_0)} - e^{ik(z\cos\varphi_0 - y\sin\varphi_0)}\right]. \tag{12.2.1}$$

First, we determine the current excited by this field on the surface of the resonator when its width $2l = \infty$. In this case, the resonator is transformed into two planes that completely shadow one another. It is evident that the current excited on the external (illuminated) sides of these plates ($y = \pm b^{\pm}$), is determined by

$$j_z^{(0)} = \frac{c}{4\pi} e^{ik(z\cos\varphi_0 - b\sin\varphi_0)}. \tag{12.2.2}$$

According to the definitions introduced in § 2.1, Eqn. (12.2.2) is the *uniform current* created by an *even excitation*. It is not difficult to show that the *uniform current* created by an *odd excitation* is equal to

$$j_z^{(0)} = \mp\frac{c}{4\pi} e^{ik(z\cos\varphi_0 - b\sin\varphi_0)} \quad \text{for} \quad y = \pm b^{\pm}. \tag{12.2.3}$$

The resulting *uniform current*, excited by Eqn. (12.1.1), is the summation of Eqns. (12.2.2) and (12.2.3). As was to be expected, it is equal to

$$j_z = \begin{cases} 0 & \text{for} \quad y = +b, \\ \dfrac{c}{2\pi} e^{ik(z\cos\varphi_0 - b\sin\varphi_0)} & \text{for} \quad y = -b. \end{cases} \tag{12.2.4}$$

It is convenient to write Eqns. (12.2.2) and (12.2.3) in the form

$$j_0 = \frac{G(u)}{L(u)} e^{iuz} \quad \text{for} \quad y = +b \tag{12.2.5}$$

as a function of $G(u)$ and $L(u)$.

Utilizing these definitions and the symmetry property of edge waves, the current being excited on the resonator at $y = +b$ may be written as

$$j_z = j_0(z,u) + \sum_{n=1}^{\infty} \left[j_n(z+l,u) + j_n(l-z,-u) \right], \tag{12.2.6}$$

where $j_0(z, u)$ is given by Eqn. (12.2.5) and j_n is the of the nth-order edge wave.

Let us assume that

$$j_n(z,u) = \int_{-\infty}^{\infty} F_n(w,u) e^{iwz} \, dw. \tag{12.2.7}$$

According to Eqns. (12.1.11) and (12.1.15), the electric field excited by this current together with the current flowing on the plate at $y = -b$ will be equal to

$$E_{n,z} = -\frac{2\pi}{ck} \int_{-\infty}^{\infty} L(w) F_n(w,u) e^{iwz} \, dw, \quad y = \pm b \tag{12.2.8}$$

for an even excitation and

$$E_{n,z} = \mp\frac{2\pi}{ck}\int_{-\infty}^{\infty} L(w)F_n(w,u)e^{iwz}\,dw\,, \quad y=\pm b \tag{12.2.9}$$

for an odd excitation.

Noting that

$$F_0(w,u) = \frac{G(u)}{L(u)}\delta(u-w)\,, \tag{12.2.10}$$

we find that for $y = \pm b$

$$E_{0,z} = -\frac{2\pi}{ck}g(z) \quad \text{for even excitation} \tag{12.2.11}$$

and

$$E_{0,z} = \mp\frac{2\pi}{ck}g(z) \quad \text{for odd excitation.} \tag{12.2.12}$$

It follows that the boundary conditions

$$E_{0,z} + E_z^0 = 0 \quad \text{for} \quad y = \pm b \tag{12.2.13}$$

and

$$E_z + E_z^0 = 0 \quad \text{for} \quad y = \pm b, \ |z| < l \tag{12.2.14}$$

lead to the requirement that

$$E_{n,z} = 0 \quad \text{for} \quad y = \pm b, \quad z > 0, \quad n = 1,2,3\ldots. \tag{12.2.15}$$

As a result, edge waves are defined by the following system of recursive functional equations

$$\left.\begin{aligned}
&\int_{-\infty}^{\infty} F_1(w,u)e^{iw\zeta}\,dw = -j_0(\zeta - l, u) \quad \text{for} \quad \zeta < 0, \\
&\int_{-\infty}^{\infty} \left[F_n(w,u) + F_{n-1}(-w,-u)e^{-i2lw}\right]e^{iw\zeta}\,dw = 0, \quad \zeta < 0, \quad n = 2,3,\ldots \\
&\int_{-\infty}^{\infty} L(w)F_n(w,u)e^{iwz}\,dw = 0, \quad \text{for} \quad \zeta > 0, \quad n = 1,2,3,\ldots
\end{aligned}\right\} \tag{12.2.16}$$

analogous to the system of Eqns. (8.3.10) derived for the case of the strip.

Applying the factorization method to this system of equations, we obtain

$$F_1(w,u) = \frac{1}{2\pi i}\frac{G(u)}{L_+(u)L_-(w)}\frac{e^{-iul}}{w - \widehat{u}}, \tag{12.2.17}$$

$$\underset{n=2,3,\ldots}{F_n(w,u)} = \frac{1}{2\pi i L_-(w)}\int_{-\infty}^{\infty} L_+(t)F_{n-1}(t,-u)\frac{e^{i2lt}}{t + \widehat{w}}\,dt\,. \tag{12.2.18}$$

Thus,

$$F_2(w,u) = \frac{1}{(2\pi i)^2}\frac{G(u)e^{iul}}{L_-(u)L_-(w)}\int_{-\infty}^{\infty}\frac{L_+(t)}{L_-(t)}\frac{e^{i2lt}\,dt}{(t+\widehat{w})(t+\widehat{u})}, \tag{12.2.19}$$

$$\underset{n=3,4,\ldots}{F_n(w,u)} = \frac{1}{(2\pi i)^n}\frac{G(u)e^{i(-1)^n ul}}{L_-\left[(-1)^n u\right]L_-(w)} \times \tag{12.2.20}$$

$$\times \int_{-\infty}^{\infty} \cdots \int_{-\infty}^{\infty} \frac{\hat{h}(t_{n-1})\hat{h}(t_{n-2})\ldots\hat{h}(t_1)dt_{n-1}dt_{n-2}\ldots dt_1}{(t_{n-1}+\overset{\frown}{w})(t_{n-2}+\overset{\frown}{t_{n-1}})\ldots(t_1+\overset{\frown}{t_2})\,(t_1+\overset{\frown}{(-1)^n u})}, \qquad (12.2.20, \text{cont.})$$

where

$$\hat{h}(w)=\frac{L_+(w)}{L_-(w)}e^{i2lw}\,. \qquad (12.2.21)$$

The factorization function, $L(w)$,

$$L(w)=L_+(w)L_-(w), \qquad (12.2.22)$$

is described by Vainshtein in references [96, 120]. The function $L_+(w)$ is holomorphic for $\mathfrak{Im}\{w\} > -\mathfrak{Im}\{k\}$ and does not have zeros there. The function $L_-(w)$ is holomorphic for $\mathfrak{Im}\{w\} < \mathfrak{Im}\{k\}$ and does not have zeros there. In [96, 120], Vainshtein shows that

$$L_+(w)=\sqrt{k+w}\,e^{\chi_+(w)}, \quad L_-(w)=L_+(-w), \qquad (12.2.23)$$

where the value of $(k+w)^{1/2}$ is determined by the branch for which $0 \le \arg\{k^{1/2}\} < \pi/2$. In addition,

$$\chi_+(u)=\frac{1}{2\pi i}\int_{\infty e^{i3\pi/4}}^{\infty e^{-i\pi/4}} \ln\left(1\pm e^{i2vb}\right)\frac{dw}{w-u}+$$

$$+\begin{cases} 0 & \text{for} \quad \mathfrak{Re}\{u\}>0 \\ \ln\left(1\pm e^{i2b\sqrt{k^2-u^2}}\right) & \text{for} \quad \underset{3\pi/4<\arg(u)<\pi}{\mathfrak{Re}\{u\}<0}\,, \end{cases} \qquad (12.2.24)$$

where the branch of the natural log must be selected such that in the region $-\mathfrak{Im}\{k\} < \mathfrak{Im}\{u\} < \mathfrak{Im}\{k\}$, the function $\chi_+(u)\cdot\chi_-(u)$ approaches zero as

$|u| \to \infty$. To do this, we must choose the principal value of the logarithm, $\ln z = \ln|z| + i\varphi$, $-\pi < \varphi < \pi$. The upper (lower) sign in this expression relates to even (odd) excitation.

The integral term can be approximated by the Vainshtein function

$$U(s,p) = \frac{1}{2\pi i}\int_{-\infty}^{\infty} \ln\left(1 - e^{i2\pi p - t^2/2}\right)\frac{dt}{t - se^{i\pi/4}} , \tag{12.2.25}$$

where,

$$s = \sqrt{\frac{2b}{k}}\, u , \tag{12.2.26}$$

$$\left.\begin{aligned} p &= \frac{kb}{\pi} + \frac{1}{2} - \left[\frac{kb}{\pi} + \frac{1}{2}\right] && \text{for even excitation} \\ p &= \frac{kb}{\pi} - \left[\frac{kb}{\pi}\right] && \text{for odd excitation} \end{aligned}\right\}. \tag{12.2.27}$$

Here, the $[x]$ denotes the greatest integer less than x.

The absolute error, shown during a derivation of $U(s,p)$ is $(kb)^{-1}$ for $s = 0$ and $(kb)^{-3/2}$ for $s = (2kb)^{1/2}$. This function was derived and studied by Vainshtein. Tables of this function for real values of s and p are provided in references [96, 120, 146]. These references provide an asymptotic formula for $U(s,p)$. In particular, for $s \gg 1$

$$U(s,p) \sim \frac{e^{-i\pi/4}}{s}\sum_{m=0}^{\infty}\frac{U_m(p)}{\left(se^{i\pi/4}\right)^{2m}} , \tag{12.2.28}$$

where

$$U_m(p) = -i\frac{(2m-1)!!}{\sqrt{2\pi}}\sum_{n=1}^{\infty}\frac{e^{i2n\pi p}}{n^{m+3/2}} , \tag{12.2.29}$$

with $(2m-1)!! = 1\cdot 3\cdot 5\cdots(2m-1)$.

Moreover, $U(s,p)$ exhibits the following properties

$$\left.\begin{aligned} &U(s,N+p)=U(s,p), \quad N=\pm1,\pm2,\ldots \\ &U(-s,p)=-U(s,p). \end{aligned}\right\} \tag{12.2.30}$$

Substituting Eqn. (12.2.23) into Eqn. (12.2.21), we obtain

$$\hat{h}(kt)=\sqrt{\frac{1+t}{1-t}}e^{iqt}e^{\chi_+(kt)-\chi_+(-kt)}, \quad q=2kl. \tag{12.2.31}$$

Here, the branch for $\sqrt{1\pm t}$ is chosen such that $\sqrt{1}\ = 1$. A comparison of Eqn. (12.2.31) with Eqn. (8.3.24) shows that they differ only by the slowly changing exponential, which possesses a finite value for $kt \neq w_m$. Note also that as $kb \rightarrow \infty$, the terms $F_n(w, u)$ transform into functions representative of a solitary strip and form a uniformly convergent series. Having made these observations, we can hypothesize that the series of $F_n(w, u)$ is uniformly convergent, and thus, term-by-term integration is possible

The conclusion of section will formulate an integral equation for

$$\hat{\Phi}(w,u)=L(w)\sum_{n=1}^{\infty}F_n(w,u). \tag{12.2.32}$$

Returning to Eqn. (12.2.18) it is not difficult to show that $\hat{\Phi}(w, u)$ satisfies the *edge wave integral equation*

$$\hat{\Phi}(w,u)=\hat{\Phi}_1(w,u)+\frac{L_+(w)}{2\pi i}\int_{-\infty}^{\infty}\frac{e^{i2lt}}{L_-(t)}\frac{\hat{\Phi}(t,-u)}{t+\widehat{w}}dt, \tag{12.2.33}$$

where

$$\hat{\Phi}_1(w,u)=L(w)F_1(w,u). \tag{12.2.34}$$

This is analogous to Eqn. (8.4.10) for strip diffraction.

§12.3 Rigorous Expressions for the Diffracted Field in the Far Zone and Interior to the Resonator

In accordance with Eqns. (12.1.11), (12.2.6), and (12.2.7) the Fourier transform of the total current can be represented as

$$F(w,u)=F_0(w,u)+\sum_{n=1}^{\infty}\left[F_n(w,u)e^{iwl}+F_n(-w,-u)e^{-iwl}\right]. \qquad (12.3.1)$$

Substituting this expression into Eqn. (12.1.9) and referring Eqn. (12.2.10), we find that for $y > b$

$$H_x=-\frac{4\pi}{c}\frac{G(u)}{L(u)}\cos v^0 b e^{i(uz+v^0 y)}-$$
$$-\frac{4\pi}{c}\int_{-\infty}^{\infty}\cos vbe^{ivy}\sum_{n=1}^{\infty}\left[F_n(w,u)e^{iw(z+l)}+F_n(-w,-u)e^{-iw(l-z)}\right]dw. \qquad (12.3.2)$$

One should observe that all the $F_n(w, u)$ are proportional to

$$\frac{1}{L_-(w)}=\frac{L_+(w)}{L(w)}=\frac{L_+(w)}{2v\cos vb}e^{-ivb}, \qquad (12.3.3)$$

which possesses poles at $w = w_m$. Since these points are the roots of $\cos(vb) = 0$, it is clear that the integrand in Eqn. (12.3.2) will not possess poles at $w = w_m$.

Applying the method of steepest descent [78] to Eqn. (12.3.2), we find that the far field is equal to

$$H_x=\Phi(w,u)\frac{e^{i(kr+\pi/4)}}{\sqrt{2\pi kr}}. \qquad (12.3.4)$$

Here,

$$r = \sqrt{y^2 + z^2}\,, \tag{12.3.5}$$

$$\Phi(w,u) = \tilde{\Phi}(w,u)e^{iwl} + \tilde{\Phi}(-w,-u)e^{-iwl}\,, \tag{12.3.6}$$

$$\tilde{\Phi}(w,u) = \frac{4\pi^2 i}{c} L(w) e^{-ikb\sin\varphi} \sum_{n=1}^{\infty} F_n(w,u), \tag{12.3.7}$$

where

$$\left. \begin{array}{c} w = k\cos\varphi, \quad u = k\cos\varphi_0, \quad v = k\sin\varphi, \quad v^0 = k\sin\varphi_0, \\ 0 < \varphi < \pi, \quad 0 < \varphi_0 < \pi. \end{array} \right\} \tag{12.3.8}$$

The two terms on the right hand side of Eqn. (12.3.6) describe cylindrical waves propagating from the left ($z = -l$) and right ($z = l$) ends of the resonator, respectively.

The scattered field for an odd excitation is determined analogously and can also be presented in the form given by Eqns. (12.3.4)–(12.3.8) with appropriate functions $L(w)$, $L_{\pm}(w)$, $F_n(w, u)$. Here, it is necessary to explain some details regarding the above calculations for the far field.

First, in Eqn. (12.3.2) we have a term that is outside the integral. An analogous term exists in the field expression for odd excitation. The sum of these two terms gives the plane wave that exists for the whole region $y > 0$ and differs in sign from the incident wave given by Eqn. (12.1.1).

Second, the terms with $m = 1$ in the integral of Eqn. (12.3.2) and in the analogous expression for the odd excitation contain a pole at $w = u$. Because of this, they create a plane wave that exists only outside the region of the geometrical optics shadow behind the resonator. It turns out that, here, this plane wave completely cancels the plane wave mentioned in the first comment.

As a result, the total scattered field contains the beam of the plane wave that cancels the incident wave given by Eqn. (12.1.1) in the shadow region behind the resonator, as expected. However, in the far field, the angle width of this beam tends to zero, which is why it does not contribute to the scattered far field given by Eqn. (12.3.4).

The field inside the resonator will now be studied. Equations (12.1.9) and (12.3.1) show that for $-b<y<b$ an even excitation creates the field

$$H_x = -\frac{i4\pi}{c}\frac{G(u)}{L(u)}\sin v^0 y e^{i(uz+v^0 b)} - \\ -\frac{4\pi i}{c}\int_{-\infty}^{\infty}\sin vye^{ivb}\sum_{n=1}^{\infty}\left[F_n(w,u)e^{iw(z+l)}+F_n(-w,-u)e^{-iw(l-z)}\right]dw. \tag{12.3.9}$$

Changing the variable of integration from w to $-w$ for the second term within the brackets, we obtain

$$H_x = -\frac{i4\pi}{c}\frac{G(u)}{L(u)}\sin v^0 y e^{i(uz+v^0 b)} - \\ -\frac{4\pi i}{c}\int_{-\infty}^{\infty}\sin vye^{ivb}\sum_{n=1}^{\infty}\left[F_n(w,u)e^{iw(z+l)}+F_n(w,-u)e^{iw(l-z)}\right]dw. \tag{12.3.10}$$

Some comments are needed, here. First, the integrand contains poles w_m that are roots of the function $L(w)$. As is clear from Eqns. (12.1.10), (12.1.19), (12.1.20) for these values of w_m, the quantities $v_m^2 = k^2 - w_m^2$ must be real. The requirement that $\mathfrak{Im}\{v_m^2\} = 0$ leads to hyperbolic equation $w_m'' = k'k''/w_m'$ that describes the location of the poles shown in Fig. 12.3.1. Here, we assume temporarily that the wave number k has a small imaginary part $0 \le \mathfrak{Im}\{k\} \ll 1$. Therefore, all poles w_m with $\mathfrak{Re}\{w_m\} > 0$ are above the integration contour while those with $\mathfrak{Re}\{w_m\} < 0$ are below it. To define the single valued function $v = (k^2 - w^2)^{1/2}$ we introduced branch cuts along the lines $k \to k + i\infty$ and $-k \to -k - i\infty$.

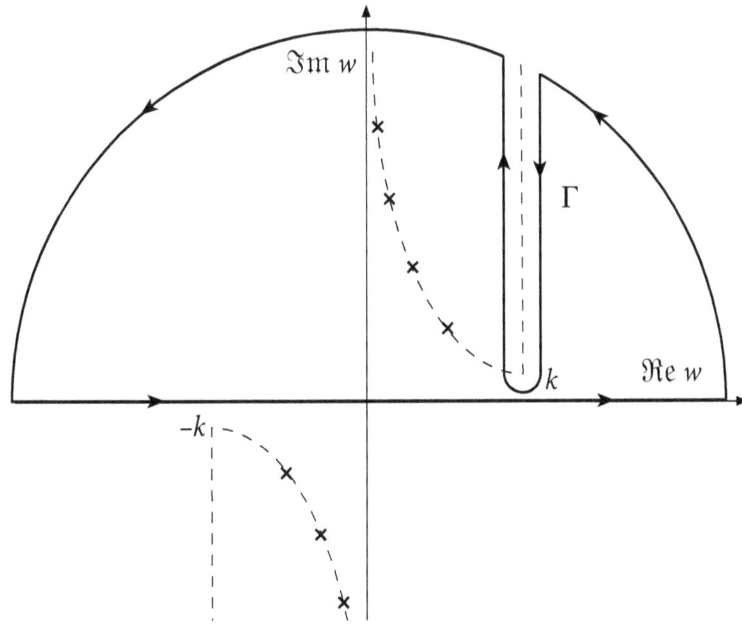

Figure 12.3.1: Contour of integration for the evaluation of Eqn. (12.3.10).

Now, we proceed to the analysis of Eqn. (12.3.10). The first term will be considered in more detail. Incorporating Eqns. (12.1.6) and (12.1.13), one can show that it is equal to

$$H_x = -i\sin(ky\sin\varphi_0)e^{ikz\cos\varphi_0} \tag{12.3.11}$$

and cancels the incident field given by Eqn. (12.1.3).

Let us now examine the integral

$$\int_\Gamma \sin\nu y e^{i\nu b} F_n(w,u)e^{iw\zeta}dw \quad \text{for} \quad \zeta > 0 \tag{12.3.12}$$

along the contour Γ, encircling the branch cut $k \to k+i\infty$ in a clockwise direction (Fig. 12.3.1). Equation (12.3.12) can be rewritten as

$$\int_{\Gamma} \frac{\sin vy}{2v\cos vb} \mathcal{H}_{+}(w,u) e^{iw\zeta}\, d\zeta, \tag{12.3.13}$$

where

$$\mathcal{H}_{+}(w,u) = L(w) F_n(w,u). \tag{12.3.14}$$

According to Eqn. (12.2.16), $\mathcal{H}_{+}(w, u)$ is holomorphic for $\Im\mathfrak{m}\{w\} > 0$ and, therefore, Eqn. (12.3.13) is equal to zero. Taking this result into account and altering the path of integration of Eqn. (12.3.10) along the contour given by Γ allows us to obtain the total field inside the resonator

$$H_x = \frac{8\pi^2}{c} \sum_{m=1}^{\infty} \frac{\sin v_m y}{L'_{-}(w_m)} e^{iv_m b} \times \\ \times \sum_{n=1}^{\infty} \left[\mathcal{F}_n(w_m, u) e^{iw_m(z+l)} + \mathcal{F}_n(w_m, -u) e^{iw_m(l-z)} \right], \tag{12.3.15}$$

in which

$$|y| < b, \quad |z| < l, \tag{12.3.16}$$

$$\left.\begin{aligned} &\mathcal{F}_n(w,u) = L_{-}(w) F_n(w,u), \\ &\mathcal{F}_1(w,u) = \frac{c}{i8\pi^2} \frac{e^{-i(ul+v^0 b)}}{w-\widehat{u}} \times \begin{cases} L_{-}(u) & \text{for even excitation} \\ -L_{-}(u) & \text{for odd excitation} \end{cases} \\ &v^0 = k\sin\varphi_0 \end{aligned}\right\} \tag{12.3.17}$$

and $L'_{-}(w)$ represents the derivative of $L_{-}(w)$ with respect to w.

For the case of odd excitation, the total field inside the resonator is expressed by an analogous expression:

$$H_x = i\frac{8\pi^2}{c}\sum_{m=0}^{\infty}\frac{\cos v_m y}{L'_-(w_m)}e^{iv_m b}\times$$
$$\times\sum_{n=1}^{\infty}\left[\mathcal{F}_n(w_m,u)e^{iw_m(z+l)}+\mathcal{F}_n(w_m,-u)e^{iw_m(l-z)}\right]. \tag{12.3.18}$$

It is possible to rewrite $L'_-(w)$ as

$$L'_-(w_m)=\epsilon_m\frac{i2w_m b}{L_+(w_m)}\quad, \tag{12.3.19}$$

where

$$\epsilon_0=2,\quad \epsilon_1=\epsilon_2=\ldots=1. \tag{12.3.20}$$

Thus, the total field inside the resonator is a superposition of waveguide modes traveling from its ends. It can be seen that beginning with some number m, the values of w_m are completely imaginary, and the corresponding terms represent evanescent modes that decay exponentially.

§12.4 Physical Interpretation and Asymptotic Expressions for $\mathcal{F}_n(w,u)$

The previous section demonstrated that $\mathcal{F}_n(k\cos\varphi, u)$ determines the scattering pattern of the nth-order edge wave radiated into free space, and that $\mathcal{F}_n(w_m, u)$ is the amplitude of the edge wave propagating inside the resonator. We will now study these functions in greater detail.

Let us begin with $\mathcal{F}_2(w, u)$. Proceeding from Eqns. (12.2.19) and (12.3.17), we can determine that it is equal to

$$\mathcal{F}_2(w,u)=\frac{G(u)e^{iul}}{2\pi i L_-(u)}\sum_m\frac{L_+^2(w_m)}{\epsilon_m 2iw_m b}\frac{e^{i2w_m l}}{(w_m+w)(w_m+u)}- \tag{12.4.1}$$

$$-\frac{1}{(2\pi i)^2}\frac{G(u)e^{iul}}{L_-(u)}\int_k^{k+i\infty}\frac{L_+^2(t)}{\sqrt{k^2-t^2}}\frac{e^{i2lt}dt}{(t+w)(t+u)}, \qquad (12.4.1, \text{cont.})$$

where the integral is evaluated along the left side of the branch cut $k \to k+i\infty$, where $\arg\{(k-t)^{1/2}\} = -\pi/4$. The first line describes edge waves created at the left end of the resonator ($z = -l$) by the diffraction of waveguide modes traveling from its right end ($z = l$). The second line is associated with the second-order edge wave created at $z = -l$ due to the diffraction of the first-order edge wave propagating along the exterior surface ($y = \pm b^{\pm}$) from the right end of the resonator.

Yet a more interesting physical picture is presented by $F_3(w, u)$. According to Eqns. (12.2.20, cont.) and (12.3.17) it is equal to

$$F_3(w,u)=\frac{G(u)e^{-iul}}{2\pi i L_-(u)}\times$$

$$\times\left\{\sum_{m_2}\frac{L_+^2(w_{m_2})}{L'(w_{m_2})}\frac{e^{i2w_{m_2}l}}{(w_{m_2}+w)}\sum_{m_1}\frac{L_+^2(w_{m_1})}{L'(w_{m_1})}\frac{e^{i2w_{m_1}l}}{(w_{m_1}+w_{m_2})(w_{m_1}-u)}-\right.$$

$$-\frac{1}{2\pi i}\int_k^{k+i\infty}\frac{L_+^2(t)}{\sqrt{k^2-t^2}}\frac{e^{i2lt}dt}{(t+w)}\sum_m\frac{L_+^2(w_m)}{L'(w_m)}\frac{e^{i2w_ml}}{(w_m+t)(w_m-u)}- \qquad (12.4.2)$$

$$-\frac{1}{2\pi i}\sum_m\frac{L_+^2(w_m)}{L'(w_m)}\frac{e^{i2w_ml}}{(w_m+w)}\int_k^{k+i\infty}\frac{L_+^2(t)}{\sqrt{k^2-t^2}}\frac{e^{i2lt}dt}{(t+w_m)(t-u)}+$$

$$\left.+\frac{1}{(2\pi i)^2}\int_k^{k+i\infty}\frac{L_+^2(t_2)}{\sqrt{k^2-t_2^{\,2}}}\frac{e^{i2lt_2}dt_2}{(t_2+w)}\int_k^{k+i\infty}\frac{L_+^2(t_1)}{\sqrt{k^2-t_1^{\,2}}}\frac{e^{i2lt_1}dt_1}{(t_1+t_2)(t_1-u)}\right\},$$

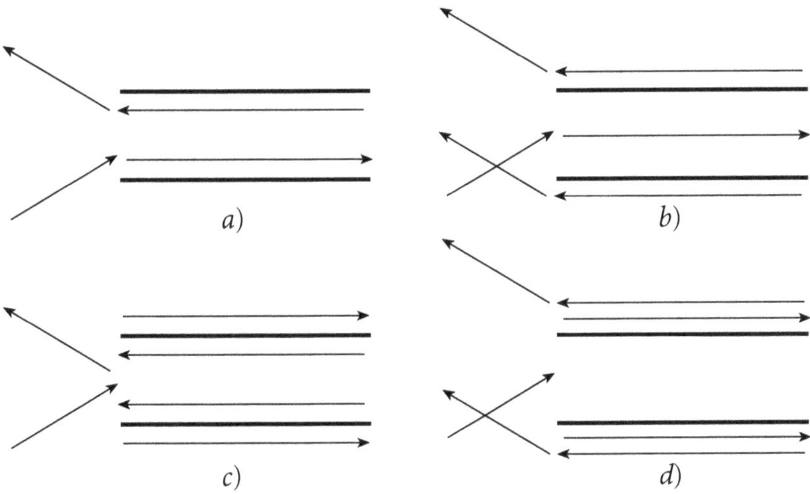

Figure 12.4.1: Progression of waves describing $\mathcal{F}_3(w,u)$.

The meaning of this equation is clear. Here, the first line in the braces describes edge waves created in the following manner. The incident wave striking the left end of the resonator ($z = -l$) excites waveguide modes (with number m_1) inside the resonator. Each of these waves, having propagated to the other end of the resonator ($z = l$) produces, in turn, a finite number of new waveguide modes (with number m_2). These new modes travel to the opposite end of the resonator and are radiated into surrounding space. This process is schematically represented in Fig. 12.4.1*a*.

The second line in Eqn. (12.4.2) is depicted in Fig. 12.4.1*b*. The first-order waveguide modes diffract at the right end of the resonator and excite second-order edge waves that propagate on the exterior surface of the resonator. These waves upon reaching the opposite end of the resonator, create third-order edge waves that radiate into free space.

The third line is depicted by Fig. 12.4.1*c*. Here, the incident wave creates a first-order edge wave on the exterior surface of the resonator. Having reached the opposite end of the resonator, they create internal waveguide modes. These modes travel to the opposite end of the resonator and are radiated into surrounding space. A detailed description of the last line of Eqn. (12.4.2) is not provided as its meaning is clearly depicted in Fig. 12.4.1*d*.

The remaining functions, $\mathcal{F}_n(w,u)$ qualitatively are no different from $\mathcal{F}_3(w,u)$. They possess an even more complicated structure, consisting of a large number of terms already introduced in $\mathcal{F}_3(w,u)$ and illustrated in Fig. 12.4.1. Also one should note that the integral along the branch cut is always connected with the edge waves flowing along the exterior of the resonator. Let us determine an asymptotic representation for them.

The function $\mathcal{F}_m(w,u)$ is composed of multiple integrals of the type

$$\underset{n=3,4,\ldots,m-1}{\hat{\varphi}_n(\alpha,\alpha_0)\,e^{iq(n-1)}} = \frac{\sqrt{1+\alpha}\sqrt{1+\alpha_0}}{(i\pi)^{n-1}}\times$$

$$\times\int_1^{1+i\infty}\cdots\int_1^{1+i\infty}\frac{\bar{\bar{h}}(t_1)\bar{\bar{h}}(t_2)\ldots\bar{\bar{h}}(t_{n-1})dt_1dt_2\ldots dt_{n-1}}{(t_1+\alpha_0)(t_2+t_1)(t_3+t_2)\ldots(t_{n-1}+t_{n-2})(t_{n-1}+\alpha)} \tag{12.4.3}$$

with $\bar{\bar{h}}(t)=h(t)\exp[2\chi_+(kt)]$ and $h(t)$ defined by Eqn. (8.3.24).

$$\underset{n=3,4,\ldots,m-1}{\hat{\varphi}_n(\alpha,\alpha_0)\,e^{iq(n-1)}} = \frac{\sqrt{1+\alpha}\sqrt{1+\alpha_0}}{(i\pi)^{n-1}}\times$$

$$\times\int_1^{1+i\infty}\cdots\int_1^{1+i\infty}\frac{h(t_1)h(t_2)\ldots h(t_{n-1})dt_1dt_2\ldots dt_{n-1}e^{2\sum_{s=1}^{n-1}\chi_+(kt_s)}}{(t_1+\alpha_0)(t_2+t_1)(t_3+t_2)\ldots(t_{n-1}+t_{n-2})(t_{n-1}+\alpha)}, \tag{12.4.4}$$

where $\alpha=\cos\varphi$, $\alpha_0=\cos\varphi_0$, $q=2kl$, and the integral is evaluated long the left side of the branch cut $1\to 1+i\infty$. A comparison of Eqn. (12.4.4) with Eqn. (10.1.7), shows that it differs from the latter by a factor of $\exp[2\Sigma\chi_+(kt_s)]$. Since this factor is a finite analytical function of t_s, the integral of Eqn. (12.4.4) is subject to Lemma 2 of § 9.1 for asymptotic expansions with $q\gg 1$. It is evident that an asymptotic expansion of $\hat{\varphi}_n(\alpha,\alpha_0)$ is easy to derive from the asymptotic expansion of $\varphi_n(\alpha,\alpha_0)$. To do this, an additional factor, $\exp[2(n-1)\chi_+(k)]$ in the first term of Eqn. (10.2.3) and the additional factor of

$$e^{2\sum_{s=1}^{n-1}\chi_+(kt_s)}$$

will be introduced on the right hand side of Eqns. (10.2.5) and (10.2.6). Using this method, the principal term of this expansion is of the form

$$\hat{\varphi}_{\substack{n\\ n=3,4,\ldots}}(\alpha,\alpha_0)\sim\varphi(q,\alpha)\varphi(q,\alpha_0)\varphi^{n-3}(q,1)e^{2(n-1)\chi_+(k)} \quad \text{for} \quad q\gg 1, \tag{12.4.5}$$

where $\varphi(q,\alpha)$ is determined by the equations of §§ 9.2 and 9.3, as before.

In principle, the introduced relationships permit the construction of an asymptotic series for $\mathcal{F}_n(w,u)$ for $kl\gg 1$. However, here, the following difficulties arise. First, $\mathcal{F}_n(w,u)$ possesses a complex compound structure, which reflects the coupling of a large number of different types of modes excited within the resonator. The second difficulty is the fact that an exact solution for $\chi_+(kt)$ has not been found. It has been estimated only approximately by $U(s,p)$. Thus,

$$e^{\chi_+(k)}=e^{U\left(\sqrt{2kb},p\right)}\left\{1+O\left[(kb)^{-3/2}\right]\right\}. \tag{12.4.6}$$

Therefore, utilizing the Vainshtein function U, we are able to determine correctly a finite number of terms of the asymptotic expansion for $\mathcal{F}_n(w,u)$. Thus, for example, assuming that

$$\chi_+(kt)=U\left(\sqrt{2kbt},p\right)$$

it is possible to obtain the following asymptotic expansion

$$\begin{aligned}\mathcal{F}_2(w,u)\sim\frac{G(u)e^{iul}}{2\pi iL_-(u)}\Bigg\{&\sum_m\frac{L_+^2(w_m)}{L'(w_m)}\frac{e^{i2w_ml}}{(w_m+w)(w_m+u)}-\\ &-\frac{1}{2(w-u)}\sum_{m=0}a_m\left[\frac{\varphi_q^{(m)}(q,\alpha_0)}{\sqrt{1+\alpha_0}}-\frac{\varphi_q^{(m)}(q,\alpha)}{\sqrt{1+\alpha}}\right]\Bigg\},\end{aligned} \tag{12.4.7}$$

for $q = 2kl \to \infty$, where $\varphi_q^{(m)}(q, \alpha)$ is a derivative of a Fresnel integral and is determined by Eqn. (9.2.6). The coefficients a_m are the Taylor series coefficients given by

$$\sqrt{1+te^{2U\left(\sqrt{2kbt},p\right)}} = \sum_{m=0}^{\infty} a_m (t-1)^m . \tag{12.4.8}$$

It is not difficult to show that for $kb \gg 1$, the coefficient a_1 contains a term on the order of $(kb)^{-1/2}$, the coefficient a_2 contains a term on the order of $(kb)^{-1}$, and the coefficient a_3 contains a term on the order of $(kb)^{-3/2}$. However, according to Eqn. (12.4.6) the coefficient a_0 is determined with an error of $(kb)^{-3/2}$. Thus, it is already incorrect to retain terms with a coefficient of a_3 in Eqn. (12.4.7).

Considering, however, that the properties of a resonator are determined principally by waveguide modes, it is possible to proceed with an analysis of $\mathcal{F}_n(w, u)$ neglecting the integrals along the branch cut, i.e., neglecting the coupling of the waves along the outside surface of the resonator.

§12.5 Approximate Expressions for the Scattering Pattern and Amplitude of Edge Waves

Omitting all terms in Eqn. (12.2.18) on the order of $(kl)^{-1/2}$, i.e., neglecting the mutual coupling of waves on the exterior of the resonator, it is possible to obtain the following relationship for $\mathcal{F}_n(w, u)$:

$$\mathcal{F}_n(w,u) = \sum_m \frac{L_+^2(w_m)}{L'(w_m)} \frac{e^{i2w_m l}}{(w_m + w)} \mathcal{F}_{n-1}(w_m, -u) . \tag{12.5.1}$$

The amplitudes of the waveguide modes are conveniently represented by

$$\mathcal{F}_n(w_{m_n}, u) = \sum_{m_{n-1}} R_{m_{n-1}, m_n} \mathcal{F}_{n-1}(w_{m_{n-1}}, -u) \tag{12.5.2}$$

or

$$\mathcal{F}_n\left(w_{m_n},u\right)=\sum_{m_{n-1},m_{n-2},\ldots,m_1} R_{m_{n-1},m_n} R_{m_{n-2},m_{n-1}}\ldots R_{m_1,m_2}\mathcal{F}_1\left[w_{m_1},-(-1)^n u\right], \tag{12.5.3}$$

where

$$R_{m,n}=\frac{L_+^2\left(w_m\right)}{L'\left(w_m\right)}\frac{e^{i2w_m l}}{\left(w_m+w_n\right)}, \tag{12.5.4}$$

$$L'\left(w_m\right)=\epsilon_m\cdot i2w_m b, \tag{12.5.5}$$

and the sum is evaluated with respect to m_v assuming the values 1,2,3,.... According to Eqn. (12.1.20), under odd excitation these indexes can also equal zero. It is natural to refer to $R_{m,n}$ as the transformation coefficient of waveguide mode m to waveguide mode n. This transformation occurs as the result of waveguide modes diffracting at the ends of the resonator. The properties of these coefficients will now be studied.

First of all, let us show that the construction of Eqn. (12.5.4) can be explained starting from simple physical relationships. From Eqn. (12.3.15) it can be seen that the waveguide mode of index m, propagating from $z=-l$ is determined by

$$\frac{\sin v_m y}{w_m b}L_+\left(w_m\right)e^{iw_m(z+l)}$$

and can be represented in the form of a sum of Brillouin waves

$$\frac{e^{iw_m l}}{2iw_m b}L_+\left(w_m\right)\left[e^{i\left(w_m z+v_m y\right)}-e^{i\left(w_m z-v_m y\right)}\right],$$

propagating at the angle φ_m ($w_m=k\cos\varphi_m$, $v_m=k\sin\varphi_m$) with the z-axis (see, Fig. 12.5.1). We will examine the diffraction of a plane wave described by the first term in this formula from a half-plane occupying the region bounded by $y=b$, $-\infty\leqslant z\leqslant l$. Diffraction from a half-plane creates an edge

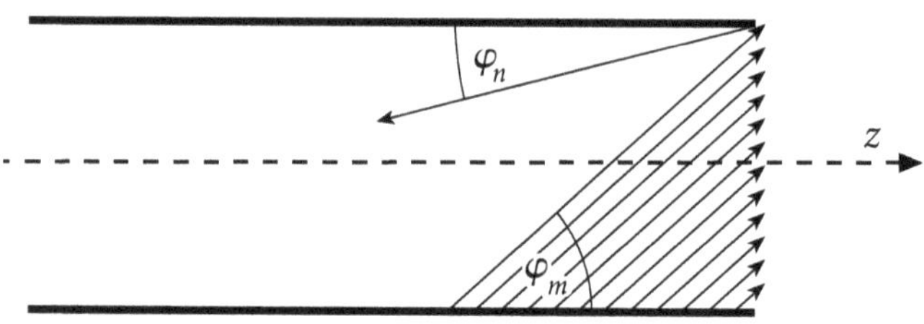

Figure 12.5.1: Diffraction of Brillouin waves from the end of a parallel plate waveguide.

cylindrical wave, the amplitude of which in the direction of φ_n is proportional to the product

$$\frac{L_+(w_m)}{2iw_m b}\frac{\sqrt{k+w_m}\sqrt{k+w_n}}{w_m+w_n}e^{i2w_m l}.$$

The root of $\sqrt{k+w_{m,n}}$ can be considered as a kernel for

$$L_+(w_{m,n})=\sqrt{k+w_{m,n}}\,e^{U(s_{m,n},p)}.$$

Moreover, for even excitation $2iw_m b = L'(w_m)$. Thus, we see that the product $R_{m,n}\cdot L_+(w_n)$ determining the amplitude of the nth wave, is proportional to the amplitude of the cylindrical wave occurring as the result of the mth wave, being diffracted from the end of the resonator, and propagating at an angle $\varphi_n = \cos^{-1}(w_n/k)$, which corresponds to the direction of the Brillouin wave for the nth waveguide mode. A quantitative estimate for $R_{m,n}$ will now be found.

Let us assign the index, j, to the propagating waveguide mode with the highest number, $m=j$. Then all the modes with index $m<j$ will be propagating, and all modes with index $m>j$ will be evanescent. Further, let the assumption be made that the frequency of the incident wave is near the critical frequency of the jth wave. In this case $w_j \approx 0$, and subsequently, referring to Eqns. (12.1.19) and (12.1.20)

$$\frac{kb}{\pi}+\frac{1}{2}=j+p \quad \text{for even excitation} \tag{12.5.6}$$

$$\frac{kb}{\pi}=j+p \quad \text{for odd excitation} \tag{12.5.7}$$

where $p < 1$. It is evident that the integers j for even and odd types of excitation are different. Moreover, it will be assumed that $j \gg 1$ for $kb \gg 1$. It can be seen that the jth mode determined in this manner exhibits specific properties which separate it from all other waveguide modes.

Actually, since

$$w_j \approx \sqrt{\frac{k}{2b}}\sqrt{4\pi p}\,, \quad w_m \approx \sqrt{\frac{k}{2b}}\sqrt{4\pi(j-m)} \quad \text{for} \quad 0<|j-m|<j \tag{12.5.8}$$

then the value of Eqn. (12.2.26) becomes

$$s_j \approx \sqrt{4\pi p}<1, \quad |s_m| \approx \sqrt{4\pi|j-m|} \gg 1 \quad \text{for} \quad 0<|j-m|<j. \tag{12.5.9}$$

Therefore, (see reference [96,120])

$$\left.\begin{aligned} &2U(s_j,p) \approx \ln\frac{2s_j^2}{i}+i(1+i)\beta_0 s_j, \quad \beta_0=0.824, \\ &U(s_m,p) \approx 0 \quad \text{for} \quad j-m \neq 0 \end{aligned}\right\} \tag{12.5.10}$$

and subsequently,

$$\left.\begin{aligned} &|R_{j,j}| \approx 1, \quad |R_{m,n}| \leqslant |R_{m,j}|, \\ &|R_{j,m}| \approx \frac{s_j}{\sqrt{\pi|j-m|}} \quad \text{for} \quad 0<|j-m|<j, \end{aligned}\right\} \tag{12.5.11}$$

$$\left|R_{m,j}\right| \approx \frac{1}{\epsilon_m 4\pi\left|j-m\right|} \quad \text{for} \quad 0<\left|j-m\right|<j. \qquad (12.5.11, \text{cont.})$$

Together with these estimates are the useful formulas

$$R_{j,j} \approx -e^{i(M+\beta_0+i\beta_0)s_j}, \quad M=\sqrt{\frac{2k}{b}}l, \qquad (12.5.12)$$

$$R_{j,m} = \frac{2w_j}{w_m+w_j} R_{j,j}, \qquad (12.5.13)$$

$$R_{m,n} \approx \frac{R_{m,j}}{1+w_n/w_m}. \qquad (12.5.14)$$

Thus, all of the coefficients, $R_{m,n}$ are small in comparison with $R_{j,j}$, the modulus of which is close to unity. On this basis, the jth mode is weakly transformed into other types of modes and weakly radiates into surrounding space. Modes of other orders, $m \neq j$, on the other hand, strongly radiate from the resonator, since they are weakly reflected from its ends and are weakly transformed into the remaining types of modes. These results permit the approximate summation of the multiple series given by Eqn. (12.5.3) for $\mathcal{F}_n(w_m, u)$.

We can begin with an estimate for $\mathcal{F}_2(w_{m_2}, u)$, by rewriting it as

$$\mathcal{F}_2\left(w_{m_2}, u\right) = R_{j,m_2}\mathcal{F}_1\left(w_j, -u\right) + \sum_{m_1}{}' R_{m_1,m_2}\mathcal{F}_1\left(w_{m_1}, -u\right), \qquad (12.5.15)$$

where the prime indicates that for the summation it is necessary to omit the term with index $m_1 = j$. Let $0 < |j-m_2| < j$, then according to Eqn. (12.5.11)

$$\mathcal{F}_2\left(w_{m_2}, u\right) = O\left(\frac{s_j}{\sqrt{\pi(j-m_2)}}\right) + \sum_{m_1}{}' O\left(\frac{1}{4\pi(j-m_1)}\right), \qquad (12.5.16)$$

Also, if $m_2 = j$ and $0 < |j-m_1| < j$, then

$$\mathcal{F}_2\left(w_j,u\right)=O(1)+\sum_{m_1}{}' O\left(\frac{1}{4\pi\left(j-m_1\right)}\right). \tag{12.5.17}$$

For sufficiently small values of s_j the terms of the series in Eqn. (12.5.16) can be on the same order as the first term. Therefore, we are not able to discard the series in Eqn. (12.5.15) for $m_2 \neq j$.

Let us further examine $\mathcal{F}_3(w_{m_3},u)$ *by* rewriting it in the form

$$\begin{aligned}\mathcal{F}_3\left(w_{m_3},u\right)&=R_{j,m_3}R_{j,j}\mathcal{F}_1\left(w_j,u\right)+R_{j,m_3}\sum_{m_1}{}' R_{m_1,j}\mathcal{F}_1\left(w_{m_1},u\right)+\\&+\sum_{m_2}{}' R_{m_2,m_3}R_{j,m_2}\mathcal{F}_1\left(w_j,u\right)+\sum_{m_2}{}' R_{m_2,m_3}\sum_{m_1}{}' R_{m_1,m_2}\mathcal{F}_1\left(w_{m_1},u\right).\end{aligned} \tag{12.5.18}$$

For $0<\left|j-m_{1,2,3}\right|<j$, retaining the same sequence of terms, this expression can be written in the form

$$\begin{aligned}\mathcal{F}_3\left(w_{m_3},u\right)&=O\left[\frac{s_j}{\sqrt{\pi\left(j-m_3\right)}}\right]+\frac{1}{4\pi^{3/2}}\left(\frac{s_j}{\sqrt{j-m_3}}\right)\sum_{m_1}{}' O\left(\frac{1}{j-m_1}\right)+\\&+\frac{1}{4\pi^{3/2}}\sum_{m_2}{}' O\left[\frac{s_j}{\left(j-m_2\right)^{3/2}}\right]+\frac{1}{\left(4\pi\right)^2}\sum_{m_2,m_1}{}'' O\left[\frac{1}{\left(j-m_2\right)\left(j-m_1\right)}\right].\end{aligned}$$

If $m_3=j$, then

$$\begin{aligned}\mathcal{F}_3\left(w_j,u\right)&=O(1)+\frac{1}{4\pi}\sum_{m_1}{}' O\left(\frac{1}{j-m_1}\right)+\\&+\frac{1}{4\pi^{3/2}}\sum_{m_2}{}' O\left[\frac{s_j}{\left(j-m_2\right)^{3/2}}\right]+\frac{1}{\left(4\pi\right)^2}\sum_{m_2,m_1}{}'' O\left[\frac{1}{\left(j-m_2\right)\left(j-m_1\right)}\right].\end{aligned}$$

Neglecting terms on the order of $(j-m_2)^{-1}\cdot(j-m_1)^{-1}$, we obtain

$$\mathcal{F}_3(w_m,u)=R_{j,m}\left[R_{j,j}\mathcal{F}_1(w_j,u)+\sum_{m_1}{}'R_{m_1,j}\mathcal{F}_1(w_{m_1},u)\right]+ \tag{12.5.19}$$

$$+\mathcal{F}_1(w_j,u)\sum_{m_1}{}'R_{m_1,m}R_{j,m_1}\,.$$

Analogous estimates can be made for $\mathcal{F}_{4,5,6\ldots}$. These estimates show that

$$\underset{n=3,4,5\ldots}{\mathcal{F}_n(w_m,u)}=R_{j,m}R_{j,j}^{n-3}\times$$

$$\times\left[R_{j,j}\mathcal{F}_1\left(w_j,-(-1)^n u\right)+\sum_{m_1}{}'R_{m_1,j}\mathcal{F}_1\left(w_{m_1},-(-1)^n u\right)\right]+ \tag{12.5.20}$$

$$+R_{j,j}^{n-3}\mathcal{F}_1\left(w_j,-(-1)^n u\right)\sum_{m_1}{}'R_{m_1,m}R_{j,m_1}\,,$$

neglecting terms of the type

$$\frac{s_j}{(j-m_1)^{5/2}},\quad \frac{s_j^2}{(j-m_1)^2},\quad \frac{1}{(j-m_1)^2},\quad m_1\neq j$$

when $m\neq j$, and also neglecting small values of higher orders.

To interpret the physical meaning of this equation, it is convenient to introduce the following notation. Here, $m^{(n)}$ is used to denote the mth waveguide mode that has passed down the length of the waveguide and undergoes n reflections from its ends. The conversion of the mth mode to the vth mode will be denoted by $m\rightarrow v$. Then, the first term of Eqn. (12.5.20) can be conveniently presented with the aid of the transformation

$$j^{(n-1)}\rightarrow m\,,$$

which shows that the jth mode having originated due to the first-order diffraction traversed the resonator $n-1$ times and upon the last reflection, excited the mth mode. The series of terms in the brackets describes the following transformation of waves:

$$m_1^{(1)} \to j^{(n-2)} \to m, \quad m_1 \neq j.$$

Finally, the last term of Eqn. (12.5.20) corresponds to the transformation

$$j^{(n-2)} \to m_1^{(1)} \to m, \quad m_1 \neq j.$$

Our attention will now be turned to the following detail. In transforming from Eqn. (12.5.3) to Eqn. (12.5.20), we specifically neglected the series

$$\sum_{m_{n-1}, m_{n-2}, \ldots, m_1}' R_{m_{n-1}, m} R_{m_{n-2}, m_{n-1}} \ldots R_{m_1, m_2} \mathcal{F}_1\left(w_{m_1}, -(-1)^n u\right), \tag{12.5.21}$$

where the sum is over all values of indices with the exception of $m_1 = m_2 = \ldots = m_{n-1} = j$. For sufficiently small values of s_j, some terms in Eqn. (12.5.21) can be comparable with some of those retained in Eqn. (12.5.20). However, as we will now see, Eqn. (12.5.21) can nevertheless be neglected in the calculation of the resulting edge wave, $\sum_{n=1} \mathcal{F}_n(w_m, u)$, which in the final calculation, is of interest.

Evidently, utilizing Eqn. (12.5.20) it is not difficult to show that

$$\sum_{n=1} \mathcal{F}_n(w_m, u) = \mathcal{F}_1(w_m, u) + \mathcal{F}_2(w_m, u) +$$

$$+ \frac{1}{\mathcal{D}_j}\left[\mathcal{F}_1(w_j, u) + R_{j,j}\mathcal{F}_1(w_j, -u)\right]\left(R_{j,m}R_{j,j} + \sum_{m_1}' R_{m_1,m}R_{j,m_1}\right) + \tag{12.5.22}$$

$$+ \frac{R_{j,m}}{\mathcal{D}_j}\sum_{m_1}' R_{m_1,j}\left[\mathcal{F}_1(w_{m_1}, u) + R_{j,j}\mathcal{F}_1(w_{m_1}, -u)\right],$$

where

$$\mathcal{D}_j = 1 - R_{j,j}^2 . \tag{12.5.23}$$

The contribution to $\sum_{n=1} \mathcal{F}_n(w_m, u)$, produced by neglecting terms of Eqn. (12.5.21) can be determined. This is easiest to do by examining the term

$$R_{m,m}^{n-1} \mathcal{F}_1\left(w_m, -(-1)^n u\right), \quad m \neq j \tag{12.5.24}$$

from Eqn. (12.5.21). During the calculation of $\sum_{n=1} \mathcal{F}_n(w_m, u)$, this term produces an additional expression

$$\frac{R_{m,m}^2}{\mathcal{D}_m}\left[\mathcal{F}_1(w_m, u) + R_{m,m}\mathcal{F}_1(w_m, -u)\right], \quad m \neq j, \tag{12.5.25}$$

where

$$\mathcal{D}_m = 1 - R_{m,m}^2 . \tag{12.5.26}$$

However, according to the conditions of the problem $|R_{j,j}| \approx 1$ and $|R_{m,m}| \ll 1$. Therefore, it is clear that the value of Eqn. (12.5.25) is negligible compared to Eqn. (12.5.22).

§12.6 Resonant Part of the Field Inside the Resonator

Equation (12.5.22) together with Eqn. (12.3.15) and (12.3.18) determine the field inside the resonator. The resonant part of the field is

$$H_x = \frac{i\alpha_j}{\mathcal{D}_j} R_{j,j} e^{-ikb\sin\varphi_0} \sum_m^j (-1)^m \frac{\sqrt{1+\alpha_m}\,\psi_+(k\alpha_m)}{\epsilon_m kb\alpha_m} \times \\ \times \begin{Bmatrix} \sin(v_m y) \\ \cos(v_m y) \end{Bmatrix} \left[A(m,\beta)e^{ik(z+l)\alpha_m} + A(m,-\beta)e^{ik(l-z)\alpha_m}\right], \tag{12.6.1}$$

where

$$\alpha_m = w_m/k = \cos\varphi_m, \qquad \nu_m = k\sin\varphi_m, \qquad \beta = u/k = \cos\varphi_0,$$
$$\epsilon_0 = 2, \qquad \epsilon_1 = \epsilon_2 = \ldots = 1 \tag{12.6.2}$$

$$A(m,\beta) = B(m,\beta) + R_{j,j}B(m,-\beta), \tag{12.6.3}$$

$$B(m,\beta) = \sqrt{1-\beta}\,\psi_-(k\beta)e^{-ikl\beta}\times$$
$$\times\left[\frac{R_{j,j}}{(\alpha_m+\alpha_j)(\alpha_j-\beta)} + \frac{1}{\alpha_m+\alpha_j}\sum_{m_1}^{j-1}\frac{R_{m_1,j}}{\alpha_{m_1}-\beta} + \frac{1}{\alpha_j-\beta}\sum_{m_1}^{j-1}\frac{R_{m_1,m}}{\alpha_{m_1}+\alpha_j}\right]. \tag{12.6.4}$$

Here, we dropped the superscripts on the quantities $\alpha_m^{(1,2)}$, $\nu_m^{(1,2)}$, $\psi_\pm^{(1,2)}$, $R_{j,j}^{(1,2)}$, $R_{m,n}^{(1,2)}$, and $D_j^{(1,2)}$, which correspond to the field components of $H_x^{(1,2)}$. In Eqn. (12.6.1), the factor $\sin(\nu_m y)$ pertains to the field $H_x^{(1)}$ and the factor $\cos(\nu_m y)$ pertains to the field $H_x^{(2)}$.

In addition, when the excitation is even, we have

$$\left.\begin{aligned} &\nu_m^{(1)} = \sqrt{k^2 - \left[w_m^{(1)}\right]^2} = \frac{\pi}{2b}(2m-1), \quad m = 1,2,3,\ldots, \\ &j^{(1)} = \left[\frac{kb}{\pi} + \frac{1}{2}\right], \quad p^{(1)} = \frac{kb}{\pi} + \frac{1}{2} - j^{(1)}, \end{aligned}\right\} \tag{12.6.5}$$

$$\left.\begin{aligned} &\psi_-^{(1)}(k\alpha) = \psi_+^{(1)}(-k\alpha), \\ &\psi_+^{(1)}(k\alpha) = e^{U\left(\sqrt{2kb\alpha},\,p^{(1)}\right)}\begin{cases} 1 & \text{for } \alpha > 0 \\ \left(1 + e^{i2kb\sqrt{1-\alpha^2}}\right) & \text{for } \alpha < 0 \end{cases} \end{aligned}\right\}. \tag{12.6.6}$$

For the odd excitation we have

$$\left.\begin{aligned} & v_m^{(2)} = \sqrt{k^2 - \left[w_m^{(2)}\right]^2} = \frac{\pi}{2b}(2m) = \frac{\pi}{b}m, \quad m = 0,1,2,3,\ldots \\ & j^{(2)} = \left[\frac{kb}{\pi}\right], \quad p^{(2)} = \frac{kb}{\pi} - j^{(2)}, \end{aligned}\right\} \tag{12.6.7}$$

$$\left.\begin{aligned} & \psi_-^{(2)}(k\alpha) = \psi_+^{(2)}(-k\alpha), \\ & \psi_+^{(2)}(k\alpha) = e^{U\left(\sqrt{2kb}\alpha,\, p^{(2)}\right)} \begin{Bmatrix} 1 & \text{for} \quad \alpha > 0 \\ \left(1 - e^{i2kb\sqrt{1-\alpha^2}}\right) & \text{for} \quad \alpha < 0 \end{Bmatrix}. \end{aligned}\right\} \tag{12.6.8}$$

Note that a simplified version of Eqn. (12.6.1) derived using the approximate relationships of Eqn. (12.5.8), was obtained previously by another method in reference [147]; however, it contains some typographical errors.

Under the conditions

$$\left.\begin{aligned} & (M + \beta_0)s_j = (2n+1)\pi, \quad s_j = \sqrt{4\pi p} = \sqrt{2kb}\alpha_j \ll 1 \\ & \beta_0 = 0.824, \quad n = 0,1,2\ldots \end{aligned}\right\} \tag{12.6.9}$$

when according to Eqn. (12.5.12),

$$R_{j,j} \approx 1 - \beta_0 s_j, \quad D_j \approx 2\beta_0 s_j, \tag{12.6.10}$$

the system becomes resonant at the jth mode, which is symmetrical with respect to the z-coordinate:

$$\begin{aligned} H_{x,j} = & \frac{(-1)^n M \cos(w_j z)}{(2n+1)\pi\beta_0} \frac{e^{i(kb+\pi/4)}}{\sqrt{kb}} \begin{Bmatrix} -\sin(v_j y) \\ i\cos(v_j y) \end{Bmatrix} e^{-ikb\sin\varphi_0} \times \\ & \times \left\{ T(\alpha_j, \beta) + T(\alpha_j, -\beta) + \sum_{m_1}^{j-1} R_{m_1,j} \left[T(\alpha_{m_1}, \beta) + T(\alpha_{m_1}, -\beta) \right] \right\}, \end{aligned} \tag{12.6.11}$$

where

$$T(\alpha,\beta)=\frac{\sqrt{1-\beta}}{\alpha-\beta}\psi_{-}(k\beta)e^{-ikl\beta},\quad T(\alpha_j,\pm\alpha_j)=O\left(\sqrt{kb}\right). \tag{12.6.12}$$

Setting here $\varphi_0 = \pi/2$, we obtain the Fialkovskii expression of [147].

$$H_x=(-1)^n\frac{M}{(n+1/2)\pi\beta_0}\begin{pmatrix}-\sin v_j y\\ i\cos v_j y\end{pmatrix}\cos w_j z,\quad n=0,1,2,\ldots \tag{12.6.13}$$

where M is a large parameter (see, Eqn. (12.5.12)) that indicates the concentration of the field in the resonator.

Under the conditions of

$$\left.\begin{array}{l}(M+\beta_0)s_j=2n\pi,\quad s_j=\sqrt{4\pi p}=\sqrt{2kb}\alpha_j\ll 1\\ \beta_0=0.824,\quad n=1,2,3\ldots\end{array}\right\} \tag{12.6.14}$$

when

$$R_{j,j}\approx-1+\beta_0 s_j,\quad D_j\approx 2\beta_0 s_j, \tag{12.6.15}$$

the system becomes resonant at the jth mode, which is antisymmetrical with respect to the z-coordinate:

$$\begin{aligned}H_{x,j}=&\frac{(-1)^n M\sin(w_j z)}{2n\pi\beta_0}\frac{e^{i(kb+\pi/4)}}{\sqrt{kb}}\left\{\begin{array}{c}-\sin(v_j y)\\ i\cos(v_j y)\end{array}\right\}e^{-ikb\sin\varphi_0}\times\\ &\times\left\{T(\alpha_j,\beta)-T(\alpha_j,-\beta)-\sum_{m_1}^{j-1}R_{m_1,j}\left[T(\alpha_{m_1},\beta)-T(\alpha_{m_1},-\beta)\right]\right\},\end{aligned} \tag{12.6.16}$$

This resonance may be observed when the illumination angle $\varphi_0 < 90°$ and $w_j l/2 \approx \pi/2$. The second condition is fulfilled when $M(\pi p)^{1/2} \approx \pi$. For $\varphi_0 = 90°$ this resonance disappears ($H_{x,j} = 0$), since

$$T(\alpha_m,+0)=T(\alpha_m,-0). \tag{12.6.17}$$

Note that the incident field $H_{0x}^{(1,2)}$ does not undergo diffraction when $\varphi_0 = \varphi_m^{(1,2)} = \cos^{-1}(w_m^{(1,2)}/k)$ with $m = 0, 1, 2,\ldots$. For this condition, the plates of the resonator are located at the nodes of the incident electric field $E_{0z}^{(1,2)}$ and do not disturb this incident field. We recall that Eqns. (12.6.1), (12.6.11), (12.6.16) were derived subject to the condition

$$u=k\cos\varphi_0\neq w_m,\quad m=0,1,2,\ldots,j-1 \tag{12.6.18}$$

and therefore, are not applicable when $\varphi_0 = \varphi_m^{(1,2)}$ or $\varphi_0 = \pi-\varphi_m^{(1,2)}$ with $m = 0, 1, 2,\ldots j-1$. This remark also applies to the results found in reference [147].

Let us now examine how the diffracted field $H_x^{(1,2)}$ vanishes when $u \to w_m^{(1,2)} \geq 0$ and $\varphi_0 \to \varphi_m^{(1,2)} \geq 0$ ($m = 0, 1, 2,\ldots j-1$). Here, we refer to the rigorous Eqns. (12.3.15), (12.3.17), (12.3.18), and to asymptotic Eqns. (12.5.1), (12.5.2), (12.5.3) which neglect all multiple edge waves running over the external faces of the resonator plates ($y = \pm b^{\pm}$).

According to Eqns. (12.3.15) and (12.3.18), the total field is

$$H_x^{total}=\frac{8\pi^2}{c}\sum_m^{\infty}\frac{L_+(w_m)}{\epsilon_m i2w_m b}\begin{Bmatrix}\sin v_m y\\ i\cos v_m y\end{Bmatrix}e^{iv_m b}\times$$
$$\times\sum_{n=1}^{\infty}\left[\mathcal{F}_n(w_m,u)e^{iw_m(z+l)}+\mathcal{F}_n(w_m,-u)e^{iw_m(l-z)}\right] \tag{12.6.19}$$

where the upper indices were omitted.

Using the relationships

$$\mathcal{F}_1(w,w_m)=0 \quad \text{for } w\neq w_m\text{, including } w\to w_m, \tag{12.6.20}$$

$$\lim_{u\to w_m}\mathcal{F}_1(w_m,u)=\begin{Bmatrix}-1\\ 1\end{Bmatrix}\frac{cL'(w_m)}{8\pi^2 iL_+(w_m)}e^{-i(w_m l+v_m b)}, \tag{12.6.21}$$

$$R_{m,n}\mathcal{F}_1(w_m,w_m) = -\mathcal{F}_1(w_n,-w_m) \tag{12.6.22}$$

and Eqns. (12.5.1), (12.5.2), and (12.5.3), one can establish that

$$\left.\begin{aligned}\mathcal{F}_n(w_m,w_v) &= -\mathcal{F}_{n-1}(w_m,w_v), \quad n=3,5,7\ldots \\ \mathcal{F}_n(w_m,-w_v) &= -\mathcal{F}_{n+1}(w_m,-w_v), \quad n=1,3,5,\ldots\end{aligned}\right\} v=0,1,2,3\ldots,j. \tag{12.6.23}$$

Consequently,

$$\left.\begin{aligned}\sum_{n=2}^{\infty}\mathcal{F}_n(w_m,w_v) &= 0, \\ \sum_{n=1}^{\infty}\mathcal{F}_n(w_m,-w_v) &= 0,\end{aligned}\right\} v=0,1,2,\ldots,j \tag{12.6.24}$$

and

$$\begin{aligned}H_x^{total} &= \frac{8\pi^2}{c}\frac{L_+(w_m)}{\epsilon_m i2w_m b}\begin{Bmatrix}\sin v_m y\\ i\cos v_m y\end{Bmatrix}e^{iv_m b}\mathcal{F}_1(w_m,w_m)e^{iw_m(z+l)} = \\ &= \begin{Bmatrix}i\sin v_m y\\ \cos v_m y\end{Bmatrix}e^{iw_m z} = \begin{Bmatrix}H_{0x}^{(1)}\\ H_{0x}^{(2)}\end{Bmatrix}.\end{aligned} \tag{12.6.25}$$

When $\varphi_0 \to \pi - \varphi_m^{(1,2)}$ and $u \to -w_m^{(1,2)}$, Eqn. (12.6.19) transforms into

$$\begin{aligned}H_x^{total} &= \frac{8\pi^2}{c}\frac{L_+(w_m)}{\epsilon_m i2w_m b}\begin{Bmatrix}\sin v_m y\\ i\cos v_m y\end{Bmatrix}e^{iv_m b}\mathcal{F}_1(w_m,w_m)e^{iw_m(l-z)} = \\ &= \begin{Bmatrix}i\sin v_m y\\ \cos v_m y\end{Bmatrix}e^{-iw_m z} = \begin{Bmatrix}H_{0x}^{(1)}\\ H_{0x}^{(2)}\end{Bmatrix}.\end{aligned} \tag{12.6.26}$$

Thus, under the incident angles $\varphi_0 = \varphi_m^{(1,2)}$ and $\varphi_0 = \pi - \varphi_m^{(1,2)}$ related to the directions of Brillouin waves, the incident field $H_{0x}^{(1,2)}$ does not undergo diffraction at the resonator. In this case, the edge waves cancel each other pairwise and the total field inside the resonator exactly equals the incident field $H_{0x}^{(1,2)}$. We emphasize that the numbers $w_m^{(1)}$ and $w_m^{(2)}$ are different. Therefore, when the $H_x^{(1)}$ part of the diffracted field disappears, its other part $H_x^{(2)}$ is present and vice versa.

§12.7 Radiation and Scattering from an Open Resonator

The field in the far zone is determined by Eqns. (12.3.4)–(12.3.8) as a function of $L(w)F_n(w, u)$. With an error of $(kl)^{-1/2}$, this last function can be presented (see Eqn. (12.5.1)) as

$$\frac{1}{2\sqrt{k}} L(w) F_n(w,u) = \sum_m \mathcal{P}(w_m, w) \mathcal{F}_{n-1}(w_m, -u), \quad {}_{n=2,3,\ldots} \tag{12.7.1}$$

where

$$\mathcal{P}(w_m, w) = \frac{L_+(w)}{2\sqrt{k}(w_m + w)} \frac{L_+^2(w_m)}{L'(w_m)} e^{i2w_m l} \tag{12.7.2}$$

is naturally referred to as the radiation pattern of the field radiated by the mth mode from a semi-infinite planar waveguide. The physical discussion, presented in Section 12.5 shows, namely, that the value of $\mathcal{P}(w_m, w)$ is proportional to the amplitude of the edge wave, which results from the diffraction of the mth Brillouin mode at the end of the resonator.

Using Eqn. (12.7.1) we can write

$$\begin{aligned} &\frac{1}{2\sqrt{k}} L(w) \sum_{n=1}^{\infty} F_n(w,u) = \\ &\quad = \frac{1}{2\sqrt{k}} L_+(w) \mathcal{F}_1(w,u) + \sum_m^{\infty} \mathcal{P}(w_m, w) \sum_{n=1}^{\infty} \mathcal{F}_n(w_m, -u). \end{aligned} \tag{12.7.3}$$

The series

$$\sum_{n=1}^{\infty} \mathcal{F}_n(w_m, -u)$$

can be approximated by Eqn. (12.5.22). It should be noted that the resonant terms in this series are separated into groups according to their values. Retaining only the most important of them, we have

$$\sum_{n=1}^{\infty} \mathcal{F}_n(w_m, -u) = \mathcal{F}_1(w_m, -u) + \mathcal{F}_2(w_m, -u) + \\ + \frac{R_{j,j}^2}{\mathcal{D}_j}\left[\mathcal{F}_1(w_j, -u) + R_{j,j}\mathcal{F}_1(w_j, u)\right]. \tag{12.7.4}$$

With the aid of this expression and Eqns. (12.7.3), (12.7.4), and (12.3.7), it is not difficult to determine that the radiation pattern of the modes radiating from the left end ($z = -l$) of the resonator into surrounding space is

$$\tilde{\Phi}(w,u) = \sum_{n=1}^{n=3} \tilde{\Phi}_n(w,u) + \tilde{\Phi}_D(w,u), \quad w = k\cos\varphi, \quad u = k\cos\varphi_0, \tag{12.7.5}$$

where

$$\tilde{\Phi}_1(w,u) = \pm\frac{L_+(w)L_-(u)}{2(w-u)} e^{-i(ul+\xi)}, \quad \xi = kb(\sin\varphi + \sin\varphi_0) \tag{12.7.6}$$

$$\tilde{\Phi}_2(w,u) = \pm\sqrt{k}L_+(u)\sum_m^j \frac{\mathcal{P}(w_m, w)}{w_m + u} e^{i(ul-\xi)}, \tag{12.7.7}$$

$$\tilde{\Phi}_3(w,u) = \pm\sqrt{k}L_-(u)\sum_m^j \sum_{m_1}^j \frac{\mathcal{P}(w_m, w)}{w_{m_1} - u} R_{m_1,m} e^{-i(ul+\xi)}, \tag{12.7.8}$$

$$\tilde{\Phi}_D(w,u) = \pm\sqrt{k}\frac{R_{j,j}^2}{D_j}\left[\frac{L_+(u)}{w_j+u}e^{iul} + R_{j,j}\frac{L_-(u)}{w_j-u}e^{-iul}\right]P(w_j,w)e^{-i\xi}. \quad (12.7.9)$$

Here, the upper sign is chosen when the resonator is excited by the field $H_{0x}^{(1)}$ and the lower sign is chosen when it is excited by the field $H_{0x}^{(2)}$. The functions $\tilde{\Phi}_{1,2,3}$ represent the radiation patterns of the first-order, second-order, and third-order edge waves, respectively. The function $\tilde{\Phi}_D$ is the resonant part of the radiation generated by the guided waves with index j and all orders $n > 3$.

One should note that Eqn. (12.7.5) leads to Eqn. (12.3.6), which gives the expression for the directivity pattern of the total scattered field

$$\Phi(w,u) = \tilde{\Phi}(w,u)e^{iwl} + \tilde{\Phi}(-w,-u)e^{-iwl}, w = k\cos\varphi, u = k\cos\varphi_0, \quad (12.7.10)$$

which exactly satisfies the reciprocity principle, i.e, $\Phi(w,u) = \Phi(u,w)$.

The physical content of Eqn. (12.7.5) will now be considered. Let us first examine $\tilde{\Phi}_1$, which corresponds to the scattered field given by

$$H_x = \pm\frac{L_-(u)L_+(w)}{2(w-u)}e^{-ik(l\cos\varphi_0+b\sin\varphi_0)}\frac{e^{i(kr+\pi/4)}}{\sqrt{2\pi kr}}e^{i(wl-vb)}. \quad (12.7.11)$$

Let $\pi/2 < \varphi < \pi$ and $\pi/2 < \varphi_0 < \pi$, then $w < 0$ and $u < 0$, and

$$H_x = \frac{\sqrt{k+w}\sqrt{k-u}}{2(w-u)}\left[e^{i(wl+vb)} \pm e^{i(wl-vb)}\right]e^{-ik(l\cos\varphi_0+b\sin\varphi_0)}\times$$
$$\times\frac{e^{i(kr+\pi/4)}}{\sqrt{2\pi kr}}e^{U\left(\sqrt{(2b/k)}w,p\right)-U\left(\sqrt{(2b/k)}u,p\right)}. \quad (12.7.12)$$

Here, $\nu = k\sin\varphi$. It is seen that the scattered field is the sum of the two edge waves radiating from the edges $y = \pm b$, $z = -l$. The directivity pattern of these waves is proportional to the directivity pattern of a cylindrical wave, created by the diffraction of a plane wave from a single half-plane. Representing the field radiated from the open end of a waveguide as a sum of two edge waves was first provided by Vainshtein [96,120]. He also emphasized the important role of edge waves play in diffraction problems.

When $kb \gg 1$ and the walls of the resonator are further separated, the functions $U(\sqrt{2b/k}\,w, p)$ and $U(\sqrt{2b/k}\,u, p)$ that represent the mutual coupling tend to zero. Therefore, summing the fields of Eqn. (12.7.12) related to the even and odd excitations yields

$$H_x \approx \frac{2\cos(\varphi/2)\sin(\varphi_0/2)}{\cos\varphi - \cos\varphi_0} e^{-ik(l\cos\varphi_0 + b\sin\varphi_0)} \frac{e^{i(k\bar{r}+\pi/4)}}{\sqrt{2\pi k\bar{r}}}, \tag{12.7.13}$$

for the total first-order edge wave, where

$$\bar{r} = \sqrt{(y+b)^2 + (z+l)^2}$$

represents the distance from the left end of the lower plate. As expected, we obtained a cylindrical wave radiating from the isolated lower half-plane, $y = -b$, $-l \leqslant z \leqslant \infty$ when illuminated by the plane wave of Eqn. (12.1.1). Under the condition $\pi/2 < \varphi_0 < \pi$, the left end of the upper plate ($z = -l$, $y = +b$) is the shadow region and does not contribute to the scattered field within the considered approximation.

Returning to Eqn. (12.7.5), note further that

$$\mathcal{P}(w_m, w) = \frac{\cos(\varphi/2)\cos^2(\varphi_m/2)e^{2ikl\cos\varphi_m}}{i\sqrt{2}\epsilon_m kb(\cos\varphi + \cos\varphi_m)\cos\varphi_m} e^{2U(s_m,p)+\chi_+(w)} \tag{12.7.14}$$

and

$$\mathcal{P}(w_m, -w_m) = -\frac{1}{\sqrt{2}}\cos(\varphi_m/2)e^{i2kl\cos\varphi_m + U(s_m,p)}, \tag{12.7.15}$$

where

$$s_m = \sqrt{\frac{2b}{k}}w_m, \quad w_m = k\cos\varphi_m, \tag{12.7.16}$$

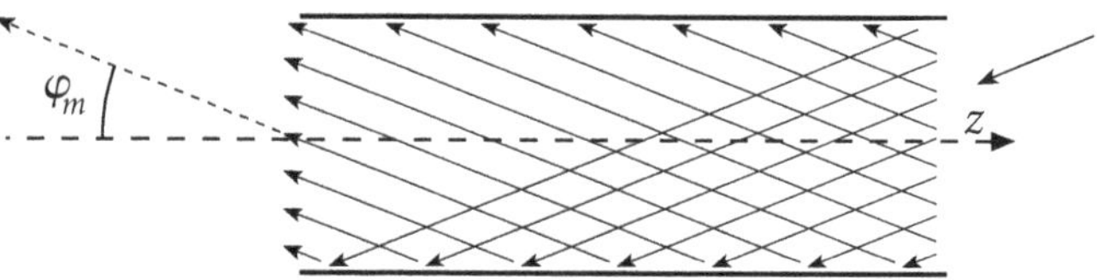

Figure 12.7.1: Diffraction of Brillouin waves from a parallel plate waveguide.

$$\Re e\{w_m\} \geqslant 0, \quad s = \sqrt{\frac{2b}{k}}w, \quad w = k\cos\varphi, \qquad (12.7.16, \text{cont.})$$

Equations (12.7.14) and (12.7.15) show that the radiation pattern of the mth waveguide mode striking the left end of the resonator possesses a maximum near $\varphi = \pi - \varphi_m$. This phenomenon is easily explained, starting from physical relationships. A waveguide mode of order m can be considered to be the superposition of two plane waves (Brillouin waves) propagating at an angle φ_m with respect to the z-axis (see Fig. 12.7.1). At the open end of the resonator the Brillouin wavefront is a planar segment of a constant phase surface which is similarly inclined with respect to the axis of the resonator. It is well understood that the maximum radiation exhibited by such a surface is in the direction of the surface normal, i.e., in the $\varphi = \pi \pm \varphi_m$ directions. At this point, we want to emphasize that this phenomenon takes place only for the radiation of propagating modes from the open end of the waveguide when the field source located between the plates.

In the problem of scattering which we study in this chapter, the field sources are external to the waveguide. These sources are the fields $H_{0x}^{(1)}$ and $H_{0x}^{(2)}$. As discussed in the previous section, these fields do not undergo diffraction at the resonator when their directions coincide with the directions of characteristic waveguide modes $\varphi_m^{(1)} = \cos^{-1}(w_m^{(1)}/k)$ and $\varphi_m^{(2)} = \cos^{-1}(w_m^{(2)}/k)$. According to the reciprocity principle, the scattered fields $H_x^{(1)}$ and $H_x^{(2)}$ will be exactly zero in the directions $\varphi = \pm\varphi_m^{(1)}, \pi\pm\varphi_m^{(1)}$ and $\varphi = \pm\varphi_m^{(2)}, \pi\pm\varphi_m^{(2)}$, respectively. Thus, it turns out that the behavior of radiation from the open end of a waveguide with an internal source and the behavior of scattering from the open section of a waveguide are completely different. Where the radiation from the waveguide has maximum, the field $H_x^{(1,2)}$ scattered from the waveguide section is exactly zero!

As a result, one can formulate the following rigorous statement

$$\left.\begin{aligned}\Phi^{(1)}\left(\pm w_m^{(1)},u\right)=\Phi^{(1)}\left(w,\pm u_m^{(1)}\right)=0,\\ \Phi^{(2)}\left(\pm w_m^{(2)},u\right)=\Phi^{(2)}\left(w,\pm u_m^{(2)}\right)=0.\end{aligned}\right\} \tag{12.7.17}$$

We will demonstrate this statement using the asymptotic expressions given by Eqns. (12.5.1)–(12.5.3). Since the functions $\Phi^{(1,2)}(w,u)$ are symmetrical, i.e., $\Phi^{(1,2)}(w,u)=\Phi^{(1,2)}(u,w)$, it is sufficient to consider the scattering only for the incident directions $\varphi_0=\varphi_N^{(1,2)}$ where $0\le\varphi_N\le\pi/2$ and $u=w_N^{(1,2)}>0, N=1,2,3,\ldots,j$.

Dropping the upper indices and using Eqn. (12.3.6), one can represent the directivity pattern given by Eqn. (12.3.7) as

$$\Phi(w,w_N)=\tilde{\Phi}(w,w_N)e^{iwl}+\tilde{\Phi}(-w,-w_N)e^{-iwl}, \tag{12.7.18}$$

where

$$\tilde{\Phi}(w,w_N)=i\frac{4\pi^2}{c}e^{-ikb\sin\varphi}L_+(w)\sum_{n=1}^{\infty}\mathcal{F}_n(w,w_N). \tag{12.7.19}$$

Referring to Eqn. (12.5.1), the series here can be written in the form

$$\begin{aligned}&\sum_{n=1}^{\infty}\mathcal{F}_n(w,w_N)=\\ &\quad-\mathcal{F}_1(w,w_N)+\sum_m\frac{L_+^2(w_m)}{L'(w_m)}\frac{e^{i2w_ml}}{w_m+w}\sum_{n=2}\mathcal{F}_{n-1}(w_m,-w_N)=\\ &\quad=\mathcal{F}_1(w,w_N)+\sum_m\frac{L_+^2(w_m)}{L'(w_m)}\frac{e^{i2w_ml}}{w_m+w}\sum_{n=1}\mathcal{F}_n(w_m,-w_N).\end{aligned} \tag{12.7.20}$$

According to Eqns. (12.6.20) and (12.6.24) this quantity is zero, and therefore, $\tilde{\Phi}(w, w_N) = 0$. One should note that possible singularities at $w = -w_m$ in Eqn. (12.7.20) are suppressed by the factor $L_+(-w_m) = 0$ in Eqn. (12.7.19).

The directivity pattern of the radiation from the right end ($z = +l$) of the resonator can be represented as

$$\tilde{\Phi}(-w,-w_N) = i\frac{4\pi^2}{c}e^{-ikb\sin\varphi}L_-(w)\times$$
$$\times\left\{\mathcal{F}_1(-w,-w_N)+\mathcal{F}_2(-w,-w_N)+\sum_{n=3}^{\infty}\mathcal{F}_n(-w,-w_N)\right\}. \quad (12.7.21)$$

Referencing Eqns. (12.5.1), (12.6.20), (12.6.24), and (12.3.17) produces,

$$\mathcal{F}_2(-w,-w_N) = \sum_m \frac{L_+^2(w_m)}{L'(w_m)}\frac{e^{i2w_m l}}{w_m - w}\mathcal{F}_1(w_m, w_N) = -\mathcal{F}_1(-w,-w_N), \quad (12.7.22)$$

$$\sum_{n=3}\mathcal{F}_n(-w_m,-w_N) = \sum_m \frac{L_+^2(w_m)}{L'(w_m)}\frac{e^{i2w_m l}}{w_m - w}\sum_{n=3}\mathcal{F}_{n-1}(w_m, w_N) =$$
$$= \sum_m \frac{L_+^2(w_m)}{L'(w_m)}\frac{e^{i2w_m l}}{w_m - w}\sum_{n=2}\mathcal{F}_n(w_m, w_N) = 0. \quad (12.7.23)$$

Possible singularities at $w = w_m$ are suppressed by the factor $L_-(w_m) = 0$ in Eqn. (12.7.21). Equations (12.7.22) and (12.7.23) show that $\tilde{\Phi}(-w,-w_m) = 0$. Thus, both the scattered field $H_x^{(1,2)}$ outside the resonator and the diffracted field $H_x^{(1,2)}$ inside the resonator are equal to zero when it is excited by the incident field $H_{0x}^{(1,2)}$ in the Brillouin angles $\varphi_m^{(1,2)}$. This happens due to the pairwise cancelation of the edge waves as is shown by Eqn. (12.6.23).

However, the asymptotic form given by Eqn. (12.7.5) together with Eqn. (12.7.9) does not provide a value for the field $H_x^{(1,2)}$ that is precisely equal to zero when the resonator is excited by the incident field $H_{0x}^{(1,2)}$ at the Brillouin angles. The reason for this is that Eqn. (12.7.5) violates the compensation conditions of Eqn. (12.6.23) for edge waves of orders $n \geq 3$. In this case, the resultant field is equal to the third- and higher-order uncompensated edge waves.

It is obvious that the amplitude of third-order edge waves is small. The magnitude of the resonant part of the field is determined by the use of the following estimations, valid under the conditions given by Eqns. (12.6.9) and (12.6.14):

$$\frac{1}{\sqrt{k}}L_+(w_j)=O(s_j),\quad \sqrt{k}\frac{L_+(w)}{w+w_j}=O\left(\sqrt{kb}\right),\quad w\approx w_j, \tag{12.7.24}$$

$$\lim_{w\to w_j}\sqrt{k}\frac{L_-(w)}{w-w_j}=\sqrt{k}\frac{i2w_jb}{L_+(w_j)}=O\left(\sqrt{kb}\right), \tag{12.7.25}$$

$$\frac{1}{D_j}\mathcal{P}(w_j,w)=\begin{cases} O(1) & \text{for } |w|\approx w_j \\ O\left(\dfrac{1}{\sqrt{kb}}\right) & \text{for } |w|\gg w_j \end{cases}. \tag{12.7.26}$$

According to these estimates,

$$\Phi_D(w,u)=O\left(\frac{1}{\sqrt{kb}}\right)\quad \text{for } |w|\gg w_j,\ |u|\gg u_j \tag{12.7.27}$$

$$\Phi_D(w,u)=O(1)\quad \text{for } |w|\gg w_j,\ |u|\approx w_j \text{ and for } |w|\approx w_j,\ |u|\gg w_j, \tag{12.7.28}$$

$$\Phi_D(w,u)=O\left(\sqrt{kb}\right)\quad \text{for } |w|\approx w_j,\ |u|\approx u_j. \tag{12.7.29}$$

Consequently, $\Phi_D(w,u)$ has the largest values at $|w|\approx w_j$, $|u|\approx w_j$. This result can be easily explained. The resonant field in the system is generated by critical modes with index j. Therefore, its most intense excitation takes place when the resonator is illuminated by plane waves at angles φ_0 close to $\pm\varphi_j$, $\pi\pm\varphi_j$. In accordance with the reciprocity principle, it is also most "luminescent" in the vicinity of angles $\varphi=\pm\varphi_j$, $\pi\pm\varphi_j$.

It should be remembered that the pattern of the total scattered field is

$$\Phi(w,u)=\Phi^{(1)}(w,u)+\Phi^{(2)}(w,u), \tag{12.7.30}$$

where the function $\Phi^{(1)}$ results from the even excitation and the function $\Phi^{(2)}$ corresponds to the odd excitation created by the incident electric field given by Eqns. (12.1.5) and (12.1.7), respectively. Here, $w = k\cos\varphi$, $u = k\cos\varphi_0$, and the functions $\Phi^{(1,2)}(w,u)$ are determined by Eqn. (12.3.6) and Eqns. (12.7.5)–(12.7.9), which are valid for $y > b$, i.e., for $0 < \varphi < \pi$. We also assume that $0 < \varphi_0 < \pi/2$. To calculate the field in the region $-\pi < \varphi < 0$, one should use the symmetry properties

$$\Phi^{(1)}(-\varphi)=-\Phi^{(1)}(\varphi),\quad \Phi^{(2)}(-\varphi)=\Phi^{(2)}(\varphi). \tag{12.7.31}$$

Note that the patterns $\tilde{\Phi}_1(w,u)$ and $\tilde{\Phi}_1(-w,-u)$ of the first-order edge waves scattered from the left and right ends of the resonator are singular in the direction of the main shadow lobe ($\varphi = \varphi_0$) where $w = u$. However, their singularities cancel each other and their sum equals

$$\begin{aligned}\Phi_1^{(1)}(u,u)&=\tilde{\Phi}_1^{(1)}(u,u)e^{iul}+\tilde{\Phi}_1^{(1)}(-u,-u)e^{-iul}=\\&=\left\{\frac{1}{2\sin\varphi_0}+\left[ikl+\frac{d}{d\beta}U\left(\sqrt{2kb}\beta,p^{(1)}\right)\right]\sin\varphi_0\right\}\times\\&\times\left(1+e^{i2kb\sin\varphi_0}\right)+ikb\cos\varphi_0 e^{i2kb\sin\varphi_0},\end{aligned} \tag{12.7.32}$$

$$\begin{aligned}\Phi_1^{(2)}(u,u)&=\tilde{\Phi}_1^{(2)}(u,u)e^{iul}+\tilde{\Phi}_1^{(2)}(-u,-u)e^{-iul}=\\&=-\left\{\frac{1}{2\sin\varphi_0}+\left[ikl+\frac{d}{d\beta}U\left(\sqrt{2kb}\beta,p^{(2)}\right)\right]\sin\varphi_0\right\}\times\\&\times\left(1-e^{i2kb\sin\varphi_0}\right)+ikb\cos\varphi_0 e^{i2kb\sin\varphi_0}.\end{aligned} \tag{12.7.33}$$

Here, $\beta = \cos\varphi_0$, and $p^{(1,2)}$ are determined by Eqns. (12.6.5), (12.6.7), and

$$\frac{1}{\sqrt{2kb}}\frac{d}{d\beta}U\left(\sqrt{2kb}\beta,p\right)=$$

$$=\frac{e^{-\pi/4}}{2\pi}U_0'(p)+\frac{s}{2\pi}\int_{-\infty}^{\infty}\frac{dt}{\left(e^{-i2\pi p+t^2/2}-1\right)\left(t-se^{i\pi/4}\right)} \tag{12.7.34}$$

where s = $\sqrt{2kb}\beta$ and

$$U_0'(p)=\int_{-\infty}^{\infty}\frac{dt}{e^{-i2\pi p+t^2/2}-1}=\sqrt{2\pi}\sum_{n=1}^{\infty}\frac{e^{i2\pi pn}}{\sqrt{n}}. \tag{12.7.35}$$

More details about the Vainshtein function $U(s,p)$ are in Appendix B of [96].

Taking into account Eqns. (12.7.32) and (12.7.33), we introduce the normalized scattering pattern

$$h(w,u)=\frac{\Phi(w,u)}{ikd}=h^{(1)}(w,u)+h^{(2)}(w,u) \tag{12.7.36}$$

where

$$h^{(1,2)}(w,u)=\tilde{h}^{(1,2)}(w,u)e^{iwl}+\tilde{h}^{(1,2)}(-w,-u)e^{-iwl}, \tag{12.7.37}$$

$$\tilde{h}^{(1,2)}(w,u)=\frac{\tilde{\Phi}^{(1,2)}(w,u)}{ikd},\quad d=2(l\sin\varphi_0+b\cos\varphi_0). \tag{12.7.38}$$

In the case when $\tan\varphi_0 > b/l$, the parameter d is the width of the incident wave beam intercepted by the resonator. The quantity $2d$ represents the first term in the high frequency asymptotic expansion for the total scattering cross section

$$S=2d\,\Re\mathrm{e}\left\{h(k\cos\varphi_0,k\cos\varphi_0)\right\}. \tag{12.7.39}$$

Referring to Eqns. (12.7.5)–(12.7.9), the function $\tilde{h}$ is equal to

$$\tilde{h}^{(1,2)}(w,u) = \sum_{n=1}^{3} \tilde{h}_n(w,u) + \tilde{h}_D^{(1,2)}(w,u). \tag{12.7.40}$$

In accordance with estimations provided by Eqns. (12.7.27)–(12.7.29), the resonant part of function $\tilde{h}^{(1,2)}(w,u)$ has maximal values in the vicinity $|w| \approx w_j$, $|u| \approx w_j$, i.e., at $\varphi \approx \varphi_j$, $\pi \pm \varphi_j$ with $\varphi_0 \approx \varphi_j$, where

$$\max|h_D(w,u)| = O\left[\frac{1}{M}\right]. \tag{12.7.41}$$

§12.8 Results of Numerical Calculations

This section presents calculations of the function $|h(w,u)|$ for normal incidence of a plane wave, when $\varphi_0 = \pi/2$ and $u = k\cos\varphi_0 = 0$.

Figures 12.8.1 and 12.8.2 show results obtained for parameters $kb = 100\pi$ and $kl = 50\pi$ for which the characteristic oscillations were investigated in [142] and [147]. In this case, $h^{(2)}(w,0) = 0$ and $h(w,0) = h^{(1)}(w,0)$. The solid line in Fig. 12.8.1 shows the function $20\log|h_1^{(1)}|$, which is the total pattern of the first-order edge waves scattered from both ends of the resonator.

In Fig. 12.8.2, the solid line depicts the field generated by the sum of the first-order and second-order edge waves. The third-order edge waves and the resonant part $h_D^{(1)}$ have practically no effect on the scattering pattern for the chosen parameters kb and kl. It turns out that in this case, $|h_3^{(1)}| \approx 10^{-4}$ and $|h_D^{(1)}| \approx 10^{-5}$. The small values of $h_D^{(1)}$ are due to the small value of $R_{j,j} \approx 0.1$. The function $|h^{(1)}|$ exhibits fast oscillations as the angle φ is varied and is equal to zero at the Brillouin angles

$$\varphi = \varphi_n^{(1)} = \sin^{-1}\left[\frac{(2n-1)\pi}{2kb}\right], \quad n = 1,2,3,\ldots,j. \tag{12.8.1}$$

For comparison, the dashed curves in Figs. 12.8.1 and 12.8.2 illustrate the function $20\log|h_0|$ which is the directivity pattern of the field scattered

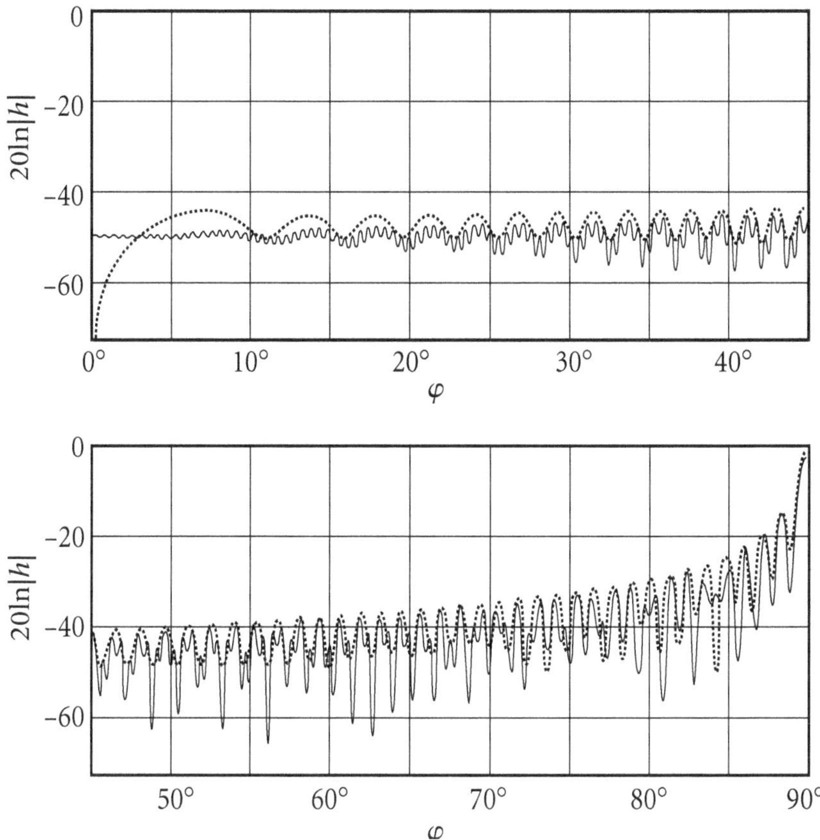

Figure 12.8.1: The solid line depicts the function $20\log|h_1^{(1)}|$ which is the total pattern of the first-order edge waves scattered from both ends of the resonator. The dashed line shows the directivity pattern $20\log|h_0|$ of the field scattered from a single strip ($y = -b$, $|z| \leq l$) excited by an incident plane wave. Reprinted from [26] with the permission of *Radiotekhnika i elektronika.*

from a single strip ($y = -b$, $|z| \leq l$) excited by the same incident plane wave. In the interval $0 \leq \varphi \leq 45°$ the curve for h_0 is close to the envelope of the maxima of the function $h^{(1)}$. Both curves merge gradually as the angle φ increases.

The second component of the scattered field, $h^{(2)}(w, 0)$, appears when $kb \neq n\pi$, $n = 1, 2, 3, \ldots$. Figure 12.8.3 shows the product $\mathrm{M}|h_D^{(2)}(0, 0)|$ for the

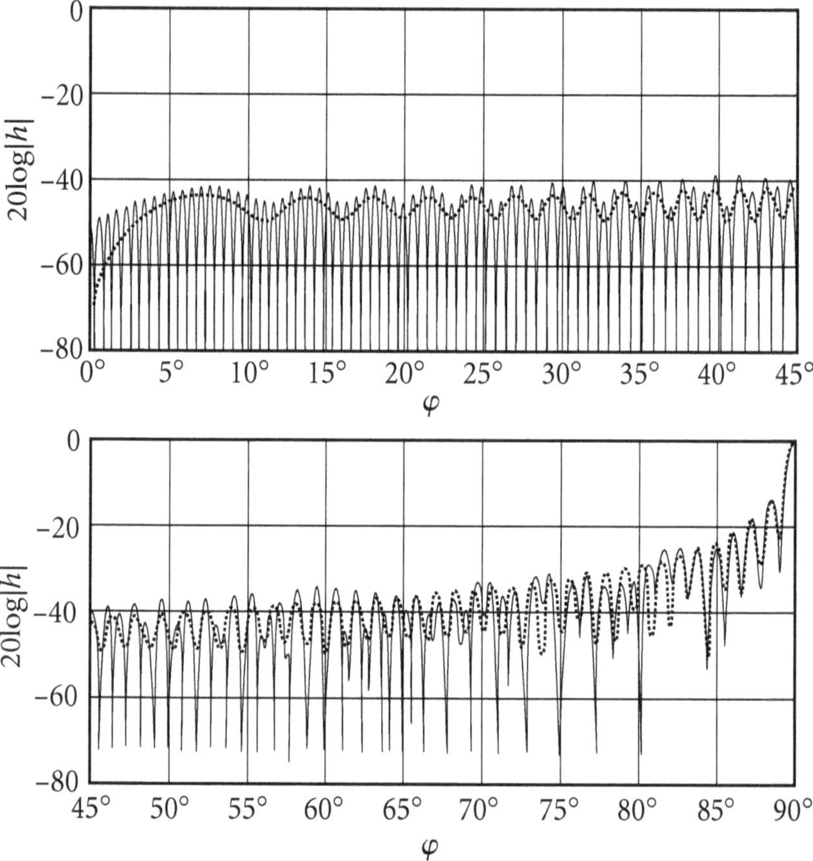

Figure 12.8.2: The solid line depicts the field generated by the sum of the first-order and second-order edge waves. The dashed line shows the directivity pattern $20\log|h_0|$ of the field scattered from a single strip ($y = -b$, $|z| \leq l$) excited by an incident plane wave. Reprinted from [26] with the permission of *Radiotekhnika i elektronika.*

values $kl = 50\pi$ and $kb = (100{+}p)\pi$ with $0 \leq p \leq 0.3$. This curve exhibits distinct resonant peaks in the vicinity of the values $p = p_n \ll 1$, which satisfy the condition [see Eqn. (12.6.9)]:

$$(M+\beta_0)\sqrt{4\pi p_n} = (2n+1)\pi, \quad n=0,1,2,3.... \tag{12.8.2}$$

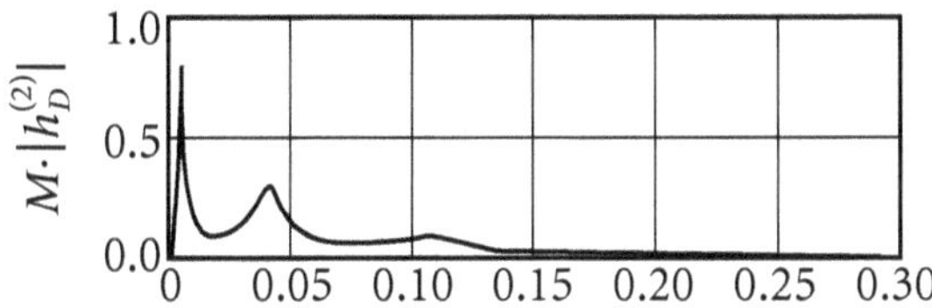

Figure 12.8.3: The product $M|h_D^{(2)}(0,0)|$ for $kl = 50\pi$ and $kb = (100+p)\pi$ with $0 \le p \le 0.3$. Reprinted from [26] with the permission of *Radiotekhnika i elektronika.*

when

$$h_D^{(2)}(0,0) \approx \frac{1}{M\beta_0}\left[R_{j,j}^{(2)}\right]^3 \approx \frac{1}{M\beta_0}, \quad R_{j,j}^{(2)} \approx 1 - \beta_0\sqrt{4\pi p}. \tag{12.8.3}$$

The total scattering cross section is determined by

$$S = 4l\Re e\left\{h^{(1)}(0,0) + h^{(2)}(0,0)\right\}. \tag{12.8.4}$$

For the above parameters $kl = 50\pi$ and $kb = (100+p)\pi$ with $p \ll 1$, the following asymptotic estimations are valid:

$$\left.\begin{aligned} &h^{(1)}(0,0) \approx 1, \quad h^{(2)}(0,0) = O(p) + O\left[\frac{\sqrt{p}}{M}\right], \\ &h_{2,3}^{(2)}(0,0) \approx O\left[\frac{\sqrt{p}}{M}\right], \quad h_D^{(2)}(0,0) = O\left[\frac{1}{M}\right]. \end{aligned}\right\} \tag{12.8.5}$$

This results in the final estimation

$$S \approx 4l\Re e\left\{h^{(1)}(0,0) + h_D^{(2)}(0,0)\right\} = 4l\left\{1 + O\left[\frac{1}{M}\right]\right\} \tag{12.8.6}$$

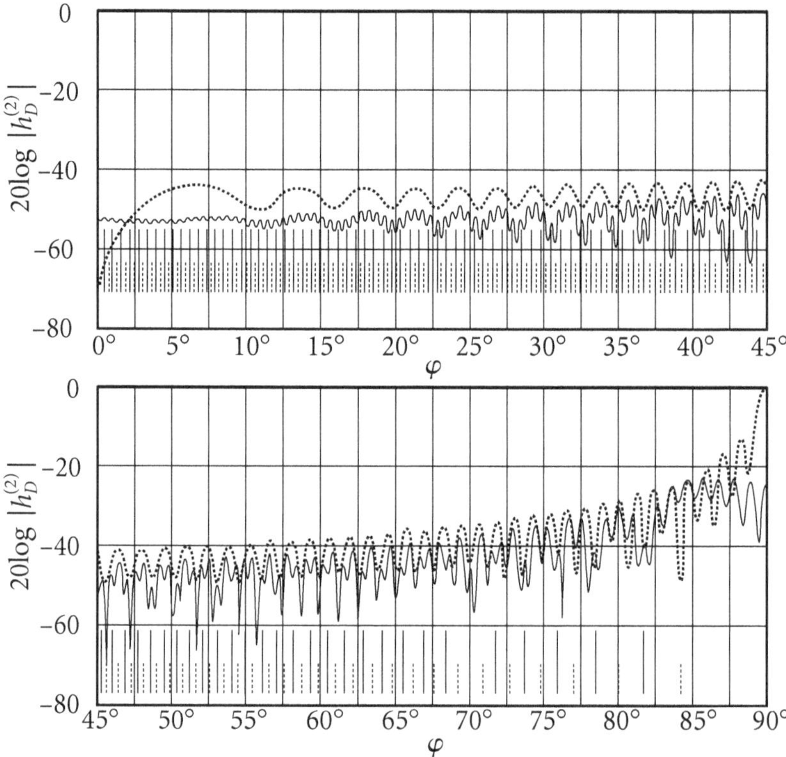

Figure 12.8.4: The solid curve depicts the spatial distribution of the resonant radiation $h_D^{(2)}(w,0)$ for the values $kl = 50\pi$ and $kb = (100+p)\pi$, where $p = 0.005$. The dashed curve presents the scattering pattern of the single lower plate ($y = -b$, $|z| \leq l$). The solid-line vertical segments denote the locations of zeros ($\varphi = \varphi_m^{(2)}$) of the function $h^{(2)}(w,0)$, and the dashed vertical lines show the locations of zeros ($\varphi = \varphi_m^{(1)}$) of the function $h^{(1)}(w,0)$. Reprinted from [26] with the permission of *Radiotekhnika i elektronika.*

even in the resonant case when $p^{(2)} = p_n$. Consequently, for open parallel plates resonators (with parameter $M \gg 1$) the resonant part of the total scattering cross section is small compared to the non-resonant part caused by scattering from the illuminated external side of the resonator.

Another interesting observation follows from Fig. 12.8.2. According to Eqn. (12.8.6) the total scattering cross section of the resonator is nearly equal to the total scattering cross section of the single lower plate. This means that the total power scattered from the resonator and the total power scattered from the single strip are practically equal. However, the resonator pattern $h^{(1)}$ has much larger number of minima than the strip pattern h_0. Thus it follows that, the resonator pattern $h^{(1)}$ must also have higher maxima than the strip pattern h_0. This is clearly seen in Fig. 12.8.2, especially in the interval $0 \leq \varphi \leq 45°$.

The spatial distribution of the resonant radiation $h_D^{(2)}(w, 0)$ is shown by the solid curve in Fig. 12.8.4 for the values $kl = 50\pi$ and $kb = (100+p)\pi$, where $p = 0.005$, which is close to the value $p = 0.0046$ that corresponds to the first resonant peak of Fig. 12.8.3. The solid-line vertical segments in Fig. 12.8.4 denote the locations of zeros ($\varphi = \varphi_m^{(2)}$) of the function $h^{(2)}(w,0)$, and the dashed vertical lines show the locations of zeros ($\varphi = \varphi_m^{(1)}$) of the function $h^{(1)}(w,0)$. As before, the dashed curve presents the scattering pattern of the single lower plate ($y = -b, |z| \leq l$). The most intense resonant radiation $h_D^{(2)}(w,0)$ is observed in the interval $\varphi_j^{(2)} \leq \varphi \leq 90°$, as was to be expected.

Within the limits of graphical accuracy, one can assume that the function $|h^{(1)}(w, 0)|$ will not change if the value $kb = 100.005\pi$ is replaced by $kb = 100\pi$. A comparison of Figs. 12.8.2 and 12.8.4 indicates that the pattern of the resonant radiation $|h_D^{(2)}(w, 0)|$ has the same qualitative character as the pattern $|h^{(1)}(w, 0)|$ almost over the entire range of angles φ, with exception of the vicinity of the main lobe. The envelope of the maxima of the pattern $|h^{(1)}(w, 0)|$ is higher than that of the pattern $|h_D^{(2)}(w, 0)|$. This difference is about 10 dB for many angles and approaches 15 dB near $\varphi = 7°$ and 25 dB near $\varphi = 90°$.

§12.9 Fundamental Results

- The field diffracted at a parallel plate resonator consists of two components, $H_x^{(1)}$ and $H_x^{(2)}$, with different type of symmetry with respect to the central plane between the resonator reflectors. Exact and asymptotic expressions for the field inside and outside the resonator are have been derived and analyzed. The following properties of the diffracted field have been established

- The resonance frequencies of the $H_x^{(1)}$ and $H_x^{(2)}$ components are different. When one is resonant the other is not.
- The resonance in the system is the result of constructive interference of critical waveguide modes reflected from the open ends.
- Scattering patterns of the $H_x^{(1)}$ and $H_x^{(2)}$ components in the far zone approach zero values in the directions of corresponding Brillouin waves, $\varphi_0 = \varphi_n^{(1,2)}$.
- $H_x^{(1)}$ or $H_x^{(2)}$ disappears completely everywhere inside and outside the resonator when the incidence direction approaches the corresponding Brillouin angles ($\varphi \to \varphi_m^{(1,2)}$). This phenomenon is the consequence of the resonator plates locations at the nodes of the electric vector related to the incident field $H_{0x}^{(1)}$ and $H_{0x}^{(2)}$.
- The influence of the resonance on the scatted field is negligible in the vicinity of the main shadow lobe, but it can be appreciable in side lobes. The main contribution to the scattered field is given by the first-order and second-order edge waves of the non-resonant component. Its maxima are higher than the maxima of the resonant component for the most of scattering directions.
- For this particular type of resonators with two parallel plates, the total scattered power associated with the resonant component is always small compared to the total power of the first-order edge waves of the non-resonant component when $kd \gg 1$ where $d = 2(l\sin\varphi_0 + b\cos\varphi_0)$ and $\tan\varphi_0 > b/l$.

§12.10 Additional Comments

- Some results of this chapter were published in [26].
- General properties of other types of open resonators were considered in the references [112, 113, and 148]. However, these books focus mainly on the study of eigen frequencies and eigen modes. In contrast, this chapter is devoted to the scattering properties of open resonators, which are more important in the study of radar cross section.
- We also refer to [148] for a resonance scattering by gratings and to [149] for diffraction from abruptly terminated open waveguides.

Conclusions

This book has examined the systematic study of edge waves and their role in the formation of the diffracted field.

Chapter 1 investigated the characteristics of the field scattered from black bodies. This field can be interpreted as the shadow radiation because it concentrates mainly in the vicinity of the shadow lobe. Central to this theory is the shadow contour theorem which states that different black bodies with the same shadow contour generate the same scattered field. The shadow contour is the boundary between illuminated and shadowed parts of the object surface.

It was shown that black bodies do not radiate in the direction of the source. In other words, no backscattering from black bodies exists. However, in the vicinity of the main shadow lobe, large black bodies create the same strong scattered field as that created by perfectly conducting bodies possessing the same shadow contour. The scattering in these directions is well known as the forward scattering and it was extensively investigated for a long tine. The theory of black bodies developed in the present book elucidates this phenomenon. It shows that the forward scattering can be actually interpreted as the shadow radiation. Another interesting observation is that in some directions a black body scatters more energy than a perfectly conducting body of the same shape and size.

It was also established in Chapter 1 that the total scattering cross-section for black bodies is half the total scattering cross-section of metallic bodies with identical shadow contours. This result is of fundamental importance. It shows that even with application of perfectly absorbing coatings on a metallic object, its total scattered power can be reduced solely by factor of two. This means that against bistatic radar, it is impossible to completely mask the scattering object by any radar absorbing materials.

This book is primarily focused on diffraction from perfectly conducting bodies with edges and vertices. Chapters 2 and 3 present some elements of the asymptotic technique known as the Physical Theory of Diffraction (PTD) [3, 4, 13, 64], which is extended here for objects with concave elements. This technique is demonstrated in the case of scattering from bodies of revolution. The scattered field is investigated in the directions along the symmetry axis which represents the focal line for edge waves. Only primary edge waves are calculated. It is assumed that objects are large in terms of wavelength, therefore the higher-order edge waves can be neglected.

The central and original idea of PTD is the separation of the surface current into the *uniform* and *nonuniform* components. The uniform component is determined by geometrical optics.The nonuniform component has a diffraction nature and is caused by any deviation of the object surface from a tangential plane, such as a smooth bending, edges, and vertices. Chapter 4, shows that this nonuniform current component is responsible for the depolarization of backscattering. A special technique is proposed for measurement of the field radiated by the nonuniform current, or in other words, by the curvature and its discontinuities (edges, vertices) in the target surface.

The parabolic equation method is widely used in diffraction problems related to smooth objects. In Chapter 6, this technique is extended to diffraction problems for objects with edges and wire ends. Interesting results are obtained in the study of diffraction at a wedge, dipole, and a strip. In particular, it is shown that the parabolic equation method leads to the same exact expressions for the current on a strip as those predicted by the elliptic wave equation.

Approximate expressions found in Chapter 6 for the current waves at dipoles and strips were used in Chapter 7 to calculate scattering patterns. A physical approach is developed here to derive scattering patterns which satisfy the reciprocity principle. Unique expressions are presented which are valid both for long and short dipoles.

The theory developed in Chapters 6 and 7 for multiple edge waves is based on the parabolic equation. A more rigorous investigation of higher-order edge waves is developed in Chapters 8–11 for the strip diffraction problem. This theory is based on the solution of functional equations and provides new exact and asymptotic expressions for the current and for the scattered field. These asymptotic expressions are applicable for arbitrary directions of incidence and observation, including grazing directions. Their error does not exceed an order of $(kl)^{-n/2}$ for $2kl \gg n$ where n is any positive integer, k is the wavenumber, and $2l$ is the width of the strip.

A mathematical theory developed in Chapters 8–11 was extended in Chapter 12 to study diffraction at an open resonator formed by two parallel plates. Here, it was shown that the resonance phenomenon is caused by the constructive interference of critical waveguide modes reflected from open ends. Simple asymptotic expressions were derived which describe the field characteristics inside and outside the resonator. The influence of the resonance in the system on the scattered field was investigated. It was shown that this influence is negligible in the vicinity of main lobes, but it can be appreciable in side lobes. New physical phenomena related to the Brillouin directions of incidence and observation have been established.

The theory developed in this book can be useful for investigation of diffraction at other, more complex scattering objects and structures.

References

1. T. Young. "On the Theory of Light and Colours." *Philosophical Transactions of the Royal Society of London* 91, Part 1. (1802): pp. 12–48.

2. T. Young. *A Course of Lectures on Natural Philosophy and Mechanical Arts*. London, 1807.

3. P. Ya. Ufimtsev. "Diffraktsiya na kline i lente" part I of "Priblizhennyi raschet diffraktsii ploskikh electromagnitnykh voln na nekotorykh metallicheskikh telakh," [Diffraction from a wedge and a strip, part I of Approximate computation of the diffraction of plane electromagnetic waves at certain metallic objects]. *Zhurnal tekhnicheskoi fiziki* 27, no. 8 (1957): pp. 1840–1849. [English translation published by: *Soviet Physics–Technical Physics.*]

4. P. Ya. Ufimtsev. "Diffraktsiya na diske i konechnom tsilindre" part II of "Priblizhennyi raschet diffraktsii ploskikh electromagnitnykh voln na nekotorykh metallicheskikh telakh," [Diffraction from a disk and a finite cylinder, part II of Approximate computation of the diffraction of plane electromagnetic waves at certain metallic objects]. *Zhurnal tekhnicheskoi fiziki* 28, no. 11 (1958): pp. 2604–2616. [English translation published by: *Soviet Physics–Technical Physics.*]

5. P. Ya. Ufimtsev. "Secondary Diffraction of Electromagnetic Waves by a Strip." *Soviet Physics–Technical Physics* 3, no. 3 (1958): pp. 535–548.

6. P. Ya. Ufimtsev. "Secondary Diffraction of Electromagnetic Waves by a Disk." *Soviet Physics–Technical Physics* 3, no. 3 (1958): pp. 549–556.

7. E. N. Maizel's, P. Ya. Ufimtsev. "Reflection of Circularly Polarized Electromagnetic Waves from Metallic Bodies." *Radio Engineering and Electronic Physics* 5, no. 12 (1960): pp. 70–76.

8. P. Ya. Ufimtsev. "Simmetrichnoe obluchenie konechnykh tel vrashcheniya" [Symmetrical illumination of finite bodies of revolution]. Odessa: *Doklad na 1-om Vsesoyuznom simpoziume no diffraktsii voln,* 1960.

9. P. Ya. Ufimtsev. "Rasseyanie ploskikh voln na tonkikh tsilindricheskikh provodnikakh" [Scattering of plane waves from thin cylindrical conductors]. Odessa: *Doklad na 1-om Vsesoyuzonom simpoziume no diffraktsii voln*, 1960.

10. E. N. Maizel's and P. Ya. Ufimtsev. "Otrazhenie elektromagnitnykh voln krugovoi polyarizatsii ot metallicheskikh tel" [Reflection of circularly polarized electromagnetic waves from metallic bodies]. Odessa: *Doklad na 1-om Vsesoyuzonom simpoziume no diffraktsii voln* [*Proceedings of the First All-Union Symposium on the Diffraction of Waves*], 1960.

11. P. Ya. Ufimtsev. "Reflections of circularly polarized radio waves from metallic bodies." *Radio Engineering and Electronic Physics* 6, no. 12 (1961): pp. 1878–1879.

12. P. Ya. Ufimtsev. Symmetrical Illumination of Finite Bodies of Revolution." *Radio Engineering and Electronic Physics* 6, no. 4 (1961): pp. 492–500.

13. P. Ya. Ufimtsev. *Metod kraevykh voln v fizicheskoi teorii diffraktsii* [*Method of edge waves in the physical theory of diffraction*]. Moscow: Sovetskoe Radio, 243 pp. 1962. [Machine translated into English by the U.S. Air Force, Foreign Technology Division (National Air Intelligence Center), Wright-Patterson AFB, OH, 1971. Technical Report AD 733203, Defense Technical Information Center of USA, Cameron Station, Alexandria, VA, 22304–6145, USA].

14. P. Ya. Ufimtsev. "Diffraction of Plane Wave Electromagnetic Waves from a Thin Cylindrical Conductor." *Radio Engineering and Electronic Physics* 7, no. 2 (1962): pp. 241–249.

15. P. Ya. Ufimtsev. "Fizicheskaya teoriya diffraktsii" [Physical Theory of Diffraction]. Gor'kii: *Doklad ha 2-m Vsesoyuznom simpoziume po diffraktsii voln* [*Proceedings of the second All-Union symposium on the diffraction of waves*], 1962.

16. P. Ya. Ufimtsev. "Transverse Diffusion for Diffraction by a Wedge." *Radio Engineering and Electronic Physics* 10, no. 6 (1965): pp. 866–875.

17. P. Ya. Ufimtsev. "Issledovanie blizhnego polya vibratora i toka v nem s pomoshch'yu parabolicheskogo uravneniya" [Research on the Near Field and Current of a Dipole Using Parabolic Equations]. Khar'kov: *Doklad na 3-om Vsesoyuznom simpoziume po diffraktsii voln* [*Proceedings of the third All-Union symposium on the diffraction of waves*], 1967.

18. P. Ya. Ufimtsev. "Asymptotic Investigation to the Problem of Diffraction on a Ribbon." *Radio Engineering and Electronic Physics* 13, no. 10 (1968): pp. 1631–1633.

19. P. Ya. Ufimtsev. "Current waves in a thin wire and in a ribbon." *U.S.S.R. Computational Mathematics and Mathematical Physics* 8, no. 6 (1968): pp. 250–261.

20. P. Ya. Ufimtsev. "Diffraction of Electromagnetic Waves at Black Bodies and Semi-transparent Plates." *Radiophysics and Quantum Electronics* 11, no. 6 (1968): pp 527–538.

21. P. Ya. Ufimtsev. "Metod fizicheskoi teorii diffraktsii" [Method of the Physical Theory of Diffraction]. From: *Lektsii pervoi Vsesoyuznoi shkoly-seminara po diffraktsii i rasprostraneniyu voln, Palanga 1965* [*Lectures of the first All-Union school-seminar on the diffraction and propagation of waves, Palanga 1965*]. Moscow-Khar'kov, 1968.

22. P. Ya. Ufimtsev. "Asymptotic Investigation of the Problem of Diffraction on a Strip." *Radio Engineering and Electronic Physics* 14, no. 7 (1969): pp. 1014–1025.

23. P. Ya. Ufimtsev. "Asimptoticheskoe reshenie zadachi o diffraktsii na lente v sluchae granichnykh uslovii Dirikhle" [Asymptotic solution to the problem of diffraction from a strip using Dirichlet boundary conditions]. *Radio Engineering and Electronic Physics* 15, no. 5 (1970): pp. 782–790.

24. P. Ya. Ufimtsev. "Asymptotic Expansion in the Theory of Diffraction of a Plane Wave from a Plate." *Soviet Physics — Doklady* 14, no. 8 (1970): pp. 755–757.

25. P. Ya. Ufimtsev. "Three Lectures on the Physical Theory of Diffraction." *The Third All-Union School-Seminar on the Diffraction and Propagation of Waves.* Leningrad-Petrodvorets, May 20–June 6, 1972.

26. P.Ya. Ufimtsev. "Diffraction Field of an Open Resonator." *Radio Engineering and Electronic Physics* 19, no. 5 (1974): pp. 43–54.

27. F. Kottler. "Zur Theorie der Beugung an schwarzen Schirmen." *Annalen der Physik* 70, no. 6 (1923): pp. 405–456.

28. F. Kottler. "Electromagnetische Theorie der Beugung an schwarzen Schirmen." *Annalen der Physik* 71, no. 15 (1923): pp. 457–508.

29. B. B. Bakker and E. T. Copson. *The Mathematical Theory of Huygen's Principle,* Oxford, 1939.

30. H. E. T. Neugebauer. "Diffraction of Electromagnetic Waves Caused by Apertures in Absorbing Plane Screens." *IRE Transaction on Antennas and Propagation* AP-4, no. 2 (1956): pp. 115–119.

31. H. E. T. Neugebauer. "Extension of Babinet's Principle to Absorbing and Transparent Materials, and Approximate Theory of Backscattering by Plane, Absorbing Disks." *Journal of Applied Physics* 28, no. 3 (1957): p. 302.

32. J. A. Stratton. *Electromagnetic Theory.* New York: McGraw-Hill Book Company, 1941.

33. Ya. N. Fel'd and L. S. Benenson. Part 1 of *Antenny santimetrovykh i detsimetrovykh voln* [*Centimeter- and decimeter-wave antennas*]. Moscow: Zhukovsky Air Force Engineering Academy, 1955.

34. L. N. Zakhar'ev and A. A. Lemanskii. "Diffraction at a 'black' body." *Radio Engineering and Electronic Physics* 14, no. 11 (1969): pp. 1673–1682.

35. L. N. Zakhar'ev and A. A. Lemanskii. *Scattering of Waves from "Black" Bodies.* Moscow: Soviet Radio, 1972.

36. W. Pauli. "On Asymptotic Series for Functions in the Theory of Diffraction of Light." *Physics Review* 54, no. 2 (1938): p. 924.

37. V. H. Weston. "Theory of Absorbers in Scattering." *IRE Transactions* AP-11, no. 5 (1963): pp. 578–584.

38. Yu. N. Maksimov. "Diffraction of a Plane Wave from a Perfectly Conducting Sphere in an Inhomogeneous Medium." *Radio Engineering and Electronic Physics* 14, no. 6 (1969): pp. 829–838.

39. P. Ya. Ufimtsev. "Black Bodies and Shadow Radiation." *Soviet Journal of Communications Technology and Electronics* 35, no. 5 (1990): pp. 108–116.

40. P.Ya. Ufimtsev. "Comments on Diffraction Principles and Limitations of RCS Reduction Techniques." *Proceedings of the IEEE* 84, no. 12 (1996): pp. 1830–1850.

41. P. Ya. Ufimtsev. "Primary Shadow Radiation and Classical Models of Black Screens." *Electromagnetics* 16, no. 2 (1996): pp. 91–112.

42. M. Macdonald. *Electric Waves*, Cambridge: The University Press, 1902.

43. F. Frank and R. V. Mizes, Eds. *Die Differential- und Integralgleichungen der Mechanik und Physik*. Vol 2, *Physical Part*. New York: Dover Publications, 1961.

44. L. B. Felsen. "Backscattering from Wide-Angle and Narrow-Angle Cones." *Journal of Applied Physics* 26, no. 3 (1955): pp. 138-151.

45. O. I. Panych. "Vliyanie neravnomernoi sostovalyayushchei toka, tekushchego vblizi ostriya konusa" [Effect of a nonuniform current, flowing near the tip of a cone]. *Radiotekhnika i elektronika* 13, no. 7 (1968): pp. 1190–1202. [English translation published by: *Radio Engineering and Electronic Physics.*]

46. K. M.Siegel, R. F. Goodrich, V. H. Weston. *"Comments on far field scattering from bodies of revolution."* *Applied Science Review* Section B, vol. 8 (1959): pp. 8–12.

47. J. B. Keller. "Backscattering from a finite cone." *IRE Transactions* AP–8, no. 2 (1960): pp. 175–182.

48. J. B. Keller. "Backscattering from a finite cone-Comparison of theory and experiment." *IRE Transactions* AP–9, no. 4 (1961): pp. 411–412.

49. C. E. Schensted. "Electromagnetic and acoustic scattering by a semi-infinite body of revolution." *Journal of Applied Physics* 26, no. 3 (1955): pp. 306–308.

50. M. G. Belkina. "Diffraktsiya elecktromagnitnykh voln na diske" [Diffraction of electromagnetic waves from a disk]. *Diffraktsiya elektromagnitnykh voln na nekotorykh telakh vrashcheniya* [*Diffraction of electromagnetic waves from some bodies of revolution*]. Moscow: Sovetskoe Radio, 1957.

51. P. Ya. Ufimtsev. "Comments on 'Comparison of Three High-Frequency Diffraction Techniques.'" *Proceedings of the IEEE* 63, no. 12 (1975): pp. 1734–1737.

52. O. Breinbjerg, Y. Rahmat-Samii, and J. Appel-Hansen. "A theoretical Examination of the Physical Theory of Diffraction and Related Equivalent Currents." Report 339, Electromagnetics Institute, Technical University of Denmark, May, 1987.

53. O. Breinbjerg. "Equivalent Edge Currents Analysis of Electromagnetic Scattering by Plane Structures." (Ph. D. diss., Electromagnetics Institute, Technical University of Denmark, March, 1991).

54. G. Pelosi, S. Selleri, and P. Ya. Ufimtsev. "Newton's Observations of Diffracted Rays." *IEEE Antennas and Propagation Magazine* 40, no. 2 (1998): pp. 7–14.

55. A. Kalashnikov. "The Gouy-Sommerfeld Diffraction." Zhurnal Russkogo Fiziko-Khimicheskogo Obshchestva, Physical Section [*Journal of the Russian Physico-Chemical Society*] 44, no. 3 (1912): pp. 137–144.

56. A. Rubinowicz. "Zur Kirchhoffschen Beugungstheorie." *Annalen der Physik* Folge 4, Band 73 (1924): pp. 339–364.

57. J. B. Keller. "Geometrical Theory of Diffraction." *Journal of the Optical Society of America* 52, no. 2 (1962): pp. 116–130.

58. J. B. Keller. "Diffraction by an Aperture." *Journal of Applied Physics* 28, no. 4 (1957): pp. 426–444.

59. T. B. A. Senior, P. L. E. Uslenghi. "Experimental Detection of the Edge-Diffraction Cone," *Proceedings of the IEEE* 60, no. 11 (1972): p. 1448.

60. P. Ya. Ufimtsev. "Rubinowicz and the Modern Theory of Diffracted Rays," *Electromagnetics* 15, no. 5 (1995): pp. 547–565.

61. D. I. Butorin and P. Ya. Ufimtsev. "Explicit Expressions for an Acoustic Edge Wave Scattered by an Infinitesimal Edge Element." *Soviet Physics–Acoustics* 32, no. 4 (1986): pp. 283–287.

62. D. I. Butorin, N. A. Martynov, and P. Ya. Ufimtsev. "Asymptotic Expressions for the Elementary Edge Wave." *Soviet Journal of Communications Technology and Electronics* 33, no. 1 (1988): pp. 17–26.

63. P.Ya. Ufimtsev. "Theory of Acoustical Edge Waves." *Journal of the Acoustical Society of America* 86, no. 2 (1989): pp. 463–474.

64. P.Ya. Ufimtsev. "Elementary Edge Waves and the Physical Theory of Diffraction." *Electromagnetics* 11, no. 2 (1991): pp. 125–160.

65. P.Ya. Ufimtsev and Y. Rahmat-Samii. "The physical theory of slope diffraction." Special Issue on Radar Cross Section of Complex Objects, *Annales des Télécommunications* 50, nos. 5–6 (1995): pp. 487–498.

66. K. M. Mitzner. "Incremental Length Diffraction Coefficients." Technical Report AFAL–TR–73–26, Northrop Corporation, Aircraft Division, April, 1974.

67. R. A. Shore and A. D. Yaghjian. "Incremental Diffraction Coefficients for Plane Conformal Strips with Application to Bistatic Scattering from the Disk." *Journal of Electromagnetic Waves and Applications* 6, no. 3 (1992): pp. 359–396.

68. P. Ya. Ufimtsev. "Comments on 'Incremental Diffraction Coefficients for Plane Conformal Strips with Application to Bistatic Scattering from the Disk.'" *Journal* of *Electromagnetic Waves and Applications,* vol. 14 (2000): pp. 873–879.

69. R. Tiberio and S. Maci. "An Incremental Theory of Diffraction: Scalar Formulation." *IEEE Transactions on Antennas and Propagation* 42, no. 5 (1994): pp. 600–612.

70. R. Tiberio, S. Maci, and A. Toccafondi. "An Incremental Theory of Diffraction: Electromagnetic Formulation." *IEEE Transactions on Antennas and Propagation* 43, no. 1 (1995): pp. 87–96.

71. A. Michaeli. "Elimination of Infinities in Equivalent Edge Currents." *IEEE Transactions on Antennas and Propagation* AP–34, no. 7 (1986): pp. 912–918.

72. A. Michaeli. "Elimination of Infinities in Equivalent Edge Currents: Physical Optics Components." *IEEE Transactions on Antennas and Propagation* AP–34, no. 8 (1986): pp. 1034–1037.

73. O. Breinbjerg, "Higher Order Equivalent Edge Currents for Fringe Wave Radar Scattering by Perfectly Conducting Polygonal Plates." *IEEE Transactions on Antennas and Propagation* 40, no. 12 (1992): pp. 1543–1554.

74. P. M. Johansen. "Uniform Physical Theory of Diffraction Equivalent Edge Currents for Truncated Wedge Strips." *IEEE Transactions on Antennas and Propagation* 44, no. 7 (1996): pp. 989–995.

75. P. Ya. Ufimtsev. "Fast Convergent Integrals for Nonuniform Currents on Wedge Faces." *Electromagnetics* 18, no. 3 (1998): pp. 289–313.

[Author's note: See corrections in *Electromagnetics* 19, no. 5: (1999). Here, the last multiplier in Eqn. (3) should correctly read as $\exp[-ikr\sin\gamma_0\cos(\alpha-\varphi_0)]$.]

76. R. G. Kouyoumjian and P. H. Pathak. "A Uniform Geometrical Theory of Diffraction for an Edge in a Perfectly Conducting Surface." *Proceedings of the IEEE* 62 (1974): pp. 1448–1461.

77. D. A. McNamara, C. W. I. Pistorius, and J. A. G. Malherbe. *Introduction to the Uniform Geometrical Theory of Diffraction*, Boston–London: Artech House, 1990.

78. C. T. Copson. *Asymptotic Expansions.* Cambridge: The University Press, 1965.

79. H. B. Tran and T. J. Kim. "Monostatic and Bistatic Radar Cross Section Analysis. Vol. 1 — High Frequency Electromagnetic Scattering Theory." Technical report NOR-82-215, Northrop Corporation, 1982.

80. V. M. Babich, *et al.* "Numerical Calculation of the Diffraction Coefficients for an Arbitrary Shaped Perfectly Conducting Cone." *IEEE Transactions on Antennas and Propagation* 44, no. 5 (1996): pp. 740–747.

81. S. Blume and V. Krebs. "Numerical Calculation of Dyadic Diffraction Coefficients and Bistatic Radar Cross Section for Perfectly Conducting Semi-Infinite Elliptic Cone." *IEEE Transactions on Antennas and Propagation* 46, no. 3 (1998): pp. 414–424.

82. B. Chytil. "Depolarization of randomly spaced scatterers." *Práce ÚRE-CSAV*, no. 20. Prague, 1961.

83. B. Chytil. "Polarisation-dependent scattering cross-sections." *Práce ÚRE-CSAV*, no. 21. Prague, 1961.

84. P. Beckmann. *The Depolarization of Electromagnetic Waves.* Boulder: The Golem Press, 1968.

85. M. A. Leontovich. "Ob odnom metode resheniya zadach o rasprostranenii elektromagnitnykh voln vdol' poverkhnosti zemli" [A Method for the Solution to the Problem of the Propagation of Electromagnetic Waves Over the Surface of the Earth]. *Izvestiya AN SSSR, Seriya Fizika* 8 (1944): pp. 16–22. [English translation published by: *Bulletin of the Academy of Sciences of the U.S.S.R., Physics.*]

86. M. A. Leontovich and V. A. Fock. "Reshenie zadachi o rasprostranenii elektromagnitnykh voln vdol' poverkhnosti zemli po metodu parabolicheskogo uravneniya" [Solution to the problem of the propagation of electromagnetic waves over the surface of the earth using parabolic equations]. *Zhurnal eksperimentalnoi i teoreticheskoi fiziki* 16, no. 7 (1946): pp. 557–573. [English translation published by: *Journal of Physics of the U.S.S.R.*]

87. V. A. Fock. "Pole ploskoi volny vblizi poverkhnosti provodyashchego tela" [Field of plane waves near the surface of a conducting body]. *Izvestiya AN SSSR, Seriya Fizika* 10, no. 2 (1946): pp. 171–186. [English translation published by: *Bulletin of the Academy of Sciences of the U.S.S.R., Physics.*]

88. G. D. Mal'uzhinets and L. A. Vainshtein. "Parabolicheskoe uravnenie v luchevykh koordinatakh" [Parabolic equation in ray coordinates]. *Radiotekhnika i elektronika* 6, no. 8 (1961): pp. 1247–1258. [English translation published by: *Radio Engineering and Electronic Physics.*]

89. L. A. Vainshtein and G. D. Mal'uzhinets. "Asimptoticheskie zakony diffraktsii v polyarnykh koordinatakh" [Asymptotic characteristics of diffraction in polar coordinates]. *Radiotekhnika i elektronika* 6, no. 9 pp. (1961): 1489–1495. [English translation published by: *Radio Engineering and Electronic Physics.*]

90. V. A. Fock. and L. A. Vainshtein. "Poperechnaya diffuziya pri diffraktsii korotkikh voln na vypuklom tsilindre s plavno menyayushcheicya kriviznoi" [Transverse diffusion due to diffraction of short waves from a convex cylinder with slowly changing curvature]. *Radiotekhnika i elektronika* 8, no. 3 (1963): pp. 363–388. [English translation published by: *Radio Engineering and Electronic Physics.*]

91. G. D. Mal'uzhinets. "Doklad na Vsesoyuznom soveschanii po voprosam elektricheskikh kolebanii i voln" [Proceedings of the All-Union Meeting on electromagnetic oscillations and waves]. Gor'kii, 1946. See also *Uspekhi fizicheskikh nauk* 69, no. 2 (1959) p. 321. [English translation published by: *Soviet Physics–Uspekhi.*]

92. V. S. Buslaev. "O formulakh korotkovolnovoi asimptotiki v zadachakh diffraktsii na vypuklykh telakh" [On the formulation of short-wave asymptotics for the problem of diffraction from a convex bodies]. *Vestnik LGU* [Leningrad State University, U.S.S.R.] no. 13, (1962): pp. 5–21.

93. V. S. Buldyrev. "Korotkovolnovaya asimptotika sobstvennykh funktsii uravneniya Gel'mgol'tsa" [Short-wave asymptotics of the characteristic functions of the Helmholtz equation]. *Doklady Akademii Nauk* 163, no. 4 (1965): pp. 853–856. [English translation published by: *Soviet Physics–Doklady.*]

94. I. A. Molotkov. "Diffraktsia na vypuklom konture c plavno izmenyayushchimcya radiusom krivizny i impedansom" [Diffraction from a convex contour with a slowly changing radius of curvature and impedance]. *Problemy mathematicheskoi fiziki* 2, Izdatel'stvo LGU [Leningrad State University, U.S.S.R.], 1967. pp. 124–132.

95. C. E. Schensted. "Approximate method for scattering problems." *IRE Transactions* AP-4, no. 3 (1956): pp. 240–242.

96. L. A. Vainshtein. *Theory of Diffraction and the Factorization Method.* Boulder: Golem Press, 1969.

97. A.V. Popov. "Numerical Solution of the Wedge Diffraction Problem by the Transverse Diffusion Method." *Soviet Physics–Acoustics* 5, no. 2 (1969): pp. 226–233.

98. V. A. Fock and L. A. Vainshtein. *Proceedings of Symposium on Electromagnetic Theory and Antennas.* Copenhagen, June 25–30, 1962. Oxford: Pergamon Press, 1963, pp. 11–25.

99. N. Y. Zhu and F. M. Landstorfer. "Numerical Study of Diffraction and Slope Diffraction at Anisotropic Impedance Wedge by the Method of Parabolic Equation: Space Waves." *IEEE Transactions on Antennas and Propagation* 45, no. 5 (1997): pp. 822–828.

100. G. Pelosi, S. Selleri, and R. D. Graglia. "Numerical Analysis of the Diffraction at an Anisotropic Impedance Wedge." *IEEE Transactions on Antennas and Propagation* 45, no. 5 (1997): pp. 767–771.

101. The Centennial of Sommerfeld's Diffraction Problem, Special Issue, *Electromagnetics* 18, no. 2 (1998).

102. G. Pelosi, G. Manara, and P. Nepa. "Electromagnetic Scattering by a Wedge with Anisotropic Impedance Faces." *IEEE Antennas and Propagation Magazine* 40, no. 6 (1998): pp. 29–34.

103. P. Ya. Ufimtsev and G. D. Yakovleva. "Paraxial Mode Beams in Regular and Irregular Waveguides." *Radio Engineering and Electronic Physics* 22, no. 3 (1977): pp. 16–28.

104. E. A. Polyansky. "About the Connection Between Solutions of the Helmholtz and Schrodinger Equations." *Akusticheskii Zhurnal* [*Soviet Physics–Acoustics*] 20, no. 1 (1974): pp. 142–143.

105. L. A. Vainshtein. "Volny toka v tonkom tsilindricheskom provodnike. 1. Tok i impedans peredayushchego vibratora" [Waves of current on a thin cylindrical conductor. 1. Current and impedance of a transmitting dipole]. *Zhurnal tekhnicheskoi fiziki* 29, no. 6 (1959): pp. 673-688. [English translation published by: *Soviet Physics–Technical Physics.*]

106. L. A. Vainshtein. "Tok v passivnom vibratore i izluchenie peredayushchego vibratora" [Current on a passive dipole and radiation of a transmitting dipole]. *Zhurnal tekhnicheskoi fiziki* 29, no. 6 (1959): pp. 689–699. [English translation published by: *Soviet Physics–Technical Physics.*]

107. L. A. Vainshtein. "Volny toka v tonkom tsilindricheskom provodnike" [Waves of current on a thin cylindrical conductor]. *Zhurnal tekhnicheskoi fiziki* 31, no. 1 (1961): pp. 29–50. [English translation published by: *Soviet Physics–Technical Physics.*]

108. E. Hallen. "Exact solution of the antenna equation." *Transactions of the Royal Institute of Technology, Stockholm, Sweden.* no. 183: (1961).

109. P. Ya. Ufimtsev and A.P. Krasnozhen. "Current Waves in a Straight Thin Wire Resonator with Finite Conductivity." *Electromagnetics* 12, no. 2 (1992): pp. 121–132.

110. P. Ya. Ufimtsev and A. P. Krasnozhen. "Scattering from a Straight Thin Wire Resonator." *Electromagnetics* 12, no. 2 (1992): pp. 133–146.

111. H. T. Shamansky, A. K. Dominek, and L. Peters, Jr. "Electromagnetic Scattering by a Straight Thin Wire." *IEEE Transactions on Antennas and Propagation* AP–37, no. 8 (1989): pp. 1019–1025.

112. L. A. Vainshtein. *Otkrytye resonatory i otkrytye volnovody* [*Open resonators and open waveguides*]. Moscow: Sovetskoe Radio, 1966

113. V. M. Babich and V. S. Buldyrev. *Asymptotic Methods in Short Wave Diffraction Theory.* Berlin–Heidelberg: Springer, 1989.

114. L. A. Vainshtein. *Elektromagnitnye volny* [*Electromagnetic waves*]. Moscow: Sovetskoe Radio, 1957.

115. A. T. Fialkovskii. "Rasseyanie ploskoi volny na tonkom tsilindricheskom provodnike" [Scattering of a plane wave from a thin cylndrical conductor]. *Zhurnal tekhnicheskoi fiziki* 36, no. 10 (1966): pp. 1744–1751. [English translation published by: *Soviet Physics–Technical Physics.*]

116. L. A. Vainshtein. "Staticheskie zadachi dlya pologo tsilindra konechnoi dliny" [Static problems for a hollow cylinder of finite length]. *Zhurnal tekhnicheskoi fiziki* 32, no. 10 (1962): pp 1165–1173. [English translation published by: *Soviet Physics–Technical Physics.*]

117. M. A. Leontovich and M. L. Levin. "K teorii vozbuzhdeniya kolebanii v vibratorakh antenn" [On the theory of the excitation of oscillations on a dipole antenna]. *Zhurnal tekhnicheskoi fiziki* 14, no. 9 (1944): pp. 481–506. [English translation published by: *Soviet Physics–Technical Physics.*]

118. L. A. Vainshtein. "Chislennye resul'taty dlya passivnogo vibratora" [Numerical results for a passive dipole]. *Zhurnal tekhnicheskoi fiziki* 37, no. 7 (1967): pp. 1181–1188. [English translation published by: *Soviet Physics–Technical Physics.*]

119. K. Lindroth. "Reflection of electromagnetic waves from thin metal strips." *Transactions of the Royal Institute of Technology, Stockholm, Sweden.* no. 91 (1955).

120. L. A. Vainshtein. *Diffraktsiya* elektromagnitnykh i zvykovykh voln na otkrytom kontse volnovoda [*Diffraction* of electromagnetic and acoustic waves from an aperture at the end of a waveguide]. Moscow: Sovetskoe Radio, 1953.

121. R. W. P. King and Tai-Tsun Wu. *The Scattering and Diffraction of Waves.* Cambridge: Harvard University Press, 1959.

122. M. D. Khaskind and L. A. Vainshtein. "Plane wave diffraction by a slit and a strip." *Radiotekhnika i elektronika* 9, no. 10. (1964): pp. 1800–1811. [English translation published by: *Radio Engineering and Electronic Physics.*]

123. R. F. Millar. "Diffraction on a Wide Slit and Complementary Strip." *Proceedings of the Cambridge Philosophical Society* 54, no. 4 (1958): pp. 476–511.

124. K. Westpfahl. "Zur Theorie einer Klasse von Beugungsproblem mittels singular Integralgleichungen" [Theory of a class of diffraction prob-

lems based on singular integral equations]. *Annalen der Physik* 4, Heft 5–6 (1959): pp. 283–351.

125. E. Lüneburg and K. Westpfahl. "Beugung am Streifen: Hochfrequenz —Asymptotik und Kleinmasche Losung" [Diffraction at a strip: high frequency asymptotics and Kleinman's solution]. *Annalen der Physik* 21, Heft 1–2 (1968): pp. 12–25.

126. R. B. Kieburtz. "Construction of asymptotic solutions to scattering problems in the Fourier transform representation." *Applied Scientific Research* 12, no. 3 (1965): pp. 221–234.

127. H. Stöckel. "Beugung am Spalt" [Diffraction at a slit]. *Annalen der Physik* 16, Heft 5–6 (1965): pp. 209–219.

128. E. B. Hansen. "Scalar Diffraction by an Infinite Strip and a Circular Disc." *Journal of Mathematics and Physics* 41, no. 3 (1962): p. 229–245.

129. G. A. Grinberg. "On a new method for the solution of the electromagnetic waves diffraction problem for a plane with an infinite rectilinear slit and for related problems." *Zhurnal tekhnicheskoi fiziki* 27, no. 11 (1957): pp. 2595–2605. [English translation published by: *Soviet Physics–Technical Physics.*]

130. G. A. Grinberg. "Method of the solution of diffraction problems for planar perfectly conducting screens, based on the study of induced shadow currents." *Zhurnal tekhnicheskoi fiziki* 28, no. 3 (1958): pp. 542–568. [English translation published by: *Soviet Physics–Technical Physics.*]

131. G. A. Grinberg. "Diffraction of electromagnetic waves at a strip of a finite width." *Doklady Akademii Nauk, SSSR* 129, no. 2 (1959): pp. 295–298. [English translation published by: *Soviet Physics–Doklady.*]

132. V. N. Kuritsyn. "On the solution of the 'key' diffraction problem for a perfectly conducting strip." *Zhurnal tekhnicheskoi fiziki* 31, no. 12 (1961): pp. 1485–1490. [English translation published by: *Soviet Physics–Technical Physics.*]

133. G. Ya. Popov. "Ob odnom priblizhennom sposobe resheniya integral'nogo uravneniya diffraktsii elektromagnytnykh voln na polose konechnoi shiriny" [On a approximate method for the solution of the integral equation for diffraction of electromagnetic waves from a strip of a finite width]. *Zhurnal tekhnicheskoi fiziki* 35, no. 3 (1965): pp. 381–389. [English translation published by: *Soviet Physics–Technical Physics.*]

134. A. T. Fialkovskii. "Diffraction of plane electromagnetic waves by a slit and a strip." *Radiotekhnika i elektronika* 11, no. 2 (1966): pp. 178–186. [English translation published by: *Radio Engineering and Electronic Physics.*]

135. V. A. Borovikov. "Plane wave diffraction by a segment." *Doklady Akademii Nauk, SSSR* 159, no. 4 (1964): pp. 711–714. [English translation published by: *Soviet Physics–Doklady.*]

136. V. A. Borovikov. *Diffraktsiya na mnogougol'nikakh i mnogogrannikakh* [*Diffraction from polygons and polyhedrons*]. Moscow: Izdatel'stvo Nauka, 1966.

137. V. A. Fock. "O nekotorykh integral'nykh uravneniyakh matematicheskoi fiziki" [On certain integral equations in the mathematical physics] in *Problemy diffraktsii i rasprostraneniya eletromagnitnykh voln* [*Problems of diffraction and propagation of electromagnetic waves*] Moscow: Sovetskoe Radio,1970.

138. V.A. Fock. "O nekotorykh integral'nykh uravneniyakh matematicheskoi fiziki" [On certain integral equations of mathematical physics], *Doklady Akademii Nauk SSSR* 37 (1942): p. 147 [English translation published by: *Soviet Physics–Doklady*]

139. K. Schwarzschild. "Die Beugung und Polarisation des Lichts durch einen Spalt." *Mathematische Annalen* 55 (1902): pp. 177–247.

140. W. Braunbek. "Neue Naherungsmethode fur die Beugung am ebenen Schirm" [New Approximate method to study diffraction at a planar screen]. *Zeitschrift für Physik* 127, no. 4 (1950): pp. 381–390.

141. W. Braunbek. "Zur Beugung an der Kreisscheibe" [Diffraction at a circular plate]. *Zeitschrift für Physik* 127, no. 4 (1950): pp. 405–415.

142. A. G. Fox and T. Li. "Resonant Modes in Maser Interferometer." *Bell Systems Technical Journal* 40, no. 2 (1961): pp. 453–458.

143. D. S. Jones. "Diffraction by a Waveguide of Finite Length." *Proceedings of the Cambridge Philosophical Society* 48 (1952): pp. 118-134.

144. W. E. Williams. "Diffraction by Two Parallel Planes of Finite Length." *Proceedings of the Cambridge Philosophical Society* 50 (1954): pp. 309–318.

145. Yu. V. Pimenov. "Ploskaya zadacha diffraktsii elektromagnitnykh voln na dvukh ideal'no provodyashchich polosakh konechnoi shiriny, raspolozhennykh odna nad drugoi" [Plane diffraction problem for

two parallel plates placed one above the other]. *Zhurnal tekhnicheskoi fiziki* 29, no. 4 (1959): p. 638. [English translation published by: *Soviet Physics-Technical Physics.*]

146. E. I. Nefedov. *Pyatiznachnye tablitsy diffraktsionnoi U(s,p)—funktsii Vainshteina* [*Five-Digit Tables of the Vainshtein U(s,p) Diffraction Function*]. Moscow: Izdatel'stvo Nauka, 1969.

147. A. T. Fialkovskii. "Diffraktsiya ploskikh voln na otkrytom rezonatore, obrazovannom parallel'nymi lentami" [Diffraction of plane waves from an open resonator formed by parallel plates]. *Zhurnal tekhnicheskoi fiziki* 39, no. 5 (1969): p. 865. [English translation published by: *Soviet Physics–Technical Physics.*]

148. V. P. Shestopalov and Y. V. Shestopalov. "Spectral Theory and Excitation of Open Structures." *The Institute of Electrical Engineers*. Norfolk, U.K.: Galliard Ltd., 1996.

149. T. Rozzi and M. Mongiardo. "Open Electromagnetic Waveguides." The Institute of Electrical Engineers. Exeter, England: Short Run Press Ltd., 1997.

150. A. G. Kyurkchan and D. B. Demin. "Modelirovanie kharakteristik rasseyaniya voln telami s pogloshchayushchim pokrytiem i 'chernymi' telami" [Simulation of Wave Scattering by Bodies with an Absorbing Coating and Black Bodies]. *Zhurnal tekhnicheskoi fiziki* 74, no. 2 (2004): p. 24–31. [English translation published by: *MAIK Nauka Interperiodika.* Distributed by the *American Institute of Physics.*]

151. A. P. Yarygin. "Primenenie metoda kraevykh voln v zadachakh difraktsii na telakh nakhodyaschikhsya v plavno neodnorodnoi srede [Application of the edge waves method to problems of diffraction from bodies placed in smoothly inhomogeneous medium]. *Radiotekhnika i elektronika* 17, no. 10 (1972): pp. 1601-1609 [English translation published by: *Radio Engineering and Electronic Physics*]

152. J. L. Guiraud. "Une approche spectrale de la theorie physique de la diffraction." *Annales des Télécommunications* 38, no. 3–4 (Mars–Avril 1983): pp. 145–157.

153. P. Balling. "Fringe-Currents Effects on Reflector Antenna Crosspolarization." *Electromagnetics* 15, no. 1 (1995): pp. 55–69.

154. D. J. Andersh, M. Hazlett, S. W. Lee, D. D. Reeves, D. P. Sullivan, and Y. Chu. "XPATCH: A High-Frequency Electromagnetic-Scattering Code and

Environment for Complex Three-Dimensional Objects." *IEEE Antennas and Propagation Magazine* 36, no. 1 (February 1994): pp. 65–69.

155. D. W. Duan, Y. Rahmat-Samii, and J. P. Mahon. "Scattering from a Circular Disk: Comparative Study of PTD and GTD Techniques." *Proceedings of the IEEE* 79, no. 10 (1991): pp. 1472–1480.

156. M. V. Vesnik. "Ispol'zovanie dvukh-mernykh reshenyi v trekh-mernykh zadachakh" [Usage of two-dimensional solutions in three-dimensional problems]. *Radiotekhnika i elektronika* 38, no. 8 (1993): pp. 1416–1423 [English translation *Journal of Communications Technology and Electronics*].

157. M. Idemen and A. Buyukaksoy. "High-Frequency Surface Currents Induced on a Perfectly Conducting Cylindrical Reflector." *IEEE Transactions on Antennas and Propagation* AP–32, no. 5 (1984): pp. 501–507.

158. R. T. Brown. "Treatment of Singularities in the Physical Theory of Diffraction." *IEEE Transactions on Antennas and Propagation* AP-32, no. 6 (1984): pp. 640–641.

159. D. P. Bouche, J. J. Bouquet, H. Manene, and R. Mittra. "Asymptotic Computation of the RCS of Low Observable Axisymmetric Objects at High Frequency." *IEEE Transactions on Antennas and Propagation* 40, no. 10 (1992): pp. 1165–1174.

160. D. P. Bouche, F. A. Molinet, and R. Mittra. "Asymptotic and Hybrid Techniques for Electromagnetic Scattering." *Proceedings of the IEEE* 81, no. 12 (1993): pp. 1658-1684.

161. D. D. Gabrielyan, O. M. Tarasenko, and V. V. Shatskyi. "Ispol'zovanie predstavleniya kraevykh voln v sochetanii s metodom integral'nykh uravnenyi pri reshenii zadach difraktsii na ideal'no provodyaschikh telakh slozhnoi formy" [Usage of the edge-wave representation combined with the method of integral equations to solve problems of diffraction by ideally conducting bodies with a complicated shape]. *Radiotekhnika i elektronika* 36, no. 6 (1991): pp. 1159–1163 [English translation published by: *J. Commun. Technology and Electronics*].

162. T. B. Hansen and R. A. Shore. "Incremental Length Diffraction Coefficients for the Shadow Boundary of a Convex Cylinder." *IEEE Transactions on Antennas and Propagation* 46, no. 10 (1998): pp. 1458–1466.

163. D. S. Wang and L. N. Medgyesi-Mitschang. "Electromagnetic Scattering from a Finite Circular and Elliptic Cone." *IEEE Transactions on Antennas and Propagation* AP–33, no. 5 (1985): pp. 488–497.

164. S. S. Skyttemyr. "Cross Polarization in Dual Reflector Antennas—A PO and PTD Analysis." *IEEE Transactions on Antennas and Propagation* AP-34, no. 6 (1986): pp. 849–853.

165. P. K. Murthy and G. A. Thiele. "Non-Uniform Currents on a Wedge Illuminated by a TE-Plane Wave." *IEEE Transactions on Antennas and Propagation* AP–34, no. 8 (1986): pp. 1038–1045.

166. M. G. Cote, M. B. Woodworth, and A. D. Yaghjian. "Scattering from Perfectly Conducting Cube." *IEEE Transactions on Antennas and Propagation* 36, no. 9 (1988): pp. 1321–1329.

167. T. Akashi, M. Ando, and T. Kinoshita. "Effects of Multiple Diffraction in PTD Analysis of Scattered Field from a Conducting Disk." *Trans. IEICE* E 72, no. 4 (1989): pp. 259–261.

168. M. Ando and T. Kinoshita. "PO and PTD Analysis in Polarization Prediction for Plane Wave Diffraction from a Large Circular Disk." Digests of 1989 *IEEE Antennas and Propagation Society Int. Symp.*, June 26–30, 1989, San Jose, California.

169. M. Ando and T. Kinoshita. "Accuracy Comparison of PTD and PO for Plane Wave Diffraction from a Large Circular Disk." *Trans. IEICE* E 72, no. 11 (1989): pp. 1212–1218.

170. N. N. Youssef. "Radar Cross Section of Complex Targets." *Proc. IEEE* 77, no. 5 (May 1989): pp. 722–734.

171. M. Ando. "Modified Physical Theory of Diffraction." in *Analysis Methods for EM Wave Problems*, E. Yamashita, Ed. Boston/London: Artech House, 1990.

172. E. N. Vasil'ev, V. V. Solodukhov and A. I. Fedorenko. "The Integral Equation Method in the Problem of Electromagnetic Waves Diffraction by Complex Bodies." *Electromagnetics* 11, no. 2 (1991): pp. 161–182.

173. S. Y. Kim, J. W. Ra, and S. Y. Shin. "Diffraction by an Arbitrary–Angled Dielectric Wedge: Part II—Corrections to Physical Optics Solution." *IEEE Transactions on Antennas and Propagation* 39, no. 9 (1991): pp. 1282–1292.

174. J. M. Ruis, M. Ferrando, and L. Jofre. "GRECO: Graphical ElectroMagnetic Computing for RCS Prediction in Real-Time." *IEEE Antennas and Propagation Magazine* 35, no. 2 (1993): pp. 7–17.

175. J. M. Ruis, M. Ferrando, and L. Jofre. "GRECO: High-Frequency RCS of Complex Radar Targets in Real-Time." *IEEE Transactions on Antennas and Propagation* 41, no. 9 (1993): pp. 1308–1319.

176. M. Martinez-Burdalo, A. Martin, and R. Villar. "Uniform PO and PTD Solution for Calculating Plane Wave Backscattering from a Finite Cylindrical Shell of Arbitrary Cross Section." *IEEE Transactions on Antennas and Propagation* 41, no. 9 (1993): pp. 1336–1339.

177. L. D. Bakhrakh and I. G. Yarmakhov. "Issledovanie difraktsii electromagnitnykh voln na schelyakh konechnoi tolschiny" [Study of electromagnetic wave diffraction at slits of finite thickness]. *Radiotekhnika i elektronika* 38, no. 2 (1993): pp. 213–218 [English translation published by: *J. Commun. Technology and Electronics*].

178. P. Corona, *et al.* "Accurate Evaluation of Backscattering by 90° Dihedral Corners." *Electromagnetics* 13, no. 1 (1993): pp. 23–36.

179. M. Domingo, R. P. Torres, and M. F. Catedra. "Calculation of the RCS from the Interaction of Edges and Faces." *IEEE Transactions on Antennas and Propagation* 42, no. 6 (1994): pp. 885–898.

180. A. Altintas, O. M. Buyukdura, and P. H. Pathak. "An Extension of the PTD Concept for Aperture Radiation Problems." *Radio Science* 29, no. 6 (1994): pp. 1403–1407.

181. M. Domingo *et al.* "Computation of the RCS of Complex Bodies Modeled Using NURBS Surfaces." *IEEE Antennas and Propagation Magazine* 37, no. 6 (1995): pp. 36–47.

182. J. S. Asvestas. "A Class of Functions with Removable Singularities and Their Application in the Physical Theory of Diffraction." *Electromagnetics* 15, no. 2 (1995): pp. 143–155.

183. S. Vermersch, M. Sesques, and D. P. Bouche. "Computation of the RCS of coated objects by a generalized PTD approach." Special Issue on Radar Cross Section of Complex Objects, *Annales des Télécommunications* 50, no. 5-6 (May- June 1995): pp. 563–572.

184. P. Ya. Ufimtsev and Y. Rahmat-Samii. "Physical theory of slope diffraction." Special Issue on Radar Cross Section of Complex Objects, *Annales des Télécommunications* 50, no. 5–6 (May-June 1995): pp. 487–498.

185. J. M. Rius, M. Vall-Lossera, and A. Cardama. "GRECO: Graphical processing methods for high-frequency RCS prediction." Special Issue on Radar Cross Section of Complex Objects, *Annales des Télécommunications* 50, no. 5–6 (May–June 1995): pp. 551–556.

186. M. Boutillier and M. A. Blondell-Fournier. "CAD based high-frequency RCS computing code for complex objects: Sermat." Special Issue on Radar Cross Section of Complex Objects, *Annales des Télécommunications* 50, no. 5–6 (May–June 1995): pp. 536–539.

187. A. C. Polycarpou, C. A. Balanis, and C. R. Bitcher. "Radar cross section of trihedral corner reflectors using PO and MEC." Special Issue on Radar Cross Section of Complex Objects, *Annales des Télécommunications* 50, no. 5–6 (May–June 1995): pp. 510–516.

188. H. H. Syed and J. L. Volakis. "PTD Analysis of Impedance Structures." *IEEE Transactions on Antennas and Propagation* 44, no. 7 (1996): pp. 983–988.

189. P. M. Johansen. "Uniform Physical Theory of Diffraction Equivalent Edge Currents for Implementation in General Computer Codes." *IEEE Antennas and Propagation Soc. Symp. Digests,* July 21–26, 1996, Baltimore, MD, vol. 2, pp. 784–787.

190. P. M. Johansen. "Time-Domain Incremental Diffraction Coefficients for Edges." *USNC/URSI Radio Science Meeting Digests*, July 21–26 1996, Baltimore, MD, p. 381.

191. J. J. Kim and O. B. Kesler. "Hybrid Scattering Analysis (PTD+UFIM) for Large Airframe with Amall Details." *USNC/URSI Radio Science Meeting Digests,* July 21–26 1996, Baltimore, MD, p. 263.

192. P. Wolf. "A new approach to edge diffraction." SIAM, *J. Appl. Math.* 15, no. 6 (1967): pp. 1434–1469.

193. M. Tran Van Nhieu. "Diffraction by plane screens." *J. Acoust. Soc. Am.* 9, no. 2 (1995): pp. 796–806.

194. M. Tran Van Nhieu. "Diffraction by the edge of a three-dimensional object." *J. Acoust. Soc. Am.* 99, no. 1 (1996): pp. 79–87.

195. P. Ya. Ufimtsev, *Fundamentals of the Physical Theory of Diffraction*, John Wiley & Sohns, Inc., Hoboken, New Jersey, 2007.

196. P. Ya. Ufimtsev, "New insight into the classical Macdonald Physical Optics approximation", *IEEE Antennas & Propagation Magazine*, vol. 50, no. 3, pp. 11–20, June 2008. Corrections in *IEEE Antennas & Propagation Magazine*, vol. 50, no. 4, p. 65, August 2008.

197. P. Ya. Ufimtsev, "On polarization coupling in the PO and PTD approximations", *IEEE Trans. Antennas & Propagation*, vol. 56, no. 12, December 2008.

198. J. M. L. Bernard, M. A. Lyalinov, N. Y. Zhu, "Analytical-numerical calculation of diffraction coefficients for a circular impedance cone", *IEEE Trans. Antennas & Propagation*, vol. 56, no. 6, pp. 1616–1623, 2008.

199. F. A. Molinet, "Edge-excited rays on convex and concave structures: A review", *IEEE Antennas & Propagation Magazine*, vol. 47, no. 5, pp. 34–46, 2005.

Appendix 1:

Relationships Between the Gaussian System (GS) and the System International (SI) for Electromagnetic Units

The Gaussian system is widely used in theoretical physics. It is often called the absolute symmetrical system. This system is based on three fundamental units: centimeter (cm), gram (g), and second (s). The units for all electromagnetic quantities are derivatives from these fundamental units and they are established using physical relationships between them. Many Gaussian units are not named.

The tables presented below show the relationships between the Gaussian System (GS) of units and the International System (SI) of units, which is based on the units of meter (m), kilogram (kg), second (s), and ampere (A).

Basic Electromagnetic Units

Symbol	Physical Quantity	Name of GS Unit	GS Unit	SI Unit
q	charge	—	$\frac{1}{3}\cdot 10^{-9}$ C	coulomb (C)
ρ	charge density	—	$\frac{1}{3}\cdot 10^{-3}$ C/m^3	C/m^3

Basic Electromagnetic Units

Symbol	Physical Quantity	Name of GS Unit	GS Unit	SI Unit
J	current	—	$\frac{1}{3}\cdot 10^{-9}$ A	ampere (A)
j	current density	—	$\frac{1}{3}\cdot 10^{-5}$ A/m^2	A/m^2
j	surface current density	—	$\frac{1}{3}\cdot 10^{-7}$ A/m	A/m
ϕ	electric potential	—	300 V	volt (V)
E	electric field intensity	—	$3\cdot 10^4$ V/m	V/m
D	electric flux density	—	$3\cdot 10^4$ V/m	C/m^2
H	magnetic field intensity	Oersted	$\frac{1}{4\pi}\cdot 10^3$ A/m	A/m
B	magnetic flux density	Gauss	10^{-4} T	tesla (T)
—	energy	erg	10^{-7} J	joule (J)
—	energy density	erg/cm^3	0.1 J/m^3	J/m^3
—	power flux	erg/s	10^{-7} W	watt (W)
—	power flux density	erg/s·cm^2	10^{-3} W/m^2	W/m^2

Freespace Harmonic Electromagnetic Fields ($e^{-i\omega t}$)

Gaussian System (GS)	System International (SI)
$\nabla\times\mathbf{H} = \frac{4\pi}{c}\mathbf{j}^e - ik\mathbf{E}$	$\nabla\times\mathbf{H} = \mathbf{j}^e - i\omega\epsilon_0\mathbf{E}$

Freespace Harmonic Electromagnetic Fields ($e^{-i\omega t}$)

Gaussian System (GS)	System International (SI)
$\nabla\times\mathbf{E}=-\frac{4\pi}{c}\mathbf{j}^m+ik\mathbf{H}$	$\nabla\times\mathbf{E}=-\mathbf{j}^m+i\omega\mu_0\mathbf{H}$
$\nabla\cdot\mathbf{E}=4\pi\rho^e$	$\nabla\cdot\mathbf{E}=\frac{\rho^e}{\epsilon_0}$
$\nabla\cdot\mathbf{H}=4\pi\rho^m$	$\nabla\cdot\mathbf{H}=\frac{\rho^m}{\mu_0}$
c — speed of light	

Free Space Constitutive Relationships

Gaussian System (GS)	System International (SI)
$\mathbf{D}=\mathbf{E}$	$\mathbf{D}=\epsilon_0\mathbf{E}$
$\mathbf{B}=\mathbf{H}$	$\mathbf{B}=\mu_0\mathbf{H}$
$\epsilon_0=(1/36\pi)\cdot 10^{-9}$ Farads/meter $\mu_0=4\pi\cdot 10^{-7}$ Henrys/meter	

Continuity Equations ($e^{-i\omega t}$)

Gaussian System (GS)	System International (SI)
$\nabla\cdot\mathbf{j}^e=i\omega\rho^e$	$\nabla\cdot\mathbf{j}^e=i\omega\rho^e$

Continuity Equations ($e^{-i\omega t}$)

Gaussian System (GS)	System International (SI)
$\nabla\cdot\mathbf{j}^m = i\omega\rho^m$	$\nabla\cdot\mathbf{j}^m = i\omega\rho^m$

Field Expressions (free space — $e^{-i\omega t}$)

Gaussian System (GS)	System International (SI)
$\mathbf{E} = -\nabla\times\mathbf{A}^m + \frac{i}{k}\left[\nabla\left(\nabla\cdot\mathbf{A}^e\right)+k^2\mathbf{A}^e\right]$	$\mathbf{E} = -\nabla\times\mathbf{A}^m + \frac{i}{k}\sqrt{\frac{\mu_0}{\epsilon_0}}\left[\nabla\left(\nabla\cdot\mathbf{A}^e\right)+k^2\mathbf{A}^e\right]$
$\mathbf{H} = \nabla\times\mathbf{A}^e + \frac{i}{k}\left[\nabla\left(\nabla\cdot\mathbf{A}^m\right)+k^2\mathbf{A}^m\right]$	$\mathbf{H} = \nabla\times\mathbf{A}^e + \frac{i}{k}\sqrt{\frac{\epsilon_0}{\mu_0}}\left[\nabla\left(\nabla\cdot\mathbf{A}^m\right)+k^2\mathbf{A}^m\right]$
$\mathbf{A}^{e,m} = \frac{1}{c}\int j^{e,m}\frac{e^{ikR}}{R}dv$	$\mathbf{A}^{e,m} = \frac{1}{4\pi}\int j^{e,m}\frac{e^{ikR}}{R}dv$
$k = \frac{\omega}{c}$	$k = \omega\sqrt{\epsilon_0\mu_0}$

Poynting Vector

Gaussian System (GS)	System International (SI)
$\mathbf{P} = \frac{c}{8\pi}\Re e\left\{\mathbf{E}\times\mathbf{H}^*\right\}$	$\mathbf{P} = \frac{1}{2}\Re e\left\{\mathbf{E}\times\mathbf{H}^*\right\}$

Equivalent Surface Currents

Gaussian System (GS)	System International (SI)
$\mathbf{j}^e = \frac{c}{4\pi}\left[\mathbf{n}\times\mathbf{H}\right]$	$\mathbf{j}^e = \left[\mathbf{n}\times\mathbf{H}\right]$
$\mathbf{j}^m = -\frac{c}{4\pi}\left[\mathbf{n}\times\mathbf{E}\right]$	$\mathbf{j}^m = -\left[\mathbf{n}\times\mathbf{E}\right]$
Here, **n** represents the unit outward normal on that side of the surface where the equivalent current is placed.	

Far Fields (in free space)

Gaussian System (GS)	System International (SI)
$\mathbf{E} = \mathbf{H}\times\hat{\mathbf{k}}$	$\mathbf{E} = \sqrt{\frac{\mu_0}{\epsilon_0}}\mathbf{H}\times\hat{\mathbf{k}}$

Far Fields (in free space)

Gaussian System (GS)	System International (SI)
$\mathbf{H} = \hat{\mathbf{k}} \times \mathbf{E}$	$\mathbf{H} = \sqrt{\frac{\epsilon_0}{\mu_0}}\hat{\mathbf{k}} \times \mathbf{E}$
Here, **k** is the unit vector in the direction of propagation.	

Appendix 2:

The Key Equivalence Theorem

Suppose there are some external electric and magnetic currents with a volume density $j^{e,m}$ located in the free space. Separate the space by an arbitrary imaginary surface S in two regions, V and W, with sources $j_v^{e,m}$ and $j_w^{e,m}$ (Fig. 1.2.1-A). Choose an observation point $\boldsymbol{P}$ inside the region W (out of V). Determine two unit vectors normal to the surface S. Direct the normal $\mathbf{n}$ into the region V and the normal $\mathbf{N} = -\mathbf{n}$ into the region W. On the internal side of the surface S we introduce surface currents

$$j_{s,n}^{e} = \frac{c}{4\pi}[\mathrm{n}\times\mathrm{H}], \quad j_{s,n}^{m} = -\frac{c}{4\pi}[\mathrm{n}\times\mathrm{E}]. \tag{A2.1}$$

Here, the vectors

$$\mathrm{E} = \mathrm{E}_v + \mathrm{E}_w, \quad \mathrm{H} = \mathrm{H}_v + \mathrm{H}_w \tag{A2.2}$$

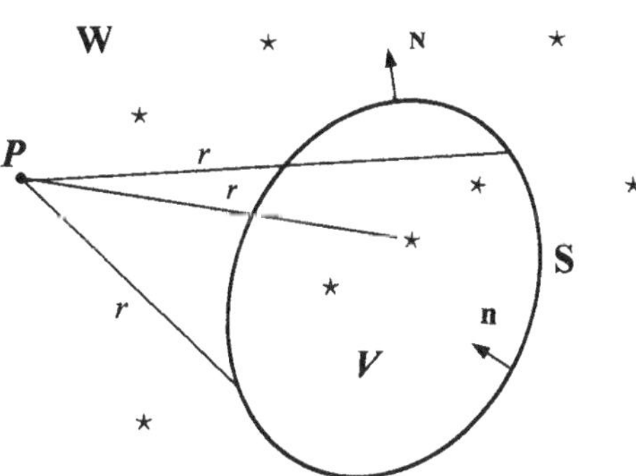

Figure 1.2.1-A

stand for the *total* electromagnetic field generated by *all* of the external currents $j_v^{e,m}$, $j_w^{e,m}$ located correspondingly in V and W.

Denote $\mathcal{E}_v$, $\mathcal{H}_v$ as the field generated at the point P by the volume currents $j_v^{e,m}$ and apply symbols $\mathcal{E}_s$, $\mathcal{H}_s$ for the field created by the surface currents $j_s^{e,m}$. Notice that according to Eqn. (A2.2) these surface currents eventually depend on the volume sources $j_v^{e,m}$ and $j_w^{e,m}$.

After that, utilizing the vector analog of Green's formula [32], one can derive the following key theorem.

Theorem 1. The sum of the fields generated at point P individually by the currents $j_v^{e,m}$ and $j_s^{e,m}$ is equal to zero

$$\mathcal{E}_v(P) + \mathcal{E}_s(P) = 0, \quad \mathcal{H}_v(P) + \mathcal{H}_s(P) = 0. \tag{A2.3}$$

Here,

$$\begin{aligned} \mathcal{E}_{v,s} &= \frac{1}{ik}(\nabla\nabla\cdot \mathrm{A}^e_{v,s} + k^2 \mathrm{A}^e_{v,s}) - \nabla\times \mathrm{A}^m_{v,s} \\ \mathcal{H}_{v,s} &= \frac{1}{ik}(\nabla\nabla\cdot \mathrm{A}^m_{v,s} + k^2 \mathrm{A}^m_{v,s}) + \nabla\times \mathrm{A}^e_{v,s}, \end{aligned}, \tag{A2.4}$$

$$\mathrm{A}_v^{e,m} = \frac{1}{c}\int_V j^{e,m}\frac{e^{ikr}}{r}dV, \quad \mathrm{A}_s^{e,m} = \frac{1}{c}\oint_S j_s^{e,m}\frac{e^{ikr}}{r}dS, \tag{A2.5}$$

In these equations, the quantity e^{ikr}/r is Green's function of the free space, $r = \sqrt{(x-\xi)^2 + (y-\eta)^2 + (z-\zeta)^2}$ is the distance between the observation and integration points. All differential operators take actions on functions of variable r with respect to coordinates of the observation point $P(x, y, z)$.

It is understood in Eqn. (A2.3) that function e^{ikr}/r and its first and second derivatives are finite and continuous. This condition is satisfied for the observation points P outside the region V. We also can bring the point P into the region V. However, in this case, one should enclose the point P with a small sphere σ and replace the surface S in Eqs. (A2.3–A2.5) with the surface $S + \sigma$. One can show that in the limiting case, when the radius of σ tends to zero, the integrals over σ lead to the following quantities

$$\mathcal{E}_\sigma(P) = -\mathrm{E}(P), \quad \mathcal{H}_\sigma(P) = -\mathrm{H}(P). \tag{A2.6}$$

Notice that while calculating these integrals one should utilize Lorentz's gauges

$$\nabla \cdot \mathrm{A}_{\sigma}^{e,m} = ik\Phi_{\sigma}^{e,m}, \tag{A2.7}$$

where

$$\Phi_{\sigma}^{e,m} = \frac{1}{4\pi}\int_{\sigma} \rho^{e,m} \frac{\mathrm{e}^{ikr}}{r} d\sigma \tag{A2.8}$$

are the scalar potentials of the fields generated by the surface charges

$$\rho_s^e = \frac{1}{4\pi}(\mathrm{n}\cdot\mathrm{E}), \quad \rho_s^m = \frac{1}{4\pi}(\mathrm{n}\cdot\mathrm{H}). \tag{A2.9}$$

In this way Eqs. (A2.3) are transformed as follows

$$\mathrm{E}(P) = \mathcal{E}_v(P) + \mathcal{E}_s(P), \quad \mathrm{H}(P) = \mathcal{H}_v(P) + \mathcal{H}_s(P). \tag{A2.10}$$

Consider now some important consequences of the key theorem (A2.3).

A. Equivalent Surface Currents

Rewrite Eqn. (A2.3) as

$$\mathcal{E}_v = -\mathcal{E}_s, \quad \mathcal{H}_v = -\mathcal{H}_s. \tag{A2.11}$$

Introduce the surface currents

$$\mathrm{j}_{s,N}^{e} = \frac{c}{4\pi}[\mathrm{N}\times\mathrm{H}] = -\mathrm{j}_{s,n}^{e}, \quad \mathrm{j}_{s,N}^{m} = -\frac{c}{4\pi}[\mathrm{N}\times\mathrm{E}] = -\mathrm{j}_{s,n}^{m} \tag{A2.12}$$

on the external side of the surface S with the normal N directed into the region W. Denote $\mathcal{E}_{s,N}$, $\mathcal{H}_{s,N}$ as the field in the region W generated by these currents. It is obvious that this field differs only in sign with the field created by the currents (A2.1),

$$\mathcal{E}_{s,N} = -\mathcal{E}_s, \quad \mathcal{H}_{s,N} = -\mathcal{H}_s. \tag{A2.13}$$

According to Eqn. (A2.11) the field $\mathcal{E}_{s,N}$, $\mathcal{H}_{s,N}$ in the region W is completely equivalent to the field radiated by the external sources $\mathrm{j}_v^{e,m}$ located inside the region V,

$$\mathcal{E}_v = \mathcal{E}_{s,N}, \quad \mathcal{H}_v = \mathcal{H}_{s,N}. \tag{A2.14}$$

This is one of the known formulations of the equivalence theorems. Its wonderful property is that the field created by the equivalent currents

(A2.12) does not depend on the shape of the surface S. It can be deformed in any way, but without change in location of the external sources $\mathrm{j}_v^{e,m}$ and $\mathrm{j}_w^{e,m}$ with respect to S. They should stay in their own regions V and W.

B. Absence of Reverse Waves in a Homogeneous Space

According to (A2.2) one can represent the field $\mathcal{E}_{s,N}(P)$, $\mathcal{H}_{s,N}(P)$ as the sum

$$\mathcal{E}_{s,N} = \mathcal{E}_{s,N}^{(v)} + \mathcal{E}_{s,N}^{(w)}, \quad \mathcal{H}_{s,N} = \mathcal{H}_{s,N}^{(v)} + \mathcal{H}_{s,N}^{(w)} \tag{A2.15}$$

Here, superscripts indicate the origination of associated quantities. The field $\mathcal{E}_{s,N}^{(v)}$, $\mathcal{H}_{s,N}^{(v)}$ is created by the equivalent surface currents

$$\mathrm{j}_{s,N}^{e} = \frac{c}{4\pi}[\mathrm{N}\times\mathrm{H}_v], \quad \mathrm{j}_{s,N}^{m} = -\frac{c}{4\pi}[\mathrm{N}\times\mathrm{E}_v], \tag{A2.16}$$

where vectors E_v, H_v represent the field in the free space generated by the external sources $\mathrm{j}_v^{e,m}$ located in the region V. The field $\mathcal{E}_{s,N}^{(w)}$, $\mathcal{H}_{s,N}^{(w)}$ is radiated by the equivalent currents

$$\mathrm{j}_{s,N}^{e} = \frac{c}{4\pi}[\mathrm{N}\times\mathrm{H}_w], \quad \mathrm{j}_{s,N}^{m} = -\frac{c}{4\pi}[\mathrm{N}\times\mathrm{E}_w], \tag{A2.17}$$

that, in their turn, are determined by the field E_w, H_w of the external sources $\mathrm{j}_w^{e,m}$ from the region W.

Taking into account (A2.15) one can rewrite the equivalence relations (A2.14) in the form

$$\mathcal{E}_v = \mathcal{E}_{s,N}^{(v)} + \mathcal{E}_{s,N}^{(w)}, \quad \mathcal{H}_v = \mathcal{H}_{s,N}^{(v)} + \mathcal{H}_{s,N}^{(w)} \tag{A2.18}$$

Now let us change the magnitude or location of sources $\mathrm{j}_w^{e,m}$. It is obvious that their field E_w, H_w on the surface S also will change, as well as the currents (A2.17) and the field $\mathcal{E}_{s,N}^{(w)}$, $\mathcal{H}_{s,N}^{(w)}$ they radiate. Denote its new values as $\tilde{\mathcal{E}}_{s,N}^{(w)}$, $\tilde{\mathcal{H}}_{s,N}^{(w)}$. Hence, the relations (A2.18) change as

$$\mathcal{E}_v = \mathcal{E}_{s,N}^{(v)} + \tilde{\mathcal{E}}_{s,N}^{(w)}, \quad \mathcal{H}_v = \mathcal{H}_{s,N}^{(v)} + \tilde{\mathcal{H}}_{s,N}^{(w)}. \tag{A2.19}$$

Now let us compare the relations (A2.18) and (A2.19). Their left sides and the first summands in the right parts are the same, while the second terms are different: $\tilde{\mathcal{E}}_{s,N}^{(w)} \neq \mathcal{E}_{s,N}^{(w)}$, $\tilde{\mathcal{H}}_{s,N}^{(w)} \neq \mathcal{H}_{s,N}^{(w)}$. However, subtracting (A2.19) and

(A2.18), we arrive to the opposite result: $\tilde{\mathcal{E}}_{s,N}^{(w)} = \mathcal{E}_{s,N}^{(w)}$, $\tilde{\mathcal{H}}_{s,N}^{(w)} = \mathcal{H}_{s,N}^{(w)}$. This contradiction leads to the conclusion that these fields are identically equal to zero,

$$\mathcal{E}_{s,N}^{(w)}(P) \equiv \tilde{\mathcal{E}}_{s,N}^{(w)}(P) \equiv 0, \quad \mathcal{H}_{s,N}^{(w)}(P) \equiv \tilde{\mathcal{H}}_{s,N}^{(w)}(P) \equiv 0. \qquad \text{(A2.20)}$$

In these equations we assume that the observation point P is located at the same region W, where the external sources are $\mathbf{j}_w^{e,m}$. From the physical point of view, one can interpret this result as absence of *reverse waves* in a homogeneous space with sources radiating electromagnetic waves.

Index

www.ingramcontent.com/pod-product-compliance
Lightning Source LLC
Chambersburg PA
CBHW050520190326
41458CB00005B/1602
9781891121661